SPEAKING FOR THE RIVER

Speaking for the River

Confronting Pollution on the Willamette,
1920s–1970s

JAMES V. HILLEGAS-ELTING

Oregon State University Press   Corvallis

Library of Congress Cataloging-in-Publication Data

Names: Hillegas-Elting, James V., author.
Title: Speaking for the river : confronting pollution on the Willamette, 1920s/1970s /
    James V. Hillegas-Elting.
Description: Corvallis : Oregon State University Press, [2018] | Includes bibliographi-
    cal references and index.
Identifiers: LCCN 2017054212 | ISBN 9780870719165 (original trade pbk. : alk.
    paper)
Subjects: LCSH: Willamette River (Or.)—Environmental conditions. | Water—
    Pollution—Oregon—Willamette River. | Environmental degradation—Oregon—
    Willamette River. | Water quality management—Oregon—Willamette River. |
    Stream conservation—Oregon—Willamette River.
Classification: LCC GE155.O7 H55 2018 | DDC 363.739/472097953—dc23
LC record available at https://lccn.loc.gov/2017054212

∞ This paper meets the requirements of ANSI/NISO Z39.48-1992
(Permanence of Paper).

First published in 2018 by Oregon State University Press
Printed in the United States of America

Oregon State University Press
121 The Valley Library
Corvallis OR 97331-4501
541-737-3166 • fax 541-737-3170
www.osupress.oregonstate.edu

Contents

Preface ix

Introduction ONE RIVER, MANY VOICES 1

Chapter 1 THE BEAUTIFUL WILLAMETTE. 19

Chapter 2 WHEN REASONABLE USE BECAME UNREASONABLE . . . 47

Chapter 3 SEWAGE AND THE CITY 83

Chapter 4 POSTWAR EFFLUENTS 117

Chapter 5 POLLUTED PARADISE. 147

Chapter 6 AT LONG LAST? 175

Chapter 7 HYDRA-HEADED POLLUTION 209

Conclusion SPEAKING FOR OURSELVES. 233

Appendix 1 Notable Willamette Valley and Lower Columbia River Pulp
and Paper Mills in the Willamette River Pollution Story. . . 239

Appendix 2 Willamette Valley Project Dams. 243

Appendix 3 North American Water Quality Entities as of 1949 245

Notes 247

Bibliographic Essay 295

Bibliography 301

Index 317

Illustrations

FIGURES

1.1   Father Dominic Waedenschwiler's 1904 cantata *Ad Willamettam* 20

1.2   Willamette watershed with major tributaries and selected cities 23

1.3   Cleveland Rockwell's sketch "Mouth of the Willamette," October 1883 24

1.4   Basalt cliffs at Willamette Falls, 1878 25

1.5   Chinookans fishing at Willamette Falls, ca. 1841 27

1.6   View from Oregon City, ca. late 1800s 29

1.7   Activity in Portland Harbor, 1899 40

2.1   Fishing for salmon at the base of Willamette Falls, 1920s 49

2.2   Downtown Portland during the Willamette River flood of June 1894 57

2.3   Derelict and dilapidated structures along Portland Harbor, January 1927 . . . 58

2.4   Construction of interceptor sewer and harbor wall, Portland Harbor, 1928 . . 59

2.5   Aerial of the Portland Harbor wall, January 1929 61

2.6   Will R. Lewis, "foe of filth," dons a gas mask, August 1936 66

2.7   Dissolved Oxygen Measurements in Portland Harbor, September 1934 67

2.8   Portland harbor pollution, January 1937 77

2.9   "Clean the Rivers!" campaign literature, fall 1938 79

2.10 Portland Mayor Joseph Carson, October 1938 80

3.1   Harold F. Wendel, Chair of the Oregon State Sanitary Authority 88

3.2   1939 map from the National Resources Committee
      Special Advisory Committee on Water Pollution 93

3.3   Untreated sewer outfall under the Burnside Bridge, Portland, 1930s 95

3.4   Citizens Sewage Disposal Committee brochure, May 1944 111

3.5   Portland's proposed sewers and treatment plant, November 1938 112

3.6   Cartoon about construction of Portland's sewage treatment plant 114

3.7   Workers excavating Portland's sewer tunnels, ca. 1950 115

4.1   December 1950 inspection of sewer line 118

4.2   Dr. David B. Charlton and L. C. Binford, November 18, 1953 121

4.3   Oregon State General Fund Biennial Appropriations,
       1939–1941 through 1953–1955. 132

4.4   North American Pulp and Paper Industry Profits, 1939–1954 133

4.5   OSSA and NCSI Funding, 1943–1949 133

4.6   Aerial view of Willamette Falls looking north, ca. 1940s 137

4.7   Barge filled with sulfite wastes, 1953 146

5.1   Willamette Valley Project Dams. 159

5.2   Columbia Slough pollution, October 1970 165

5.3   Columbia Boulevard Wastewater Treatment Plant dedication,
       October 26, 1952. 167

5.4   Columbia Boulevard Wastewater Treatment Plant, July 1954 168

5.5   Editorial cartoon on air and water pollution problems, late 1950s 171

6.1   Reductions in biochemical oxygen demand (BOD) discharges to the
       main stem Willamette River, 1957–1974 180

6.2   Lowest annual recorded Willamette River flows, 1927–1974. 183

6.3   Coming to agreement: OSSA biennial budget requests
       and legislative appropriations, 1939–1941 through 1947–1949.. 187

6.4   Pollution control on a treadmill: OSSA biennial budget requests
       and legislative appropriations, 1949–1951 through 1957–1959 189

6.5   OSSA/DEQ biennial appropriations and federal allocations for
       municipal waste treatment investments, 1939–1941 through 1969–1971 . . .190

6.6   Addressing pollution in paradise: OSSA biennial budget requests
       and legislative appropriations, 1959–1961 through 1969–1971.. 191

6.7   Changes in OSSA/DEQ budgets and state general fund
       appropriations, 1941–1943 through 1969–1971 192

6.8   Rate of change of general fund appropriations to selected state agencies
       relative to overall general fund, 1939–1940 through 1969–1971. 193

6.9   Percent change in OSSA/DEQ general fund appropriations
       per biennia compared to inflation, 1939–1941 through 1969–1971. 194

6.10   Oregon Pulp and Paper mill in Salem, 1949 .203

7.1   Lowest dissolved oxygen levels recorded in Portland harbor during
       annual June-October low-flow periods, 1950–1974. 210

TABLES

4.1   OSSA and NCSI Funding, 1943–1949 134

5.1   OSSA Data on Willamette River Water Quality, 1957 169

Preface

Spoiler alert: This book concludes that the Willamette River pollution abatement story has not ended. The book ends, but the story will continue. It will continue along the forward flow of time as events unfold and coalesce into periods that will lend themselves to additional chapters, just as the complex geological, hydrological, and climatic dynamics of the watershed itself continue to evolve.

The story will also not end because—try as I might!—I cannot possibly have uncovered all the relevant sources and pertinent details. My version of the narrative is not an immutable interpretation: a fundamental tenet of the historical discipline is that published works will be interrogated, challenged, qualified, and superseded, all in the light of advancing scholarship, new sources, and different needs.

This book began as a 30-page graduate school seminar paper in 2008, grew into a hefty 140-page master's thesis in 2009, and exists now as about 90,000 words of text. I have long been a student of human relationships to the environment and how human-wrought environmental change leads to changes in cultural values and practices. By narrowing my focus to Oregon's Willamette River watershed, I have been able to pursue this research where environmental changes are often most pronounced and widespread—within intensively built and populated landscapes, to include the effects of urbanization and industrialization upon the planet's most precious resource: water. As I got deeper into my research, I found that while the topic had been covered to some extent, it was deserving of a book-length treatment because it is an important regional story with national and international relevance.

In the mid-nineteenth century, paintings of and literature about the Willamette Valley in Oregon Country were critical in shaping a mythological framework that helped draw European American settlers to the region. This imagery established in the minds of immigrants traveling west a utopic vision of a landscape perfectly suited to farmsteading, agriculture, and community building. In much the same way as New York landscape paintings by members of the Hudson River School exemplified an ideal that was both primeval and domesticated and established the Hudson as America's river of empire, the Willamette became the iconic Pacific Northwest river. In the 1960s, when advocates in New York organized to preserve Hudson Valley landscapes and address riverine pollution, they drew upon these historic and mythological associations. In Oregon, as well, advocates readily referenced the Willamette River's central role in regional history to justify the investment of time and resources into abating pollution. In many other ways, as well, the Willamette River pollution abatement story intersects with events outside the region to provide important context. For example, a federal New Deal agency conducted the first national survey of water quality in the late 1930s, concluding that the Willamette and Lower Columbia watersheds were as polluted as heavily industrialized watersheds in the Great Lakes and Northeast regions. Thus, delving into the details of Willamette River pollution provides some depth on an important local topic with regional and national relevance.

I also chose to study this topic because David B. Charlton suggested I do so. He and I never met—he died before I attended graduate school—but in a 1975 letter he wrote that the popular version of the Willamette River clean-up story that had recently gained national attention was unduly focused on a single person—Tom McCall—to the exclusion of many others who had been working for decades on the issue. I wanted to know more about these other people, how they set about their tasks, and why they cared.

An environmental history of Willamette River water quality can be written as a tale of decline and a "fall from grace"—what academics call "declensionist" history. It could be written as a narrative of gradual progress and eventual triumph. Both of these interpretations would follow standard frameworks. I have tried to avoid these. My goal is not to frame evidence in a well-worn, mythological structure that contains the essence of its conclusion in the structure itself—either decline or progress—but to present a story that best reflects contemporary context and options. I have attempted this because both Charlton and the evidence have led me in this direction. This complex and ever-evolving story is not one that can fit into declensionist or triumphal framework and still sustain the integrity of the primary sources. One might rightly ask, however: If the story of Willamette River pollution abatement is so complex, how can one possibly hope to write a relatively succinct book on the topic?

One of the primary tasks of the historian is to organize research and analysis in a way that makes sense to a broader audience. In reviewing a series of events and stacks of materials, the historian's charge is to make conscious choices about how to arrange the evidence into a coherent narrative articulated in chapters, subsections within chapters, and so forth. One might deem this practice as arbitrary, because someone else could review these same sources, interpret them differently, and craft a strikingly different narrative. Doing this work is not arbitrary, however. The process of selecting evidence and devising a chronology provides the fundamental structure upon which the entire narrative is based. This communicates to the reader what the historian deems to be most important about the topic. To put it in another way: it is the historian's role to venture into the deep forest, examine the trees in some detail, and come back to communicate something compelling about the forest.

The reader will find the narrative centered on pivotal *political* events—events relating to the body politic, democratic governance, and public law and administration: meetings, election results, administrative decisions, governors' tenures, and legislative successes and failures. Natural systems themselves are critical agents in many historical narratives, including this one. Nonetheless, after years of research I have chosen to focus on pivotal political events not because I consider myself a policy wonk but because this is where the evidence led. A degraded natural system provided advocates with practical, economic, and moral reasons to fight pollution, but they had to express their grievances and seek solutions in the public arena and push for changes in local, state, and federal law. These activities were nested in a web of complex environmental, economic, cultural, demographic, technological, and scientific issues and opportunities. Cultural values European American settlers brought with them directly influenced their legal and social relationship to the environment; this, in turn, led directly to degraded water quality. In response, advocates struggled to reformulate the state's legal and administrative structure to clean up the river. Cleaning the river required not just regulatory changes but also infrastructure funding, technology development, and scientific advancements to quantify and address pollution.

Pollution abatement—as with so many other pressing environmental issues—has been an iterative process, with one generation's solutions contributing to the next generation's opportunities and challenges. With this realization in mind, the answer I give when asked if the river is cleaner now than before is: in some significant ways it is drastically cleaner (i.e., mats of floating feces and detritus no longer clog Portland Harbor), but in some notable ways it is more polluted (i.e., dioxins and other persistent chemicals befoul sediments in Portland Harbor and will remain a concern for centuries to come).

I want to thank my vast network of friends, family, and colleagues who, in one way or another, have made this book possible. While many have provided content or editorial input, I accept full responsibility for any remaining errors and omissions.

A historian could only write polemics or fiction if it were not for the expertise and assistance of professional archivists and librarians. These include: Geoff Wexler, Scott Daniels, and other current and former staff at the Oregon Historical Society Research Library; Brian Johnson, Mary Hansen, and their colleagues at the City of Portland Archives and Records Center; Larry Landis at the Oregon State University Archives; and staffs at the University of Oregon Archives, Environmental Protection Agency Region 10 Library, National Archives and Records Administration Pacific Alaska Region facility, Oregon State Archives, and League of Oregon Cities.

Brian White, Don Yon, and Day Marshall at the Oregon Department of Environmental Quality in Portland helped me gain access to important sources, such as the original meeting minutes of the Oregon State Sanitary Authority (which had not yet been converted into PDF files), internal historical overviews of the DEQ's work, and historic images.

I am grateful to my colleagues at the American Society for Environmental History (ASEH) and the Urban History Association who granted permission to present sections of my research at their annual conferences. Matthew Klingle and Jay Taylor, in particular, provided substantive, constructive feedback to my 2010 ASEH paper, and for this I am grateful.

I am indebted to William L. Lang for quite a number of reasons, not the least of which is the fact that he put into my head the daft idea of writing this book. As my major field adviser in graduate school at Portland State University, Bill's feedback and suggestions helped me learn how to research, think, and write as a *real*, professional historian. He also had the patience to help me work through all major iterations of this narrative, and saw it grow from a seminar paper in 2008 into what it is today.

Two other Portland State University professors helped me make the most of my graduate school experience and learn the profession: David Johnson and Carl Abbott. David, Carl, Bill, and Oregon State University Professor Emeritus William G. Robbins agreed to serve as my thesis committee; I was truly appreciative that Bill Robbins did so as he has written extensively on environmental and political topics in Oregon.

I am indebted to advisors and mentors during my undergraduate years at Western Washington University's Fairhaven College, including Kathryn Anderson, Len Helfgott, Chris Friday, Kevin Leonard, Midori Takagi, Stan Tag, and Dan First Scout Rowe.

I appreciated the constructive comments from all who have reviewed this manuscript, including Oregon State University Press's anonymous reviewers, Bill Lang, Martin V. Melosi, and my hardworking editor/historian aunt, Grace Elting Castle.

Earning a Geographic Information Systems (GIS) certificate while also working on this topic afforded me the opportunity to georeference water quality data from the 1930s and develop some of the images in this book. Portland Community College instructors Christina Friedle, Steven Jett, and Nadia Jones educated me in GIS and provided constructive feedback on the maps I created, and I thank them for it.

I am indebted to three accommodating and generous people for providing me secluded spaces from which to work on this project during key periods over the years: Sherry Palmiter (in Chelan, Washington) and Michelle Bobowick and Chuck Veneklase (in Carson, Washington).

The foundation enabling all the work I have put into this book is my family. My wife Jennifer has supported me unconditionally through this process and has allowed me the time and solitude necessary to complete it. Max and Maizy have generously tolerated the long stretches of time I have been sequestered in my office beavering away at this project and the frustration that has arisen when I was not able to meet deadlines. Luna was born a few months after I completed the first manuscript draft in 2012, and since then has been a more than worthwhile distraction during the long revision process. My love to you all!

Introduction
One River, Many Voices

Twenty-six miles from its confluence with the Columbia, Oregon's Willamette River pours over a forty-foot basalt cliff. The river roils and foams as it pools at the cliff base before continuing downstream through the city of Portland. Astride this impressive waterfall, on the west and east banks of the Willamette, two pulp and paper mills cling to the basalt, one in the town of West Linn and the other in Oregon City. The large and imposing structures exist as sprawling industrial sculptures perched above the churning river.

In viewing the natural and built landscape at this location, one is witness to more than cascading water, basalt cliffs, and aging industrial architecture. This vista affords a panoramic view of millions of years of history: Oregon's deep geologic past, thousands of years of Chinookan and Kalapuyan experience, the cultural practices of nineteenth-century European American settlers, and Oregonians' evolving cultural values of the twentieth and twenty-first centuries. It might seem an exaggeration to perceive such deep meaning at one location, but Willamette Falls is a powerful place, formed from cataclysmic geological forces, sculpted over millions of years through the river's flow and Ice Age floods, and long a center of human habitation, industry, and sustenance.

A little more than two decades after Meriwether Lewis, William Clark, and their intrepid team completed their trip to the Pacific Ocean in 1804–1805, European American newcomers to the Willamette Valley sought to take advantage of the hydraulic power potential at the falls. Hudson's Bay Company Chief Factor Dr. John McLoughlin established Oregon City in 1829 and used the falling water to power a sawmill. A short-lived paper mill replaced the sawmill from 1866 to 1867, and in 1908 the Hawley Pulp and Paper Company built the core of a mill that would operate for more than a century. Across the river in West Linn, the Crown Paper Company and Willamette Falls Pulp and

Paper Company both established pulp and paper mills in 1889 prior to combining operations under a single company. The mill shut down in 2017.

Taking advantage of hydropower at Willamette Falls, European Americans built an industry that contributed significantly to Oregon's economy for decades, provided jobs in the local metropolis and regional hinterlands, and produced a range of wood pulp–based products for consumption throughout the world. The mill facilities at Oregon City and West Linn exemplify these positive facets of the technologies and values that European Americans brought with them to the Pacific Northwest in the nineteenth century. Less apparent in the twenty-first century, but no less an integral part of the European American story in the Pacific Northwest, are the vast quantities of wastes produced by industries and concentrated urban populations. What occurred in Oregon beginning in the mid-1800s occurred everywhere Europeans and European Americans introduced industrialization and urbanization: fouled waters. Until well into the second half of the twentieth century, the mills at Willamette Falls were prominent artifacts representing extreme water pollution as much as they represented jobs, communities, economic growth, and plentiful consumer goods.

The panoramic view of all that now exists at Willamette Falls encompasses the story of water pollution not only in Oregon and the Pacific Northwest but across the globe. It includes the geological forces, climatic changes, and hydrological systems that define the natural environment; the ways in which different societies made use of the natural environment; and the many ways technological systems exacerbate or help alleviate the effects of pollution. Other aspects of the story of water pollution and its abatement are less apparent at Willamette Falls, but as important. These include the interplay of scientific understanding, cultural values, and public policy, and how these change over time. With this perspective, one can understand Willamette Falls as both a literal and figurative place where more than a century of human activities degrading the natural environment had, by 2016, shifted decidedly to something else.

This book will investigate three general areas of inquiry. First, a critical mass of Oregonians shifted their values in relation to water pollution drastically over 150 years. In the mid-nineteenth century they viewed pollution as a necessary component of progress and modernization. By the mid-twentieth century, they agreed to fund expensive sewage treatment infrastructure and supported policies that put pressure on industries to abate their pollution. By the late twentieth century, they supported even more expensive, proactive, and wide-ranging technological and policy measures to address even more complex and confounding types of water pollution. Why?

Secondly, this book looks into how the human intellect—cultural values, economic systems, scientific knowledge, and political ideologies—manifested in the physical realities of the natural world, first to pollute the environment and then to address this pollution. In what condition did European Americans find the Willamette Valley environment when they began settling there and how did they consciously and unconsciously degrade it? When they decided to address environmental damage, how did they do so?

Finally, this book uses a place-based example to shed light on an international pattern: pollution abatement and "clean streams" efforts were (and are) a shifting target. With changes over the decades in technology, scientific understanding, and cultural values, water pollution abatement advocates in Oregon and elsewhere learned firsthand that any given "solution" was, at best, context-specific and provisional. That water pollution abatement was complex and required significant effort did not deter advocates in Oregon in the 1920s, and it does not deter them now, in the second decade of the twenty-first century.

SPEAKING FOR THE RIVER

Oregon's Willamette Valley is located in the western half of the state between the Cascade Mountains on the east and Coast Range on the west. The river draining it runs south to north and intersects with the Columbia River approximately one hundred miles from where the Columbia meets the Pacific Ocean. The valley has been the most heavily populated, urbanized, and industrialized region within the present-day State of Oregon since European American settlement began in the mid-nineteenth century. For most of this time, Portland has been the most heavily populated, urbanized, and industrialized city in the valley. In draining the valley, the Willamette River has carried wastes from farms, towns, mills, factories, and other sources through Portland and onward to the Columbia. Not long after European American settlement commenced the Willamette River became the most polluted river in the state.

European American settlers began migrating to Oregon Country en masse in the 1840s. Most settlement concentrated in the lush and verdant Willamette River watershed where Chinooks and Kalapuyans had already been living for thousands of years. As with many other large-scale human migrations, a combination of "push" and "pull" factors influenced settlement along the west coast of North America. Among the environmental factors pulling European Americans to the Willamette watershed was a landscape of open grasslands interspersed with oaks; a climate favorable to agriculture; and the meandering river and its tributaries offering the prospect of transportation, irrigation, and power. Complex geological and climatic processes over tens of millions of years created this landscape. Active management of the oak savannah grasslands over thousands of years through controlled burning helped support thriving game populations and plant life that sustained Kalapuyans for many generations. Inadvertently, this activity also helped create a landscape that appealed to mid-nineteenth century migrants seeking land for agriculture and community building. Boosters, poets, and painters romanticized the Willamette Valley and the river running through it, in much the same way as the Hudson River School of painters portrayed New York's Hudson River Valley. Drawn by environmental attributes that nature had forged and Indians had cultivated, the settlers, town builders, industrialists, and other newcomers nevertheless immediately set about the task of remaking it in their own image.

It did not take long for Willamette River water quality to reflect the effects of European American settlement. Oregon poet Samuel L. Simpson wrote in the late 1860s of a "Beautiful Willamette" with "crystal depths . . . Wreathing sunshine upon the morrow," but by the 1880s Oregonians characterized it as a sewer.[1] Untreated sewage, pulp and paper mill discharges, and effluent from flax retting, food canning, and animal processing flowed unabated into the Willamette and its tributaries. Industrial society expanded and evolved in the late nineteenth and early twentieth centuries to provide Americans with more consumer products, higher wages, and an increased standard of living. The outcome of this growth was more—and more complex—wastes. The dominant cultural ethic in the United States well into the 1930s held that the water and air pollution were unavoidable consequences of progress and prosperity. The extent to which businesses and municipalities had a responsibility to decrease wastes that harmed fisheries or threatened public health was determined by how such wastes might profitably be turned into useful and saleable products—for example, adhesive or artificial vanilla flavoring, in the case of pulp and paper wastes, or fertilizer, in the case of treated sewage. In the absence of such remunerative options, this ethic concluded, the benefits of clean streams were not worth the expense.

Lest it be implied otherwise, the history of the United States clearly disproves the existence of a monolithic set of cultural values remaining unchanged over time. Challenges to dominant values in any given era, however, have often taken decades to bear fruit. The framers of the Declaration of Independence and Constitution did not deal directly with the contradictions of slavery, but for decades prior to the Civil War a growing minority of highly committed Americans worked for abolition. Similarly, it took decades after the 1848 women's rights convention in Seneca Falls, New York, for individual states to begin to allow women to vote, and it was not until 1920 that the Nineteenth Amendment to the US Constitution granted this right nationally. Similarly, the dominant cultural assumption that degraded water quality was not only an unavoidable consequence of modern life, but an inalienable right, took decades of hard work to overcome.

Until President Franklin D. Roosevelt's New Deal of the 1930s, the vast majority of cities in the United States did not have facilities to treat sewage. The majority of industrial plants had neither waste treatment systems nor any local, state, or federal requirement compelling them to build such systems. State fish and game officials and sporting groups such as the Izaak Walton League of America had become increasingly active in the first decades of the twentieth century, arguing that alternatives must be found to the use of rivers, lakes, estuaries, and other bodies of water as "waste sinks" receiving untreated effluent discharges.[2] New Deal programs offered an unprecedented amount of federal grants and subsidized loans for cities to build sewage treatment systems, and they also produced the first comprehensive national survey of water pollution.

The emergency of World War II drastically increased industrialization and urban densification in many areas of the United States, while at the same time causing most federal and state leaders to put water quality issues on the back burner. Momentum built by the Works Progress Administration, Public Works Administration, National Resources Planning Board, and other New Deal programs stalled. Pollution abatement advocates and—where applicable—state water quality agencies resumed their push as soon as they could after the war. These efforts, coupled with the 1948 and 1956 Federal Water Pollution Control Acts, increased pressure upon polluting cities and industries while offering federal funds to help build sewage treatment systems. The result was that by the late 1950s control of water pollution from point sources—such as municipal sewer outfalls and industrial drain pipes routed directly into waterways—was almost keeping pace with the rate of

industrialization and urban growth. As the *Oregonian* newspaper editorialized in 1957, water pollution control was on a treadmill.

Achieving such a stasis—in which water quality was not necessarily
improving over time but it was not getting worse—was insufficient for advocates who had been working since the 1920s and 1930s. It also was insufficient
for a growing number of Americans in the post–World War II era who found
themselves with increased disposable income and leisure time and wanted to
enjoy nearby rivers and lakes for fishing, boating, and swimming. Mainstream
American cultural values that had largely tolerated degraded water quality in
the first half of the twentieth century began to change. Advances in science
enabled more accurate and comprehensive water quality measurement and
quantification, which, in turn, supported increasingly more effective technologies and policy approaches.

In the 1960s and early 1970s this momentum led to strengthened state
and federal laws to address not only water pollution but air and land pollution as well. In cases where states had begun to improve surface water quality in the 1960s, they had done so primarily against municipal and industrial
point source polluters. Actions taken under federal laws such as the National
Environmental Policy Act (NEPA) of 1970 and Clean Water Act of 1972 went
a long way toward addressing point source pollution. Once the "low hanging
fruit" of this relatively simple pollution source had been addressed to a significant degree, what remained was far more complex. Addressing pesticide and
herbicide runoff from agriculture and forestry operations, water tainted with
radioactive runoff, petroleum, asbestos, and heavy metals washed from paved
surfaces required a coordinated federal and state approach. Research and
experience showed that this kind of pollution was much harder to address
and potentially much more harmful to human life and the environment.

The pattern outlined above applies generally to the history in the
Willamette Valley of Oregon and to many other regions of the country. Not
surprisingly, localized attempts to address degraded water quality occurred
earliest in the most intensively populated and industrialized areas. Two of the
nation's most populous cities in the second half of the nineteenth century—
Brooklyn and Boston—built the nation's pioneering first-generation sewage treatment systems; Massachusetts established its Lawrence Experiment
Station in the 1880s to increase scientific knowledge of public health threats
from polluted water, and Pennsylvania established a Sanitary Water Board in
the early 1920s. An analysis of the Oregon case uncovers insights about how

and why it followed or differed from the broader national pattern, what this tells us about Oregonians' contributions to an issue still of pressing concern, and how one might extract lessons applicable to present or future efforts to address important environmental issues. A primary lesson is the complexity of the water pollution issue: water quality becomes degraded through the addition of diverse pollutants requiring increasingly more refined technologies to detect and advanced scientific methods to quantify; this empirical evidence helps inform the creation of more comprehensive and proactive regulatory responses; and as one era's solutions alleviate certain conditions, they also tend to result in problems and challenges future generations must confront.

A RIVER RESTORED. A RIVER RESTORED?

Degraded water quality has been a central characteristic of the Willamette River since European Americans arrived in the valley in the mid-nineteenth century. Over the years, Oregonians have called the Willamette an "open sewer," described stalled cleanup efforts, and claimed that the success or failure would shape the region's future. Empirical data on the types of pollution and its effects on natural systems, however, was lacking until the 1920s and 1930s. The first comprehensive report on water quality throughout the United States was a 1939 survey from the New Deal National Resources Committee. This report found the Willamette and lower Columbia rivers as polluted as watersheds in the heavily industrialized Great Lakes and Northeast regions.

By the early 1970s, things had changed. Journalist Ethel Starbird's June 1972 *National Geographic* cover story touted the progress Oregonians—embodied by then-governor Thomas L. McCall—had made in cleaning up the Willamette River. This progress had become a nationally recognized success story, and Congress had included some elements of Oregon's administrative framework into the federal Clean Water Act. Reflecting real achievements in Oregon and coming in the midst of growing public intolerance for environmental degradation, Starbird titled her article "A River Restored" to position Oregon as an exemplar for the nation.

About twenty-five years later, Neil Mullane, water quality division administrator with the Oregon Department of Environmental Quality, wrote a chapter on the Willamette River titled—with a nod to Starbird—"A River Restored?" Mullane recast Starbird's declarative statement into a question because, as he wrote, "restoration is constant, not static," for "time brings new challenges and information on potential threats." Restoring water quality, therefore, "must mean constant vigilance."[3] The shift from a statement into a question reflected fundamental changes over the intervening years. Water pollution control that had seemed to be at hand in the early 1970s was, by the late 1990s, understood to be a much more complex, ongoing process.

Speaking for the River identifies some of the primary topics, events, people, and organizations central to Oregon's water pollution control narrative through the twentieth century. It helps explain the transition from Starbird's celebratory optimism of the early 1970s and Mullane's qualified observations of the 1990s. A question arises about whether or not the river is "clean" now, in light of the recently resolved sewer overflow problems in Portland and ongoing work to address pollution within the Portland Harbor Superfund Site. The historically informed response is, "*Yes, but…*" *Yes*, the river is cleaner now than at any time since European Americans first came to the valley in large numbers, *but* only measured in terms of point source pollutants. Degraded water quality from untreated municipal sewage and industrial waste outflows have largely been resolved.

More difficult and potentially harmful kinds of pollutants still exist within river sediments. These include toxic substances that become concentrated within the tissues of living organisms—such as fish and the people who catch and eat them—through a process called *bioaccumulation*. Dioxins, heavy metals, and other such pollutants are not resolved simply by routing a sewer line to the treatment plant or directing pulp and paper mill effluent to aeration lagoons.

A historically informed look at the Willamette River water quality story also helps provide a more accurate and functional model for actions in the present. Reviewing newspaper accounts, government records, specialists' reports, meeting minutes, correspondence, and other sources brings much-needed specificity to a narrative that is all too often overgeneralized. This helps us understand the actions and reactions of real people who faced real and complex circumstances; through this understanding, we may not only gain insight into our current array of challenges but discern options that we might pursue today to help bring about the future we desire for our descendants.

Humans have the tendency to simplify—and, often, mythologize—historical accounts. Historian Coll Thrush provides one example of this propensity in his book *Native Seattle*. Thrush explains how the earliest written narrative describing the founding of Seattle has since served as its central creation story. This story begins with the arrival of white settlers in the midst of

Indian lands in November 1851, and, after a short period of coexistence, the narrative writes Indians out of Seattle's urban history. Like the "Mayflowers, Lexingtons, and Fort Sumters" of the broader narrative of American history, these "discrete moments" serve as founding myths, "chosen out of the complexity of the past and designated as the place where one thing is said to end and another to begin."[4]

The story of a visibly cleaner Willamette River also has a simplified point of beginning, a founding myth, encapsulated in the actions of the charismatic and much-beloved journalist-turned-governor Tom McCall. Ethel Starbird presented this version in her 1972 article:

> Fishermen and conservationists were the first to call for reforms . . .
> while polluters kept insisting the corrective costs were prohibitive.
> But the cleanup crusaders kept plugging away. Tom McCall, who was
> then a popular television newscaster, was one of them. In 1962, he
> made a real shocker of a documentary film that showed how rotten
> the river had become . . . .

She continued: "Protectionist sentiment snowballed" with positive reception of the documentary, and in 1966 Oregonians voted McCall in to his first of two terms as governor.[5]

This version of the story is compelling in its simplicity. However, as Thrush finds in the mythologized narration of Seattle's founding, "beginnings and endings are rarely clear in history, and the events that we call history were rarely as deliberate or discrete as we imagine them to be." Thrush's conclusion about the role of myth in Seattle's Indian history applies equally well to the Willamette River pollution abatement story: the McCall-centric version "obscures more about actual historical events than it reveals."[6] Chemist David B. Charlton recognized this in 1975, providing the vantage point of someone actively involved in Willamette River cleanup since the 1920s. Reflecting on this firsthand experience, he posed important questions in response to "the much reported and commented clean-up of the Willamette River"—or, "what might be called the McCall version that appeared in the *National Geographic*." He asked:

> Who were really concerned and took action leading to the first
> investigations of the problem and then to take corrective action?
> Was it the State Board of Health; the Game and Fish Commissions?

Did organized sportsmen, such as the Multnomah Anglers and
Hunters Club, which was founded before the turn of the century,
play an important role in initiating action? Just how active were the
Waltonians[7] of the Portland Chapter and in other chapters . . . ?[8]

Charlton's questions were rhetorical, because he knew the answers. As
an advocate for decades, he would have told Starbird a different version of
the story. It would have encompassed activities that occurred before McCall
became involved, such as the persistent work of citizen groups and state officials, water quality research from university professors and students, and support from New Deal–era and other federal programs. It would have included
the fact that when voters approved the November 1938 citizen's initiative
creating a state sanitary authority, Oregon became the first and only state in
the nation to centralize water quality in a single body through the process of
direct democracy. It would have conveyed the frustrations involved in getting
Portland city government and industry officials to change their waste disposal
practices. Uncovering answers to Charlton's questions enables us to understand our world and our present environmental concerns in a more thorough
and accurate way.

THE CONTENTIOUS PROCESS OF ADDRESSING WILLAMETTE RIVER POLLUTION

Conflict defines the central dynamic of the Willamette River pollution abatement story. Simply put, pollution in the Willamette would not have been
abated to the extent that it has been if not for the active and persistent advocacy on the part of citizen groups and city, state, and federal government bodies. These individuals and groups were in a constant struggle with polluters.
The two primary polluters in the Willamette watershed until the early 1970s
were the City of Portland and the pulp and paper industry. These parties,
respectively, discharged sewage and papermaking wastes in vast quantities
that soured the river and choked aquatic life. People had to push Portland city
officials and the pulp and paper companies for *decades* before they began to
see some results—and then they had to keep pushing.

While the three groups central to this struggle are identifiable, they
did not represent monolithic entities. Many abatement advocates saw their
work in both practical and symbolic terms. A cleaner Willamette would pose
less of a public health risk and would support highly desirable fish such as

salmon and trout, and both of these outcomes could help sustain economic opportunities in tourism and recreation. From a symbolic perspective, some advocates argued that a polluted Willamette was a moral outrage because it fouled one of Oregon's natural gems. Many also asserted that Portland, as the largest city in the watershed, was obligated to lead by example. Cleaning the river was thus an important way for some Oregonians to reestablish a link with the cleaner river of the past while securing economic and recreational opportunities into the future. Almost without exception, however, abatement advocates through the 1960s did not press for the removal of *all* pollution from the river, nor for the curtailment of urban growth or industrial development. They were not pushing to restore habitat or reestablish the pre-contact braided river system. In Portland Harbor, advocates were not attempting to limit transportation and shipping in an important West Coast port. Advocates instead espoused a traditional conservationist wise-use approach to the river, and in doing so they asserted strongly that using the river as a waste sink to the exclusion of all other present and potential future uses was not wise. They were happy that the Willamette was a "working river," but did not interpret this to mean it had to be a filthy river.

Abatement advocates with these diverse views often came together in spite of their differences to confront Portland city officials and pulp and paper company representatives. City leaders also were not of a unified mind when it came to water pollution. City Engineer Olaf Laurgaard, for example, was the driving force behind Portland's west side harbor wall and interceptor sewer project in the late 1920s, but he opposed the development of a comprehensive citywide sewer interceptor and treatment system. Commissioner William Bowes helped write the long-awaited sewer funding measure that voters approved in November 1938, but then resisted implementing the law for nearly six years. On the other hand, city commissioner and future mayor Dorothy McCullough Lee and two other commissioners took up the clean stream movement in early 1944 and were pivotal in the passage of a costly sewer-funding scheme that led to the construction of the city's sewage treatment plant.

The pulp and paper industry, as well, was not monolithic. One generalization largely holds true: because private sector businesses are motivated by profit rather than public service and are not required to make their records and correspondence widely available to researchers, direct evidence of their internal processes and deliberations is often not available. One must try to discern industry's point of view based on news accounts, minutes from government entities such as the Oregon State Sanitary Authority (OSSA), and some limited archived materials. Even within this constrained set of sources, however, there is sufficient evidence to conclude that Willamette Valley mills did not always act as one voice when it came to their willingness to implement abatement measures—along a spectrum, some mills were more or less cooperative than others. The North American pulp and paper industry also funded substantive and scientifically sound water pollution control research beginning in the 1940s that contributed to understanding the complex Willamette River system.

This book approaches the topic predominately from the perspective of the abatement process, and seeks to specify and evaluate barriers in the way of a cleaner river. Some barriers might be deemed active resistance to the clean stream cause, but many other hindrances reflected the complexities of water pollution. Many instances suggest that arguments against pollution abatement from Portland city officials and industry representatives were not without merit. In the case of municipal sewage treatment infrastructure, some city officials argued the costs were prohibitively high, or that the city was not legally empowered to incur such debt. These were very real issues of economics and law that had to be resolved. Similarly, the pulp and paper industry accurately asserted that the science and technology involved in treating or reusing some of their wastes was extremely complex and varied from mill to mill (often day to day at each mill). Nevertheless, if abatement advocates had not continued to press Portland city leaders and industry officials to change, available evidence suggests these polluters would not have done so.

The arena of this conflict was not the open, tossing seas of naval battles nor the broad, sweeping plains of cavalry charges, but the much more mundane venues of state agency hearing rooms, water quality laboratories, engineer workspaces, the soap box, and the ballot box. What was at stake, however, was something more tangible and lasting than campaign medals and romantic odes to heroes lost: it was the health of one of Oregon's most resonant symbols and historically important economic corridors in the midst of the state's most heavily populated, urbanized, and industrialized region. Abatement advocates saw their work in symbolic *and* practical terms, and both of these motivations sustained their efforts against severe pollution and decades of indifference and intransigence. These conflicts echo environmental disputes in the early twenty-first century, centering on science, law, and economics. Other topics might make for more compelling headlines or raise

passions more predictably, but in the matter of Willamette River water pollution, there was no escaping the fundamental importance of arguments played
out in the realm of politics and policy. Environmental degradation, and people's reactions to it, heavily influenced political discussions. The evolving state
of science and technology put clear bounds around what was knowable and
implementable, and economic cycles heavily influenced decisions about how
to fund and administer abatement solutions. This book identifies the occasions during which the environment, citizen advocacy, science, technology,
and the economy were critical in determining if, and how, pollution abatement measures were defined and implemented.

THE RIVER AS CENTRAL CHARACTER

In the midst of contentious debates centered on science, law, and economics,
it is essential not to forget that the subject of this story—the reason people
cared enough to advocate for change—is the river itself. The Willamette River
watershed is a dynamic, evolving system that drains approximately 11,400
square miles of variegated terrain. For millennia, the Willamette's main channel was highly braided and meandering. Beginning in the second half of the
nineteenth century, European American settlers, with help from the US Army
Corps of Engineers, removed obstructions, reinforced banks, constructed
dams, and dredged the channel so that the river today has more in common
with a canal than Samuel Simpson's romanticized "Beautiful Willamette."
They made these changes to connect to broader national and international
commodities markets and establish urban centers that would support and
attract commerce, industry, and the arts. Central to these efforts was the
modification of the Willamette River to serve more efficiently for transportation, waste disposal, and irrigation, and to minimize costly flood damage.
European Americans rationalized and regularized the river so that it could
"work" in the way that they preferred.

In some ways, the channelized, dammed, and dredged Willamette River
system can be understood as a component of the same kind of "organic
machine" that is the broader Columbia River drainage. As illustrative as
this metaphor can be, historian Richard White, in coining this phrase, also
reminds us that it is important not to fixate on metaphors and interpretive
models to the detriment of natural systems themselves. European Americans
wrought changes within the watershed that resulted in "reductions of the
natural world to property, the reduction of action to discourse, of life to the

market" in attempts to modify "the great changing and multifarious planet to a stable and harmonious Nature."[9] However, even in the early twenty-first century, it is important to recognize that whereas the Willamette River of 2018 is no longer "natural" in the sense of not being highly modified by humans, it is still, in spite of this, *organic.*

Richard White's organic machine metaphor is one example of a defining characteristic of the sub-discipline of environmental history: explicitly including the environment itself within the historical narrative. Doing so recognizes that "nature" is, to quote White again, "at once a cultural construct and a set of actual things outside of us and not fully contained by our constructions."[10] Historian Matthew Klingle succinctly identifies this dynamic in his environmental history of Seattle when he identifies nature as an "agent"—meaning that nature is an active contributor to historical events. "Nature is not an agent in the way historians traditionally define the term," as Klingle observes, "nor is it a moral force, although some humans grant it such standing through religion or philosophy, but it is a physical agent. Nature is an integral part of the messy planet on which human action unfolds. It may not be an actor in the strictest sense, yet its actions force people to make moral choices—to dam or let flow running rivers, to protect or exterminate imperiled creatures, to clean or befoul poor neighborhoods. Nature, sometimes masked and disguised, constantly intrudes upon their actions, often because humans have underestimated their environments."[11]

Just as the Willamette River can still overrun its banks in spite of more than a century of human actions to prevent this, so too will the complex dynamic between natural systems and human artifice not be neatly contained within any simplified metaphor or model.

THE WILLAMETTE RIVER POLLUTION ABATEMENT STORY

Willamette River pollution abatement narratives already exist.[12] Many of these follow the narrative structure Ethel Starbird laid out in her 1972 *National Geographic* cover story—the "so-called Tom McCall version." To paraphrase this narrative: "Before McCall produced *Pollution in Paradise* and then served as Governor, nothing much happened to abate water pollution in Oregon."[13] Another key reference is George W. Gleeson's *Return of a River.* Gleeson understood his report was not a complex, nuanced historical analysis. Rather, in his own words, he offered a "semi-historical description" in which it was "impossible to present anything but a fraction of the data" or "to give proper credit to persons, agencies, and organizations whose collective efforts effected river improvement."[14] Because Gleeson documented some of the most important events and studies in the Willamette River water pollution story from the mid-1920s, and fully cited his work, he was successful in producing a valuable resource, of much use to all subsequent researchers. Both Starbird's and Gleeson's works appeared in 1972, and the majority of subsequent published narratives rely on these sources either explicitly or implicitly. If Starbird or Gleason are not cited in these other publications, there are telltale narrative clues. For instance, one prominent element from Gleeson's work many authors use is originally from a 1927 Portland City Club study calling the river "'ugly and filthy'" and "'intolerable'" and of such a state that "workmen refused to work on riverside construction."[15] The most readily discernable influence from Starbird's work is perpetuation of the McCall-centric narrative itself.

This book builds upon these and many other works to show how abatement advocates made use of science, technology, public administration, and economics to address Willamette River pollution. Advocates repeatedly found the broader voting public was not amenable to changing the status quo until pushed to do so by environmental conditions that impeded accustomed activities. This dynamic has been repeated in other areas involving a host of environmental concerns: *Both* significant environmental changes *and* a core set of advocates were often required to force the broader American society to adjust its accustomed ways of living. Throughout this process, the changing environment aided the abatement advocates' case. Advocates found plenty of filth, slime, dead fish, and other *qualitative* sensory evidence, but they needed to translate this into *quantifiable* data that could form the basis for technological, legal, and administrative solutions.

Considering how complex water pollution is now, during the second decade of the twenty-first century, it may strike some observers that the concerns of people fifty or more years ago are quaint, and that perhaps there are no correlations to draw between their times and ours. How does abating point source pollution from raw sewage outfalls, for instance, compare with toxic dioxins and furans saturating river sediments? In the former case, the answer could be as simple as shutting a valve to stop the flow; in the latter case, questions arise about whether to cap the sediment or to dredge it, and, if dredged, what are the most effective dredging technologies? How to time this dredging not to interfere with salmon spawning season? What to do with the sediments once dredged—is it best to process the dredged material onsite or offsite, or transport it to a landfill? In considering the kinds of contaminants that are to go in a landfill, the site may need to be built far from the influence of geologic faults and lined and covered with specially designed materials. Once contaminants at a given site have been addressed, how does one ensure that upriver sources do not recontaminate it?

The fact that the river is drastically freer from gross, visible pollution from municipal sewage and pulp and paper wastes in 2018 than it was in 1968 or 1928 suggests the first generation of abatement advocates "won" their decades-long struggle. If water pollution were such a simple matter as to be resolved with the introduction of sewage treatment systems and ways to treat, reuse, and otherwise decrease mill effluents, then this story could be relegated to the mists of time. Unfortunately, however, the pollution sources that advocates appeared to have under control by the 1970s were those most amenable to alleviation. As anyone knowledgeable with the Willamette River today can tell you, the river is far from "clean." Superfund sites, toxic runoff from city streets and farmers' fields, airborne particulates, pharmaceuticals flushed down toilets, and other realities of our modern consumer-oriented capitalist society continue to pollute the Willamette and many of the world's other rivers, lakes, and estuaries in ways that are much more complex and long lasting. Humans are only in the early stages of developing abatement and remediation solutions and changing their values to address such pollution.

Nevertheless, the people who dedicated some portion of their personal and professional time to clean up the Willamette provide valuable models of ways to address contemporary water pollution and other environmental issues. Their achievements show the value of organizing collectively to achieve important social goals. They show the value in persisting over time in the face of intransigence from civic and business leaders. They show how essential it is to approach a complex environmental topic with a diverse coalition of people offering expertise in policy, public health, chemistry, civil engineering, wildlife biology, public relations, and an array of other skills, and with a range of moral, aesthetic, economic, political, and professional motivations. They show the importance of reaching out to the broader public at key moments: not all members of society have the time to be intimately engaged in a given topic, but for truly important issues a critical mass is needed to provide support at key moments. They also show the importance of going beyond impassioned critiques and emotional jeremiads to offer scientific, technological, and policy solutions to complex environmental issues.

Before abatement advocates could provide lessons to future generations, however, first the river had to become fouled.

Chapter 1
The Beautiful Willamette

> From the Cascades' frozen gorges,
> Leaping like a child at play,
> Winding, widening through the valley,
> Bright Willamette glides away;
> Onward ever,
> Lovely river,
> Softly calling to the sea,
> Time, that scars us,
> Maims and mars us,
> Leaves no track or trench on thee.[1]

With these lines poet Samuel L. Simpson opened his 1868 paean, "The Beautiful Willamette." His verse exemplifies the highly romanticized vision many European American newcomers projected onto Oregon's Willamette Valley and the river draining it. Simpson was uncommon among his contemporaries in that he was able to earn recognition and a livelihood from his writings, but he was not unique among fellow immigrants immersed in social values, mythological foundations, and economic and political systems that fostered such a vision. Within the complex web of cultural values and practices introduced to the Willamette Valley in the early and mid-nineteenth century coexisted elements that both romanticized and exploited nature—that both sought refuge in pristine, untrammeled landscapes and that voraciously altered landscapes to extract and commodify resources. Within Simpson's "The Beautiful Willamette" is both the preservation and destruction of the river.

The Simpson family arrived in the Willamette Valley amidst a wave of immigrants that decisively shifted the demographic balance away from the

AD WILLAMETTAM
OR
BEAUTIFUL WILLAMETTE
Poem by Sam.L.Simpson
CANTATA
FOR
Mixed Choir, Soli and Orchestra
BY
F. DOMINIC, O.S.B.

RESPECTFULLY DEDICATED TO THE
WILLAMETTE VALLEY CHORAL UNION

Fig. 1.1. Cover to Father Dominic Waedenschwiler's 1904 cantata *Ad Willamettam or Beautiful Willamette*, based on Samuel Simpson's poem (University of Oregon Libraries Special Collections and University Archives, SCA OrColl 782.8 D713).

region's Indian population. Young Samuel was born in Missouri in 1845 and his parents moved the family to the Willamette Valley the following year— just two years before Oregon Country became Oregon Territory. In 1865 he graduated from Willamette University and three years later married Julia Humphrey. After a brief period in the field of law, he began his career with the *Corvallis Gazette* in 1870 and would be a journalist and poet until his death in 1899. Many European American settlers to Oregon lauded his work, particularly his poem "The Beautiful Willamette," as they built cities, harvested the forest, worked in industries, and engaged in other activities that drastically altered the landscape and river.[2]

Since the 1830s, when Europeans and European Americans first began establishing religious missions and permanent settlements in what is now the state of Oregon, the Willamette Valley has been the center of the state's population and industry. The landscape beckoned these newcomers because it reflected their idealized vision of homesteading, agriculture, and resource extraction. Set between the Coast Range to the west and the Cascade Mountains to the east, the valley these settlers laid eyes upon seemed to be the earthly manifestation of Eden itself. While traveling through the Willamette Valley under the employ of the Hudson's Bay Company in 1834, for example,

John Work observed rolling grasslands with no visible stones and "scarcely a shrub to interrupt the process of the plow which might be employed in many places with little more difficulty than in a stubble field." Seven years later George Emmons of the US Exploring Expedition described a scene in this "great Walamat Valley": an extensive prairie "from 60 to [1]00 miles either way" with "prairie to the south as far as the view extends" and "streams being easily traced by a border of trees that grew up on either bank."[3]

Mid-nineteenth century explorers, boosters, and immigrants were unaware of the complex geologic and cultural history that had molded this landscape into forms that appealed to their aesthetic, economic, and spiritual sensibilities. In part this was because the tools of geology, hydrology, and related sciences necessary to interpret the landscape were only just beginning to explain these processes. Another barrier to this knowledge was that these newcomers understood their world through the filters of mores, beliefs, and expectations that made it nearly impossible to perceive value in Indian lifeways or in an ecosystem that served as anything but an exploitable resource.

That the Willamette Valley landscape and climate conformed to preferences settlers brought with them is an essential part of the story of polluting and, decades later, trying to clean up the Willamette River. Their views influenced how they perceived the world, in turn setting the parameters for how they interacted with the environment. The region had existed for millions of years before the introduction of cultural, moral, and technological systems that resulted in severe water pollution within just a few decades. It would take nearly a century after the first waves of mass immigration to the Willamette Valley for cultural values to begin to change and for later generations to create new laws, administrative structures, and technologies that reflected a view of the Willamette River as more than a waste repository.

THE WILLAMETTE VALLEY PRIOR TO EUROPEAN AMERICAN SETTLEMENT: LANDSCAPE, CLIMATE, AND PEOPLE

The first white person to document the Willamette River was Royal Navy Lieutenant William R. Broughton of Captain George Vancouver's 1792 expedition. Broughton's party charted a course up the Columbia, noting the confluence of the Willamette River before continuing farther eastward to the entrance of the Columbia River Gorge.[4] In early April 1806 members of Meriwether Lewis and William Clark's Corps of Discovery were the first people of European descent to navigate any part of the Willamette. As the

Corps was making its return trip east after spending the winter at Fort Clatsop near the Columbia's mouth, Clark took a party a few miles south to present-day Swan Island. They called this river the *Multnomah*, the Chinookan name for a village on Sauvie Island (*måånumaå*, "those toward the water"). After speaking with the area's inhabitants, Clark's party interpreted the river's source to be "near the head waters of some of the rivers, which fall into the gulph of California." They traveled no farther up its course because they were headed east after a long and grueling winter.[5]

Had the Corps the time and inclination to explore the watershed, they would have traveled about 187 miles farther south before the river forked. One of these branches (the Middle Fork) continued south, while the other (the Coast Fork) headed southwest. The explorers would have risen just four hundred feet in elevation to this point along a meandering, braided course with many channels and obstacles. Beyond the fork, the river's branches had higher-velocity flows as runoff tumbled out of the surrounding mountains. Along their southward journey, Clark and his party would have noted ten major tributaries, later to be named the Clackamas, Tualatin, Molalla, Yamhill, Luckiamute, Santiam, Calapooia, Marys, Long Tom, and McKenzie rivers. Later measurements established that the Willamette carried the twelfth-largest water volume in the United States.[6]

The valley containing this watershed is roughly rectangular, approximately 75 miles wide and 150 miles long and encompassing about 12,045 square miles. The Coast Range to the west rises up to an average of 1,500 feet and is significant enough to block some of the coldest and wettest weather fronts moving in from the Pacific Ocean. The much more imposing Cascade Mountains to the east contain five peaks of more than 10,000 feet in elevation and provide a barrier to moisture moving farther east. These peaks are conducive to significant snowpack and year-round glaciation that provide melt water to the valley. Because of this topographical layout, the Willamette Valley is not as cold nor as wet as the Oregon coast and not nearly as dry as the Great Basin extending east of the Cascades to the Rocky Mountains.

This landscape evolved over the course of 60 million years as the dynamics of plate tectonics gradually elevated the formerly submerged transition zone between thin oceanic crust and thicker continental crust (the "continental margin"). Two much more recent events also helped form the watershed. First, Columbia River flood basalts significantly altered the landscape in the entire Pacific Northwest from 6 to 18 million years ago, including creating the

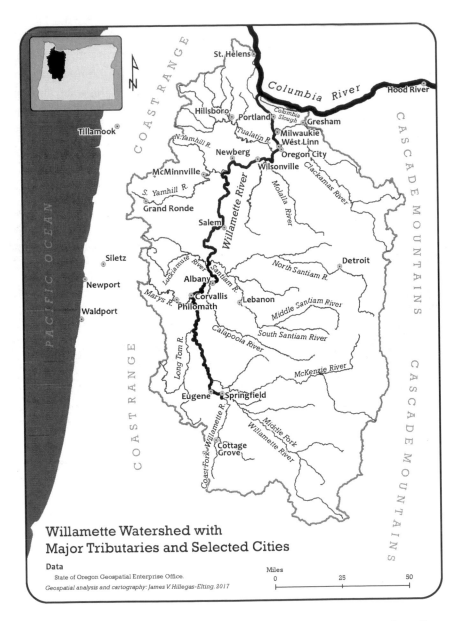

Fig 1.2. Willamette watershed with major tributaries and selected cities (James V. Hillegas-Elting, 2017).

forty-foot tall basalt cliffs of Willamette Falls. The falls separate the river into two distinct sections: the lowest twenty-six miles from the river's confluence with the Columbia is within reach of Pacific Ocean tidal fluctuations, while the river above the falls is not.[7] A second significant period of geologic activity

Fig. 1.3. Cleveland Rockwell's sketch "Mouth of the Willamette" for *West Shore* magazine, Oct. 1883 (Oregon Historical Society Research Library, bb014845).

occurred between fifteen thousand and thirteen thousand years ago. At least forty times during this period, flood waters from an extensive lake system in western Montana—Glacial Lake Missoula—poured over eastern Washington and down the Columbia River drainage to spread repeatedly into the Willamette Valley as far south as present-day Eugene.[8] These Missoula Floods resulted in both erosion and deposition throughout the valley. They spread silt and sediment from eastern and central Washington to help create highly fertile soils supporting the camas and wapato Native Americans valued for food and providing farmland European American would find exceptionally productive.[9]

By about four thousand years ago, the watershed's topography and hydrology that existed at the time of contact between Europeans and Indians was in place. The Willamette River moved slowly over relatively flat, poorly drained land, resulting in a meandering, braided, river system. The river's course changed often, with side channels diverging from the main channel to intersect downriver. Freshets inundated floodplains, scoured and deposited silt, sand, gravel, and parts of trees; these events often redirected the river to create or abandon channels. Oxbow lakes marked areas where the river had previously flowed. Sand and gravel deposits sprouted trees and shrubs and developed into small islands, until floodwaters moved these deposits farther downstream. The river made its way north to plummet over Willamette Falls. Below the falls, the river flowed through the tidal reach of the Portland Basin and into the Columbia River by way of the main channel or through Multnomah Channel separating Sauvie Island from the mainland.

Since Lewis and Clark's Corps of Discovery did not venture far up their "Multnomah River," they could not report on the characteristics of topography and climate that would, in a few decades, attract a flood of immigrants. Not long after Lewis and Clark's visit, other Europeans and European Americans began exploring the valley. The Pacific Fur Company founded Fort Astoria in 1811 near the Columbia's mouth on the Pacific Coast, and this location served as a base of operations to extract furs from throughout the region. Fur trapping expeditions from Astoria began recording their impressions of the Willamette Valley in 1812. By this time, both trappers and Indians were applying the name "Willamette" (*wálamt*) to the river. *Wálamt* was the name of the Chinookan village at the base of Willamette Falls, and extension of this name to the entire river likely occurred by way of the lingua franca of the Columbia basin, Chinuk Wawa. There is no definitive record of what the name "Willamette" meant to Chinooks.[10]

Fig. 1.4. Basalt cliffs at Willamette Falls, 1878 (Oregon Historical Society Research Library, bb015680).

In 1824, the Hudson's Bay Company (HBC) established Fort Vancouver on the north side of the Columbia east of its confluence with the Willamette as the regional center of the HBC's fur-trapping operations. Not long after, HBC Chief Factor John McLoughlin allowed some former trappers to settle with their Indian wives in a prairie at a bend in the Willamette between the present-day cities of Salem and Newberg, an area now known as French Prairie. These *Métis* were of French and Indian extraction from the Great Lakes region. Their homesteads—numbering 322 by 1841—constituted the first permanent settlements of European descendants in the Willamette Valley.[11] Other newcomers were soon to follow in the form of Christian missionaries: Jason Lee's Methodist Mission in 1834 and Norbert Blanchet's Catholic Mission in 1838.[12]

Even before Samuel Simpson's poem "The Beautiful Willamette," booster accounts and romanticized literature, poetry, and art were essential components of European American migrations west. Politicians such as Missouri Senator Thomas Hart Benton, writers such as Hall Jackson Kelley, and romantic landscape painters such as Cleveland Rockwell helped entice settlers to Oregon Country and the "Edenic" Willamette Valley.[13] Such romanticizing was a manifestation of the same cultural values and perceptions that spurred European Americans to mythologize New York's Hudson River Valley. Writers and painters characterized the Hudson as the nation's "river of empire" both because of its role as a commercial corridor fueling growth in New York City and its aesthetic beauty, captured on canvas by members of the influential Hudson River School of landscape painting from the 1820s through the 1840s.[14] The Willamette River served this same mythological role for the first waves of European American settlers in the mid-nineteenth century. Spurred by representations that struck such a deep chord, immigrant floodgates opened in the 1840s. The first wagon train to Oregon Territory arrived in 1840, and the Great Migration of 1843 brought 875 additional non-Indian residents to the valley. McLoughlin himself established a home in Oregon City on the east side of Willamette Falls in 1846. By 1845 there were about five thousand settlers in the valley.[15] For European Americans imbued with Jeffersonian agrarian ideals, romantic visions, and a sense of Manifest Destiny, their new home was—to use Goldilocks' phrasing—"just right."

Humans had long considered the Willamette Valley "just right," of course. At least eight thousand years ago Native Americans began living in the hills surrounding the valley, inhabiting land exposed by the receding Cordilleran

Fig. 1.5. Chinookans fishing at Willamette Falls, ca. 1841. Engraving by Joseph Drayton of the Wilkes Expedition (Oregon Historical Society Research Library, bb008331).

ice sheet. They established villages in the valley itself as rivers and creeks drained marshy lowlands. In this warmer climate and at higher elevations grew Douglas fir, Sitka spruce, and western hemlock. As the marshlands continued to drain, new plants spread into the valley floor, including a variety of grasses and western white oak, while forests of ash, cottonwood, maple, and alder grew along streams.[16] By the time of contact with Europeans and European Americans in the late eighteenth century, there were two distinct peoples living in the watershed. The Chinook-speaking Multnomah and Clackamas resided in villages clustered in and around the Portland Basin, Sauvie Island, and at Willamette Falls in the northern portion of the watershed. From their village *wálamt* on the west side of Willamette Falls the Clackamas controlled access to the abundant fishery.[17] Farther south at least ten Kalapuyan-speaking groups lived throughout the Tualatin and Willamette Valleys.[18] They used annual controlled burns to help propagate plants essential for food and fiber—such as camas, wapato, tarweed, and hazel nuts—and to create habitat amenable to deer and other game animals. Over thousands of years this management regime sustained the prototypical Willamette Valley landscape of open prairie interspersed with oaks and stands of Douglas fir, while more dense vegetation lined creeks, rivers, and lakes.[19]

The Willamette Valley landscape conformed to the cultural preferences of both Indians and the newly arrived European Americans. By the mid-nineteenth century, however, conflict, disease, and dispossession made life in the valley much less Edenic for Chinooks and Kalapuyans than it appeared to be for Simpson and the other newcomers. In the 1850s, a series of bloody conflicts known as the Rogue River Wars raged in southern Oregon and Northern California. A predominately single, young, and male population of whites flocked to the gold-bearing regions of the Rogue Valley and Siskiyou Mountains to make their fortunes. On the whole they showed blatant disregard for Indian communities, while many of them initiated and escalated violent acts. The bloodshed in southern Oregon did not spread north, primarily because newcomers to the Willamette Valley tended to be homesteading families drawn by the prospect of establishing communities centered on agriculture and industry, rather than belligerent, single men focused on extracting a quick fortune before moving on.[20]

Though spared the violence and conflict that Klamath, Modoc, Shasta, and other tribes to the south endured, a series of recurring epidemic diseases ravaged Chinooks and Kalapuyans beginning in the 1770s. Between 1774 and 1874 the entire native population of the northwest coast decreased about 80 percent, predominately through newly introduced diseases such as influenza, small pox, and measles. During the decade of the 1830s alone, the population of Chinooks in the Portland Basin and Kalapuyans upriver fell about 92 percent.[21] With such significant demographic changes these communities were much less capable of challenging white encroachment. This, coupled with the violence of the Rogue River Wars, led the US government to initiate a series of treaties in 1855 authorizing the army to relocate Willamette Valley tribes (and others throughout western Oregon and northern California) to the Grand Ronde Reservation in the foothills of the Coast Range at the western edge of the Willamette Valley.[22]

CHANGING LANDSCAPE, CHANGING RIVER

"... the great sewer of this great Willamette Valley"

That Samuel Simpson's Willamette River was beautiful to European American settlers in the mid-nineteenth century was a direct result of both natural forces and thousands of years of active management from Chinooks, Kalapuyans, and their forebears. Nearly 5,000 white settlers relocated to the Willamette Valley by 1845, compared to an 1841 estimate of 575 Chinooks in the Portland Basin and 600 Kalapuyans south of Willamette Falls.[23] These shifting demographics began to be reflected in the river itself. Within just a few decades of European American settlement, the Willamette was becoming much less than beautiful. European Americans brought with them an intertwined set of religious beliefs, cultural values, and technologies that soon drastically transformed the landscape and river through industrialization and urbanization.

The HBC established the first industrial facility on the Willamette River in 1831 when the company built a water-powered lumber mill at Willamette Falls in Oregon City. Lumber mills and flax processing, meat packing, vegetable canning, and related facilities proliferated in the Willamette Valley after the 1860s. By the 1880s there were nearly forty salmon canneries on the entire Columbia River system—including the Willamette—and by 1890 there were two pulp mills in the valley, one at West Linn and the other at Lebanon.[24] Here and at many other industrialized sites the newcomers converted flora and fauna into commodities, and then eagerly injected these unitized goods into regional and global markets. While waterborne shipping had been the most efficient transportation mode since the time of contact and would continue to be important into the early twenty-first century, rail connections to the Willamette Valley after the Civil War accelerated both industrialization and urban growth.[25]

Fig. 1.6. View from Oregon City, ca. late 1800s, looking southwest at Imperial Flour Mills in the foreground, with Willamette Falls and West Linn beyond (Oregon Historical Society Research Library, bb000056).

The valley's European American population expanded as well. In 1860, a year after Oregon Territory became the State of Oregon, there were 52,465 residents, most in the Willamette Valley. Statewide population exploded over the next forty years to number 413,536 in 1900, and most of this growth occurred along the Willamette River and its tributaries. Oregon's most populous cities and towns in the nineteenth century were founded along the banks of these waterways. By 1900, the population of Willamette Valley cities that would figure importantly in later pollution abatement efforts were: Portland (founded 1845), 90,426; Newberg (founded 1869), 915; Oregon City (founded 1829), 3,494; Salem (founded 1842), 4,258; Corvallis (founded 1849), 1,819; and Eugene (founded 1862), 3,236.[26]

Willamette River water quality reflected these drastic changes. One state senator opined in 1888 that the river had become "the great sewer of this great Willamette Valley."[27] Waste from practically every community and industry in the valley made its way directly into the Willamette or its tributaries. Communities that built sewer lines—Portland commenced this work in the 1880s—did so to collect wastes and route them to the nearest stream; waste treatment technologies did not yet exist. In addition to municipal sewage, untreated industrial wastes were also poured into watercourses. Key Oregon industries providing employment, civic pride, and tax revenues—such as wool cleaning, flax retting, vegetable canning, lumber production, and paper making—each used thousands of gallons of clean water daily and returned to the Willamette and other streams unfiltered and untreated effluents highly polluted with fibers, chemicals, dyes, and other substances.

By no means were Oregonians the only people to be experiencing negative environmental effects from the expansion of extractive industries and growth of increasingly concentrated populations. Unsurprisingly, the communities facing these effects most acutely were among the first to attempt to address them by adopting new approaches to urban sanitation and environmental regulation. In the late 1850s, Brooklyn, New York, built the nation's first effective system to collect human wastes and storm water from the urban landscape and pipe them to the nearest watercourse. Twenty years later, Boston followed with the nation's second extensive urban sewer project. In 1886 the State of Massachusetts established its Lawrence Experiment Station under the board of health to conduct research on the public health effects of degraded water quality. Outside the United States, engineers in London had been building sewers in a piecemeal fashion since the seventeenth century, but in 1855 city leaders overhauled and modernized the sewage system in response to typhoid epidemics and threats to economic stability. London's achievement significantly improved public health at the expense of an increasingly filthy River Thames.[28]

Water quality laws in the nineteenth century addressed the most egregious and threatening effects of pollution, but were limited in their application. One reason for this was that societies of the late nineteenth century had never before experienced the combined effects of the scientific, engineering, and technological advances that enabled more people to live more densely in highly industrialized and interconnected communities. Both new kinds of wastes and new and old kinds of pollutants were being produced in increasing amounts. The accustomed ways of dealing with these no longer worked: no longer was it always feasible to find the solution to pollution through dilution.

For centuries, humans have perceived rivers, lakes, bays, and other large bodies of water as integral components of waste removal systems—or, as historian Joel Tarr labels this approach, as "waste sinks." When these wastes are in sufficiently small doses and not made up of substances that are difficult to break down, natural processes of aeration and decomposition suffice. As the nineteenth century progressed, medical professionals, scientists, and engineers throughout North America and Europe integrated this concept into their respective approaches. They came to a consensus on how best to deal with concentrated industrial and municipal effluents: direct them to the nearest flowing stream or large body of water and allow nature to do its work. This approach functions best when the receiving waters act in accord with human desires. When wastes are sufficiently concentrated, however, or when they include more complex chemicals and substances, or when seasonal fluctuations change the dynamic of the receiving water, dilution turns out *not* to be a solution.[29] When natural systems no longer behaved the way people needed them to behave, the receiving waters became fouled, which threatened human health and constrained other potential uses of the water. These developments required changes to the status quo and compelled communities to evolve new technological, scientific, legal, and administrative solutions.

In the late nineteenth century, civic leaders in Oregon's largest city took one approach to an increasingly fouled Willamette River when they established a new water source about twenty-five miles east in the Bull Run watershed. Since the 1850s Portland residents had been taking their water from the Willamette River and an unregulated and haphazard scattering of wells. Governor Sylvester Pennoyer inaugurated the first flow from the Bull Run reservoir into Portland on January 2, 1895. With this reliable supply of clean water far removed from potential contamination, the city could continue to grow.[30] Other Willamette Valley communities followed Portland's lead in the first decades of the twentieth century and secured alternate sources of municipal water supplies.[31] The benefits of this approach became apparent in the drastic decline in public health threats from typhoid and other bacteriological infection. One consequence in the shift away from using the Willamette as a potable water source, however, was that with the decreased threat to public health Oregonians could more easily ignore or discount other effects of water pollution.

Another communal response to water pollution is to enact laws that prohibit it or that establish administrative departments to study it and devise solutions. Governments in London, Brooklyn, and Massachusetts were at the forefront of these approaches. State government in Oregon followed in 1889 by enacting its first two water quality laws to foster economic development and respond to public health threats. One law promoted drainage for both agriculture and sanitary reasons, and the other prohibited disposal of animal carcasses and other decaying matter into domestic water sources.[32]

To be effective, of course, statutory prohibitions require administrative systems and enforcement mechanisms. Oregon state leaders initially assigned these tasks to the fish commission and game commission. Legislators created a permanent board of fish commissioners in 1887 and a game board twelve years later. Water quality oversight was but one of many areas of responsibility for these entities, and of secondary importance in the overall management of commercial fisheries (in the case of the fish commission) and sport fisheries (in the case of the game commission).[33] For these commissions, degraded water conditions only motivated action if economic returns were at risk by pollution that directly threatened fish populations important to commercial operations and recreationists. Legislators expanded the state's active role in ensuring clean waters beyond this limited purview when they created the Oregon State Board of Health in 1903. As with similar agencies in other states, the mission of Oregon's board of health was to address an array of public health issues such as smallpox, typhoid, bubonic plague, and tuberculosis; provide health education; and ensure the cleanliness of domestic water supplies.[34]

A deputy warden registered the fish commission's first response against polluters in 1903 when he levied a $50 fine on the Rainier Mill and Lumber Company north of Portland for dumping sawdust into the Columbia.[35] Beyond this modest example, evidence is lacking that Oregon's initial administrative and enforcement mechanisms were having any appreciable effect on water quality. A typhoid epidemic in February 1906 linked to sewer outfalls from the city of Eugene shows the ineffectiveness of Oregon's 1889 law to protect domestic water sources. In response to this public health emergency, the *Oregonian* called the river a "common sewer for the entire valley between the Cascade and Coast Ranges of mountains from Cottage Grove to the Columbia."[36] The board of health found typhoid in the Columbia River for the first time later that year and traced the bacteria to the Willamette and streams in eastern Oregon. The secretary of the board of health was not surprised and noted the Willamette "has not been free from typhoid germs for years." The *Oregonian* advised that "it evidently behooves the many swimmers about Portland to cultivate the gentle art of keeping their mouths closed while in the water."[37]

By the 1910s, with the valley's growing population and industrial expansion degrading Willamette River water quality at an unprecedented rate, state leaders had allocated different aspects of water quality oversight to three government agencies. The board of health responded to pollution when it threatened public health; the fish commission became concerned when pollution threatened lucrative salmon runs; and the game commission was authorized to act when recreational fishing opportunities had been harmed. This three-pronged approach proved to be less than effective, so through the 1900s and 1910s legislators enacted incremental changes. Under provisions of the state's 1919 laws, for example, the game commission had the prerogative to disapprove municipal waste disposal plans that included raw sewage outfalls.[38] However, in the aggregate, Oregon's laws continued to distribute responsibility among three separate state agencies while granting them almost no enforcement powers.

WATER POLLUTION AND THE FEDERAL GOVERNMENT DURING THE PROGRESSIVE ERA

Throughout the United States well into the twentieth century, water quality laws were limited in number and scope and uniformly lacking in robust administration, monitoring, and enforcement provisions. The progression of federal water pollution control laws was similar to Oregon's experience: incremental changes through laws passed in reaction to threats to public health or the economy, with minimal enforcement powers and responsibility diffused among various agencies. National political leaders further viewed water pollution solely as a matter for local and state government. In instances where pollution involved multiple states—such as in the Great Lakes region or Hudson River watershed—the federal government preferred establishing regional compacts among affected states.[39]

The first national law addressing water quality—albeit indirectly—was the 1899 Rivers and Harbors Act. It empowered the US Army Corps of Engineers to maintain access to navigable waterways by removing debris and forbidding construction into navigable channels. The law's indirect effect on water quality pertained neither to public nor aquatic health: it empowered the corps of engineers to require the removal of mining tailings, mill debris, and other wastes that interfered with navigation.[40]

Congress established the US Public Health Service (USPHS) in 1911. It had much in common with state health boards in Oregon and elsewhere in its focus on public health and sanitation; it also conducted research into the health effects of water pollution in navigable streams. The agency's research group was based at the Office of Industrial Hygiene and Sanitation in Cincinnati. Here engineers and scientists worked in collaboration with the Bureau of Mines on an array of pioneering research that included establishing some of the earliest water quality measurement standards.[41] Much like the Oregon Board of Health, the USPHS also provided research, analysis, and consulting services and responded to public health emergencies; it had little or no powers to enforce water quality standards, and it was not concerned with the health of aquatic life.

Another development in water quality oversight at the national level during the first decades of the twentieth century was the establishment of the International Joint Commission (IJC) between the United States and Canada. The 1909 Boundary Waters Treaty created the IJC and tasked it with managing the shared resources of the Great Lakes area. This charge included addressing severe pollution from large population centers and dense industrialization. Illustrating the same focus on public health shared by the USPHS and Oregon Board of Health, a typhoid outbreak in 1910 prompted Canadian and American officials to ask the commission to develop a multinational solution to untreated sewage flowing into the lakes. IJC staff commenced work on the issue but by the early 1920s the emergency had eclipsed: municipalities drawing water supplies from the lakes had installed chlorine treatment systems that killed bacteria, thereby extinguishing the threat without the expenditure of millions of dollars in sewage treatment. Thus, the application of chlorine thwarted the first attempt to develop any kind of joint United States–Canadian approach to municipal waste treatment. It was also a significant reason why typhoid fever outbreaks in the United States fell from 35.8 to 2.1 cases per 100,000 people between 1900 and 1930.[42] With this threat dissipated, so waned Canadian and US interest in abating pollution in the Great Lakes.

While water pollution significantly worsened in the 1910s and 1920s, federal and state approaches continued to be decentralized and incremental. Beyond USPHS research that soon was to become foundational to quantifying pollution's effects on water quality, the federal government did not take a leading role in water pollution abatement. Through the 1920s, however, one unlikely national figure was to help set the stage for later abatement work: Secretary of Commerce (and future president) Herbert Hoover.

Historians have characterized the 1920s as a decade that roared with laissez-faire capitalist excess. Though the nation had fallen into a recession almost immediately after World War I ended in 1918, within a few years much of the nation was in recovery.[43] Republican presidents William G. Harding (1920–1924) and Calvin Coolidge (1924–1928) both fostered minimal business regulation. This applied as much to water quality and other environmental issues as it did to banking, finance, and the stock market. Both presidents left it up to the states to develop, administer, and enforce regulations relating to municipal sewage, industrial wastes, and water quality.[44]

Secretary of commerce during both the Harding and Coolidge administrations was a man who, in most respects, shared his bosses' approach to business regulation and environmental oversight. Herbert Hoover—who would succeed Coolidge as president from 1928 to 1932—differed from his predecessors in at least one significant way, however. Hoover spent formative years of his youth in Oregon and developed a passion for fishing and other outdoor activities. He came of age politically and professionally in the midst of the refinement and application of a range of Progressive Era conservation policies exemplified in the work of Gifford Pinchot's US Forest Service. From these experiences he formed the belief that resources can and should be conserved in a way that fostered present economic growth but did not undermine future benefits.[45]

An avid fisher, Hoover was also an active member of the Izaak Walton League of America (IWLA). Fifty-four men meeting at the Chicago Athletic

Club formed the League in January 1922, and named it for a seventeenth-century English writer and angler. Sportsmen's groups had been a feature of American society since at least the mid-nineteenth century, and while membership increased markedly by the early 1920s, these groups were widely dispersed and focused primarily on their own local or regional interests. The IWLA's explicit goal was to bring together the membership and interests of angling and hunting groups within a national framework.[46] Foremost among the group's priorities was water pollution abatement, with degradation of the Illinois River a key motivating factor for founding members.[47] Hoover became a member himself that year and contributed to the first issue of the IWLA's magazine *Outdoor America*. The league, in turn, enthusiastically supported Hoover's efforts as secretary of commerce to protect fisheries in Alaska and to address oil pollution in coastal waters.[48] Hoover served as the IWLA's honorary president from 1926 to 1932.[49]

As secretary of commerce and later as president, Hoover espoused a philosophy directly linking business success to individual prosperity and societal advancement. His objective was to encourage businesses to expand and widen American consumers' range of options; this would, in turn, increase standards of living, which would then lead to more leisure time. With more leisure available, Americans would spend more time in the out-of-doors; this would improve individual health, enhance society, and help lead to natural resource conservation. As historian Kendrick Clements writes, Hoover believed that the "goals of conservation could be achieved through local initiatives and voluntary programs of cooperation between federal authorities and private interest." Over the course of the 1920s, Hoover's ideas came more and more to define the Republican Party's approach to a range of conservation issues, including water quality.[50]

One example is the Federal Oil Pollution Control Act of 1924. Secretary Hoover called a conference of state fish commissioners in June 1921 in reaction to pressure from commercial fishers and beach resort owners along the Northeast coast who complained that oily wastes were ruining their livelihoods. At this conference, representatives of the US Army Corps of Engineers advocated for an increased role in pollution abatement to build upon their work since passage of the 1888 New York Harbor Act and 1899 Rivers and Harbors Act. Coming out of this conference, the Senate passed a strong oil control bill in 1922, but the measure died in the House.[51] In the midst of these events, the Manufacturing Chemists' Association and American Petroleum

Institute had been funding US Bureau of Mines water pollution studies. Key recommendations from these studies echoed the petroleum industry's point of view, including warning against singling out any one industry, asserting that petroleum processing facilities had already implemented sufficient pollution control devices, and suggesting that further studies be made. These recommendations served as the basis for a weaker oil pollution bill that President Coolidge signed into law on June 7, 1924.[52]

In 1927, Hoover delivered an address before the IWLA in Chicago in which his perspective on conservation is clear. He told the league the department of commerce's increasing involvement in efforts to improve recreational fishing were due to a variety of interrelated reasons, "some of them economic in their nature, some moral, and some spiritual." Fishing was "good for the soul of man" and also helped to keep "the population from moral turpitude." However, such a program would require more than increased funding for fisheries research and hatcheries, but would also mean addressing pollution, "the poison cup which we give to eggs, fry, fingerlings, adolescents, and adult fish alike." Though "there are as many opinions about pollution as there are minds concerning it," he saw the commerce department's role as helping states conduct water quality surveys that would enable state officials to develop their own stream classification systems and enforcement mechanisms.[53] His desire was not to involve the federal government in setting standards and holding polluters accountable, but in equipping states with the data they needed to do this work on their own.

Though change had been incremental and subtle, twelve years of Republican presidencies and a decade of consumer and business excesses brought about a recognition at the federal level that a more systematic and collaborative approach to water quality was required. These changes are evident at the state level as well. Legislators in Pennsylvania (1923) and Wisconsin (1927) were among the first to establish agencies within their respective boards of health that integrated various water quality concerns: public health, commercial and sports fisheries, economic interests of other businesses using the resource, and, to some extent, recreation.[54] Some among Oregon's growing number of pollution abatement advocates actively kept abreast of these changes. The Izaak Walton League was critical in this communication through the Waltonian's periodical *Outdoor America*, annual conferences, and growing network throughout North America. One Oregonian exemplified the value provided by this network of professionals and concerned citizens: nationally renowned photographer and conservationist William Lovell Finley.

Finley was born in California August 9, 1876, and his parents moved the family to Portland, Oregon, eleven years later. In his youth he and his friend Herman Bohlman avidly collected bird eggs, feathers, and skins to sell, but in the 1890s they ceased collecting to photograph and film birds instead. In 1902, Finley helped found the Oregon Audubon Society.[55] He and his wife Irene were married in 1906, and they worked together over the following decades to publish books and articles in national magazines such as *National Geographic, Atlantic Monthly,* and *Colliers.*[56] Finley became a founding member of the Portland Chapter of the IWLA in December 1922, a week after the Chicago Chapter honored him for his contributions to conservation causes.[57] Finley would later work with other prominent conservationists such as Ira N. Gabrielson and Jay N. "Ding" Darling.[58] In addition to being a skilled photographer, prolific author, and highly respected conservationist with connections among national leaders, Finley was also actively involved in changing state government policies and practices. In 1903 he helped pass the state's Model Bird Law to put an end to hunting seabirds. He and Bohlman led efforts to convince President Theodore Roosevelt to designate wildlife refuges at Three Arch Rocks, Klamath Lake, and in Malheur County. In 1911, Finley helped reorganize the Oregon Fish and Game Commissions, and shortly thereafter became state game warden. In this role he managed game and fish restocking, oversaw hatchery construction, and helped ensure hunter safety—and he also responded to periodic fish die-offs in the Willamette River.[59]

WILLAMETTE RIVER WATER QUALITY AND PORTLAND SEWAGE

"... not primarily a decorative river"

Fish kills and typhoid threats spurred the federal government's interest in Willamette River pollution. In late July 1925, the Oregon State Game Commission convened the first conference on the topic, inviting representatives from the state board of health, six Willamette Valley cities, and various civic clubs and industries. Joining this discussion was a representative of the US Bureau of Fisheries, Dr. Henry B. Ward, a zoology professor at the University of Illinois and member of the Izaak Walton League.[60] Reflecting the purview of his role with the bureau of fisheries and the priorities of his boss, Commerce Secretary Herbert Hoover, Dr. Ward expressed the issue in economic terms. He estimated that the commercial salmon fishery in the

Willamette and lower Columbia rivers contributed 4 to 6 million dollars per year to Oregon's economy. This resource was threatened, however, by pollution: "I cannot emphasize too strongly," he told conferees, "that sewage dumping means the complete destruction of fish life."[61] Ward offered federal support in the form of research services to help state officials gather the data required to generate their own responses.

USPHS Sanitary Engineer Harry B. Hommon visited Portland less than a month later to discuss river pollution. Hommon announced the federal government would provide technical assistance to the City of Portland in conducting a survey of the Willamette within municipal limits if the city council provided staffing and funding. The goal of this study would be to generate the first quantifiable data on pollution affecting public health. While specific abatement steps would be left to Portland city engineers, the study would also offer abatement suggestions.[62] The survey was to take up to three years to gather a full range of water flow and pollution concentration measurements.[63] This relatively long duration was necessary due to significant seasonal fluctuations, with annual low-water periods of late summer through early fall compounding the effects of pollution. Hommon commenced the survey in February 1926 by producing a nine-page preliminary report outlining equipment and staffing requirements needed.[64]

While federal experts and state officials were conferring on water quality problems affecting fisheries and public health, legislators added an important enhancement to state water quality laws. A 1925 law created a sanitary engineering division within the board of health tasked with approving sewer system plans.[65] The *Oregonian* supported these federal and state actions by contrasting this approach to Chicago's recently completed Sanitary and Ship Canal. Chicago officials had the canal built to reverse Illinois River flow so wastes would drain away from Lake Michigan—the city's water supply—to the southwest and into the Mississippi. Rather than attempting to quantify and address pollution at its source, the city opted to send wastes to downstream communities. Such treatment of waterways was, in the *Oregonian's* view, "evil."[66]

Concurrently, Portland leaders were planning to upgrade the city's sewerage infrastructure. City Engineer Olaf Laurgaard proposed a comprehensive harbor redevelopment plan to rid the shoreline of derelict structures and protect the downtown core from floods by building a wall along the river's western bank.[67] This work was part of a broader effort to reclaim this section

Fig. 1.7. Activity in Portland Harbor, 1899 (Oregon Historical Society Research Library, ba018579).

of Portland's waterfront for the public; what had been a bustling area of docks, warehouses, and businesses a few decades previously had, by the early 1920s, become inaccessible, unsightly, and economically unproductive.[68] In addition to the harbor wall itself, the plan included an interceptor sewer line running parallel to the waterfront to connect outfalls and a pumping station to maintain positive pressure in the interceptor line. These elements would address potential health threats and reverse depreciated values of downtown real estate caused by sewage backing up into the city's business district during high water periods.[69] The project would also help reconnect city residents to the river. *Oregon Journal* editors called upon the symbolism of Samuel Simpson's "Beautiful Willamette" in early 1921 when they urged city leaders and business interests not to allow the entirety of Portland Harbor to be developed solely for commercial purposes. They critiqued city leaders for ignoring the river's "romantic flavor" associated with Simpson and other pioneer European Americans. "Portland's commercial necessities will never demand the entire utilization of the Willamette River frontage within the city for docks," the editors claimed, and asserted that "it would be no more unreasonable to use the public funds for waterside recreation spots than for parks far from the opportunity of boating and swimming."[70] This editorial reflected what was becoming an increasingly more popular use of the Willamette: swimming, pleasure boating, and other kinds of activities.

While the Willamette was becoming more polluted from untreated and unregulated industrial and municipal wastes, Oregonians found themselves with more time and interest in being on or in the river. Leaders of the Apostolic Faith Mission of Portland conducted large-scale baptisms in the Willamette River in July 1914, and a month later the first annual marathon swim took place, from Oaks Park north to the Morrison Street Bridge.[71] In 1915 City Commissioner (soon to be Mayor) George Baker proposed increasing amenities along the river such as designated swimming zones and dressing rooms. Aware of the many sewer outfalls within city limits, Baker proposed these zones in and near Oaks Park, at Portland's south end, above most outfalls.[72] Oaks Park hosted outdoor swimming and diving competitions while raw sewage volumes in the lower Willamette increased.[73] In 1924 the Portland City Council considered forbidding recreation due to bacterial infection from raw sewage regularly carried *upstream* by backflow from the Columbia River.[74]

With significant growth in Portland's population during the first decades of the twentieth century, more residents were using the river as both a waste sink and for recreation—purposes increasingly coming into conflict. Between 1910 and 1930, population increased 46 percent, from 207,214 to 301,815 residents.[75] While city officials had authorized sewer line construction since the 1880s, these lines did not convey the sewage to a treatment plant of any kind. Rather, engineers built individual sewer lines as they were needed and routed outfalls directly to the Willamette River bisecting Portland, or to the Columbia Slough at the northern edge of the city. These lines carried sewage, industrial wastes, storm water, and—in some cases—rerouted creeks in a single conduit. Sanitary engineers and public health experts throughout Europe and North America had debated the efficacy of combined versus separated sewer systems. Combined sewers, as the name implies, relay all liquids in a single conduit, whereas separated systems dedicate one line solely for storm water runoff and another to sewage and industrial wastes. The consensus that developed by the late nineteenth century privileged combined systems. One compelling reason was cost: because combined systems required half the excavation and half the pipe of separate systems, they were much less expensive to build and maintain. Scientific understanding of how diseases spread supported this economic rationale. In the early twentieth century, the understanding that bacteria caused disease became the dominant scientific paradigm, but prior to this expert consensus held that miasmas—or bad odors—caused disease.

With this theory of disease etiology, what mattered most was carrying urban wastes away as quickly and inexpensively as possible, not separating or treating wastes. Further reinforcing the decision to build combined systems was agreement among experts that rivers, streams, and lakes had the inherent ability to cleanse themselves, so treatment was not necessary.[76]

Considering rivers as an integral element of the waste treatment system was one among many functions that Americans and Europeans characterized as a "reasonable use." A reasonably used body of water meant that it would serve the community-at-large in many ways: transportation, recreation, commerce, food, and waste disposal. Prevailing scientific and engineering orthodoxies held that not only *could* bodies of water be used to dispose of wastes, but that this was one use among others to which receiving waters *should* be used.[77] The *Portland Telegram* spoke eloquently in support of the "reasonable use" doctrine when in late 1922 it noted that, from the beginning of European American settlement, the Willamette had been a "working river," and "not primarily a decorative river." According to the paper, the Willamette

> has always dressed in working clothes. It is habitually unkempt.
> Its banks bristle with the rough rubbish of visionless, grubbing
> commerce. . . . Like most rivers in American cities, [the Willamette] is
> a sewer, a back-door area, a clutter-yard, a dump, a workshop . . . .

However, the editorial continued, the river was also "the greatest scenic possibility in this city of great scenic possibilities," and, "by taking thought it can be made beautiful without in the least impairing its utility."[78] Certainly, this image of the river contrasts with Simpson's romanticized "Beautiful Willamette," but this editorial illustrates that Willamette Valley residents perceived the river in a variety of ways.[79] Even when used "reasonably," however, water quality in and near Portland deteriorated.

As reasonable uses of the Willamette River came into conflict, there was no clear-cut and inexpensive way to deal with municipal wastes. Primary sewage treatment techniques from the nineteenth century involved screening solids and allowing the particulates to settle out before sending the remaining liquid to a nearby river or lake. Practical and effective secondary sewage treatment technology had only been developed in the 1910s. This "activated sludge" process involved the participation of microbes that digested wastes in highly aerated liquid environments and produced their own, inert, wastes that

then settled out of suspension; the process was somewhat similar to ferment-
ing beer, though involving different microbes and nutrients. The resulting
"sludge" (the beer brewing equivalent would be "trub") was free of harmful
bacteria and retained significantly less oxygen-depleting organic content.

Since the 1840s, chemists and sanitary engineers in the United Kingdom
had been trying to find ways to concentrate the nitrogen available in human
waste and make a profit from the sale of sewage sludge as agricultural fertilizer. After seventy years of attempts, the consensus was that fertilizer sales
did not cover the costs of production and transportation, so common practice in both the United Kingdom and United States was to deposit the sludge
into landfills or waterways. The effectiveness of the activated sludge process,
however, renewed interest in converting wastes into fertilizer. As increasingly
more cities in the United States began taking steps to address their municipal
wastes in the 1910s and 1920s, a common element in the deliberations was
whether or not to design a fertilizer production facility in addition to a waste
treatment facility. In the vast majority of cases the expense of these operations
far exceeded meager returns realized from fertilizer sales. A notable exception
was the city of Milwaukee, Wisconsin, where political pressures and opportunities aligned with available technologies to result in a successful sludge-to-
fertilizer operation.[80] More commonly, the novelty and expense of activated
sludge systems—plus the ongoing debate about whether it was not just as
effective to chlorinate water sources, rather than treat sewage at all—meant
that, in the 1920s, very few cities in the United States, Canada, or Europe had
built such facilities.[81] Portland was not alone in dumping the full load of its
municipal wastes directly into the nearest body of water.

Notwithstanding these technical challenges, in 1925 William Finley and
other pollution abatement advocates argued that Laurgaard's proposed harbor wall and sewer interceptor project should go further. Rather than integrate a number of raw sewage outfalls into a single outfall, they urged city
leaders to build a treatment plant and develop a more comprehensive citywide sewer system.[82] USPHS Sanitary Engineer Hommon advised against
this comprehensive approach, however. His view was that Willamette Valley
cities would not benefit from these kinds of expensive and complex urban
infrastructure projects before seeing results of USPHS-sponsored water
quality surveys. These surveys would include calculations of the assimilative
capacity of the Willamette and Columbia rivers, and with this information
Hommon asserted that it might not even be necessary for municipalities to

treat their sewage at all.[83] Laurgaard echoed this perspective. He found that Portland's existing system of combined sewers would make treatment costly because of the large amount of mixed water and sewage. He also noted the option of building an entirely new system of pipes dedicated strictly to sewage would be prohibitively expensive. Laurgaard did agree, however, with both federal officials and state "fish and game men" such as Finley that treatment of Portland's sewage would be just one step in a broader program of cleaning up the river. Unlike Finley, however, he concluded that taking this step was not worth the expense.[84]

Sustaining the growth of Portland and other communities were the many industries of the Willamette Valley, the vast majority of which routed their untreated liquid wastes into the river or its tributaries. Meat processing, flax retting, wool processing, and vegetable canning each contributed wastes that concerned abatement advocates, but the effluents of no single industry was remotely as detrimental to water quality as those from the valley's five pulp and paper mills. These mills were located at Salem, Lebanon, Newberg, Oregon City, and West Linn; though outside the Willamette River watershed, three additional regional pulp and paper mills also played a role in Oregon's water pollution abatement story: one downriver from Portland at St. Helens, Oregon, and two across the Columbia River from Portland at Camas and Vancouver, Washington. Over the course of the twentieth century, as the result of industry buy-outs and consolidation, each went under different names over the years.[85] As individual and corporate profitability waxed and waned, and as ownership changed, abatement advocates found that some mills were relatively more amenable than others to dealing with their wastes; even so, advocates found themselves constantly struggling to change industry practices in the interest of improving water quality.

THE DAWN OF ABATEMENT ADVOCACY

Federal involvement in Willamette River water quality in the mid-1920s came as the contending parties were coming into focus. Pushing for changes to current practices were the abatement advocates, composed of two overlapping groups. The first were those with some manner of state imprimatur, such as members of the state game commission, state fish commission, and state board of health. The second were members of sporting clubs, the foremost being the Izaak Walton League. These advocates sought solutions to municipal and industrial wastes. The City of Portland, in particular, with its concentrated and growing population, expelled more raw sewage than all of the valley's other communities combined. Among the valley's polluting industries, pulp and paper manufacturers stood out as the most egregious polluter. Advocates' efforts focused on these two primary polluters.

The degraded water quality Oregonians faced by the 1920s had developed over less than a century of European American settlement and development. Within a few short years of Samuel Simpson writing "The Beautiful Willamette," the river was already changed from the effects of pollution. The same values, mores, and technologies that spurred him to characterize the river as "Limpid, volatile, and free," replete with "crystal deeps" and "roseate ripples . . . Breathing fragrance round to-day," and communicate this vision to others, were also the sources of the river's degradation. The river had not been able to absorb such wide-ranging alterations to the region's landscape, economy, and society and remain white settlers' romantic symbol.[86]

Chapter 2
When Reasonable Use Became Unreasonable

Oregonians who sipped their coffee while perusing the *Morning Oregonian* of May 5, 1926, would have read on page four of section two:

FISH KILLED BY SEWAGE

EUGENE FACES CHARGE OF POLLUTING RIVER

Willamette for 12 Miles Below City Said to Be Covered with Coat of Oil

State Game Warden Edgar F. Averill was investigating "a black viscous coat" visible on the Willamette for twelve miles downstream from Eugene that was claimed to have "killed numerous fish and kept other fish from running up the river to spawn."[1] Oregon's water pollution laws and state agencies such as the fish and game commissions were supposed to have established a legal and administrative framework that made such pollution impossible. Unfortunately for fish and clean stream advocates, this was not the case.

Averill and game commission staff investigations of Eugene's sewer mishap would have occurred well after the effluent had dissipated and traveled downstream toward the Columbia. They would have been hard-pressed to collect evidence because there were no sampling stations in the vicinity of the spill, nor any game commission staff on hand at the time to view the pollution directly. The ever-flowing water would have carried the "black viscous coat" of pollutants downstream and diluted it. The most apparent signs of remaining pollution would have been dead fish, but it is not apparent that anyone collected or tested fish. A few days later the *Morning Oregonian* reported, "Stream Pollution Rouses Game Body," but this state of arousal seems only to have led to promises from Eugene city officials to make "some financial provision for the erection of proper sewage disposal plants."[2]

Farther downstream, Milwaukie city officials were moving forward with plans to build a raw sewage outlet to the Willamette River, in spite of failing to secure prior approval from the state board of health as required in Oregon law. If they moved forward, their project would also be in violation of recent enhancement of game commission powers to pursue those who "put or deposit in any of the rivers, streams, lakes or waters of Oregon . . . any dead animal carcass, or part thereof, manure, sewage, putrid, decaying or deleterious substance . . . which injures fish or will corrupt or impair the quality of the waters of said rivers, streams or lakes. . . ."[3] By the mid-1920s, in spite of this legal framework, state officials had only been able to confront polluters successfully in a few minor cases, such as logging companies and individual steamships. They had yet to succeed in cases against large-scale polluters such as municipalities and heavy industries.[4]

ORGANIZING ON THE RIVER'S BEHALF

". . . any reasonable plan"

Decades of incremental changes to the state's legal framework and some recent assistance from the US Bureau of Fisheries and US Public Health Service had not changed the status quo. Clean stream advocates sought a new approach, with abatement efforts thus far having been episodic and unorganized. Beginning in 1926, they began to coalesce the energies of various groups and individuals within a focused effort aligned by common goals and methods. State Game Warden Averill took a leadership role in this effort. Averill had both a personal and professional interest in this topic as his long and active membership in the Portland Chapter of the Izaak Walton League would testify. He was born along the Chetco River on the southern Oregon coast in 1881, and graduated from Willamette University with a bachelor of science in 1905. He worked for a few years as a reporter for the *Salem Capital Journal* and editor at both the *Pendleton Tribune* and the *East Oregonian*. In 1911 he became a district game warden and from 1914 to 1919 worked for the US Biological Survey out of Pendleton. He entered the insurance business in 1921 and, after serving as state game warden from 1925 to 1927, he would open his own insurance office in Portland. In addition to being a Waltonian, he also belonged to the Portland City Club and Audubon Society; in the 1930s he would serve as president of the Oregon Wildlife Federation.[5] Averill, like his colleague and contemporary William Finley, combined personal and professional interests with both training and experience over the course of what would be decades of involvement in Oregon's clean stream movement.

In the immediate aftermath of the Eugene fish kill and Milwaukie city leaders' disregard for state law, Averill hosted two meetings with the purpose of taking substantive action against both municipal and industrial violators of Oregon's water quality statutes. The first meeting, on May 13, 1926, included representatives of the fish commission and board of health, as well as the State Sportsmen's Association, Valley City Engineer's Association, and the Salmon Protective League. Attendees called upon the attorney general to take action against polluters. The second meeting, on May 20, was broadened to include representatives of the pulp and paper industry as well as officials from Milwaukie, Salem, Albany, and Eugene.[6] The *Oregon Journal* succinctly characterized the result of these two meetings, born out of frustration, as "River Pollution Ban Backed, But Solution Lacking." City officials again promised to remedy their sewer deficiencies. Industry executives, for their part, admitted to releasing "sulphurous acid and pulp into the Willamette," but claimed this was not detrimental to aquatic life because no dead fish had been found downstream. These executives continued, however, that even if wastes were

Fig. 2.1. Fishing for salmon at the base of Willamette Falls, 1920s (Oregon Historical Society Research Library, bb015676).

harmful, "the only fish below the mills worthwhile are salmon, and they avoid the sulphurous acid." Even though, according to the executives, water pollution "is a big problem" in the watershed generally, their wastes were either released in negligible amounts or were unavoidable. In either case, they said, "the mills should not be put to great expense unless the fish are hurt.[7] Neither municipalities nor industries had any immediate solutions.

Averill and his colleagues in the fish commission and board of health were not satisfied. William Finley attended these meetings, and he, at least, could have provided some useful comparative historical context. During his service with the state game commission overseeing recreational fisheries, he responded to a large fish kill at Willamette Falls in July 1911. Since this die-off occurred within a stretch of the river designated a commercial fishery, his colleague R. E. Clanton, the state's master fish warden, held jurisdiction for enforcing water quality regulations. Clanton filed charges against owners of the Willamette Pulp and Paper mill at Oregon City. He cited the company for discharging excessive amounts of waste pulp, dyes, and other effluents, and for not allowing sufficient water to flow through their mill race to enable fish and eel passage. This case was to be a test of water quality laws that would lead to "prosecutions of others who violate the state laws in the same manner." Later in 1911, the secretary of the state board of health attended a conference on sewage treatment technologies hosted by the Municipal League of Eugene. The aim of the conference was to move toward "the ultimate purification of the Willamette River."[8] Fifteen years later, with the benefit of hindsight on these and other events, the fact that advocates needed to convene meetings on these same issues illustrates that lofty aims expressed in 1911 had come to naught, and something different was now necessary.

The preliminary meetings in May 1926 led to a larger meeting hosted by the Salem Chamber of Commerce in September, at which even more city and county officials were present. Attendees looked beyond the Willamette Valley to water pollution and waste disposal in other areas of the state. Organizers sought to develop a definite pollution control program that included drafting a legislative proposal and educating the public. Attendees formed a committee, named officers, and elected Portland Mayor George Baker president of their new coalition, the Anti-Stream Pollution League (A-SPL). Governor Walter Pierce welcomed the group's creation by observing that "the demand for elimination of stream pollution was general throughout the state" and said he "would support any reasonable plan looking to elimination of the menace."[9] The A-SPL included representatives from the state fish and game commissions, municipalities, industries, chambers of commerce, and conservation groups. In addition to Mayor Baker, other key members included Portland City Engineer Olaf Laurgaard, State Game Warden Edgar F. Averill, and Portland attorney John C. Veatch; Veatch, like Averill, was active in the Portland Chapter of the Izaak Walton League.[10] Creating the A-SPL signified a new era in Willamette River pollution abatement. Previously, representatives of the state fish or game commissions or the board of health had some authority to act, but only within their limited jurisdictions. As the outcomes of events in 1911 had shown, this arrangement had not led to a cleaner river. Advocates outside of government were also involved, but either as individuals or as members of conservation or other civic groups with localized membership and interests. The two conferences in May 1926 brought these and other groups together, and formation of the A-SPL later that year began to unite the interests of abatement advocates and concentrate their energies on tangible efforts and outcomes. With this momentum, league members worked throughout late 1926 both to increase public awareness and to develop a bill for the 1927 legislative session that would commence mid-January.

The league's executive committee reviewed its legislative proposal with the organization's full membership in early 1927. Pennsylvania's 1923 law centralizing water pollution abatement within a state agency influenced the A-SPL's proposal (as well as a similar proposal being considered in Wisconsin).[11] The A-SPL pushed for creating a sanitary water board within the board of health composed of representatives from the board of health, game commission, and other departments empowered by law with ensuring one or another aspect of water quality. This board would have the authority to establish water quality standards, investigate infractions, pursue violators, and approve municipal waste treatment plans. Reflecting frustration in the status quo and a desire for immediate results, the A-SPL further proposed strict compliance dates that offered significant economic and engineering challenges. The group called for cities and industries to submit complete engineering plans to the sanitary water board within two years, and have their waste treatment systems completed by 1935.[12]

The A-SPL faced stiff opposition in the state's most politically powerful city and most economically significant industry. Portland City Engineer Olaf Laurgaard of the A-SPL's executive committee supported the compliance deadlines.[13] However, his boss, and league president, Mayor Baker, was strongly opposed. He declared "it would cost Portland alone $35,000,000 to build intercepting sewers and treating plants to prevent pollution." With such a high cost, "he would not be a party to any plan to rush through the legislature a measure that would slap onto the taxpayers of the city such a huge debt without giving them a voice in the matter."[14] Baker's opposition was pivotal, and he was not alone. C. C. Chapman editorialized in his influential political weekly the *Oregon Voter* that the newspapers in Albany, Corvallis, and Roseburg were justified in expressing dire concern about the financial burden on municipalities. Though he praised the league for "its insistence that the problem must be faced before the trouble becomes too aggravated . . . [and] a virulent menace to public health," he also concluded the proposal would place an undue financial burden on industries, particularly pulp and paper manufacturers.[15]

Thus, the league's proposal failed to garner support from the valley's leading polluters, who also wielded significant influence in state politics. Arguments from these polluters appeared to be based on quantitative economic and scientific data. Mayor Baker referenced high costs for sewage treatment; industry representatives asserted mill wastes were not harmful to aquatic life but, even if they were, it was not worth the abatement costs. Objectively evaluating the accuracy of these claims would not have been possible, however, because no evidence existed to support them. Mayor Baker claimed that sewage treatment infrastructure would cost Portland residents $35 million. In truth, however, he could not have known with any certainty what the costs would be. In the early 1920s, City Engineer C. H. Smith had developed preliminary sewage collection and treatment plans, but the city had not yet formulated comprehensive engineering specifications from which accurate estimates could be made.[16] Like Mayor Baker, the pulp and paper industry put forth an economic rationale to justify inaction, plus an argument that appeared to address the biological effects of mill wastes. As with Baker's pronouncements, industry executives' arguments do not withstand scrutiny. First, even considering some limited scientific research that had been done through the 1920s, no one had yet completed comprehensive studies to understand the varieties of mill wastes and their effects on fish and benthic organisms.[17] Second, the pulp and paper industry was only just beginning to conduct basic research in waste treatment and the production of marketable materials from wastes—thus, they themselves did not know what might be both technically feasible and economically viable.[18]

Therefore, mill representatives' remarks from a May 1926 meeting clearly show their primary motivation was economic: "mills should not be put to great expense'" in the interest of water pollution abatement.[19]

The A-SPL set its goals as formulating a draft bill for state legislators to consider and increasing public interest in addressing water pollution. If evaluated in relation to the goal of delivering a water pollution control bill for the 1927 legislative session, the A-SPL's work must be judged a failure. Its draft proposal never made it out of the group's own executive committee. Quantifying the extent to which the A-SPL was able to mold public opinion in favor of its plan is difficult, if not impossible, since public polling was not commonplace in the 1920s. Another method of evaluating voter interest in a proposed law would be to read letters to legislators, but this line of investigation is not possible since the proposal did not make it that far. The A-SPL's efforts also fell short of its goals in another way. Members hoped even if their initial bill was not passed, the legislature would authorize funds for a more comprehensive study of the issue and potential remedies. Under Salem Representative I. B. Giesy's sponsorship, this proposal made it to the legislature in February 1927, but it "went overboard" when his colleagues postponed it indefinitely.[20] In spite of these roadblocks, the A-SPL's efforts were not in vain, however. In providing an example of substantive, proactive, state-level organizing and advocacy, A-SPL members established a precedent for future work. They also achieved another important outcome. Those opposed to pollution control lacked data to support their arguments—but abatement advocates lacked it as well. By bringing to the foreground the critical need for clear and specific information about the causes and effects of water pollution and potential solutions, the A-SPL's work was an important catalyst for change.[21]

BROADENING THE CONSTITUENCY

"... the greatest menace to public health and economic development"

Following the A-SPL's lead, in April 1927 the Portland City Club conducted its own study. Founded in 1916 and modeled after similar organizations in the East, the City Club was a voluntary group of citizens motivated to improve municipal institutions.[22] Its report observed the nuisance that degraded water quality had become by noting the filth and stench that caused some people to refuse to work along the riverbank.[23] The river's putrescence could be measured in *qualitative* terms by sight, smell, touch, and taste—and it certainly killed fish and repelled citizens—but the extreme pollution was not yet *quantified.*

After the failed campaign to draft a proposal for the 1927 legislative session, abatement advocates achieved a notable success at the March 1929 Commonwealth Conference at the University of Oregon in Eugene. The university brought together city, county, and state officials, subject area experts, and concerned citizens at these annual events to foster a "forward moving procession of improvement . . . directed to eliciting concern for common and lasting interests and for those conditions most vital in their relation to the welfare of the whole people."[24] Organizers focused the March 1929 conference on water pollution abatement, indicating both that the issue was becoming increasingly important to Oregonians and that the A-SPL had realized some success in its goal of raising public awareness. Attendees discussed results from the City of Portland's survey of the lower Willamette published shortly before the conference. Between July 1926 and January 1929, the city staff regularly collected water samples at various depths at seven stations to

> determine whether or not the water contains a sufficient percentage
> of dissolved oxygen to properly sustain fish life in the river, and to
> discover the extent of contamination on account of bacteria present
> which may cause a detrimental effect upon the health of persons in the
> vicinity of the river or those using the river for recreational purposes.[25]

In so doing, the survey provided the first empirical data on dissolved oxygen levels and bacteria counts in the Willamette. Based on this, Portland City Engineer Laurgaard concluded that "the Willamette River flow is of an amount sufficient to dilute the sewage so that there is no detrimental effect upon fish life"—*except* for a period of about two months each year during the low-flow period of August through September, when "it appears that fish life is endangered." Though the river is adequate to support fish life for ten months out of twelve, Laurgaard continued in his report, "the water is probably not sufficiently pure to justify the use of the river for recreational purposes." The health of Portland residents was not threatened "on account of sewers having their outfalls extending into the Willamette River," however, "since Portland does not use the river water for domestic purposes." USPHS Sanitary Engineer Hommon applied a standard technique to analyze a river's ability to assimilate wastes when he compared Portland's sewage volume with the river's flow gauged in cubic feet per second. With this metric he concluded that the river flow was sufficient to carry wastes downstream and "prevent obnoxious conditions." Merely preventing the visible accumulation of "obnoxious" floating detritus and sludge was not the only metric for a reasonably used river, as Laurgaard wrote: "The sustenance of fish life is, however, a function of the dissolved oxygen present in the water."[26]

Other experts at the conference expressed stronger views, characterizing the Willamette as "the greatest menace to public health and economic development of any stream in this section, due to its badly polluted state." Another noted that "fish cannot live in polluted waters and that steps have been necessary in some places to counteract unclean waters." Dr. Frederick Stricker of the Oregon Board of Health made the case that the state "cannot hope to make a bid for tourist travel if it does not offer advantages" such as recreational fishing. Washington's state sanitary engineer was on hand to convey his experience that "one of the most serious puzzles facing experts today . . . is the problem of disposal of industrial refuses from factories and mills." He also urged Oregonians to press municipalities to build sewage treatment facilities because—in spite of Hommon's comparison of the complex hydrology of the Willamette system with a generalized model of assimilative capacity—"natural purification of the Willamette takes place slowly because of the sluggishness of the flow."[27]

Attendees agreed generally that stream pollution problems were a concern for cities and concurred that "some very definite measures must be taken in the near future for the solution of these problems." They identified two areas of focus for follow-up work: first, gathering more empirical data about the biological and hydrological systems influencing water quality; second, determining effective ways to finance the expensive waste treatment infrastructure likely to be required. Attendees deemed it important "that there should be made a study not only of the engineering problem facing the cities, but also of finances involved." Such a focus would help address the core arguments Mayor Baker and mill officials had levied against the A-SPL's legislative proposal two years earlier. To carry out this work, they unanimously established a committee under the leadership of the state board of health composed of representatives from the University of Oregon, Oregon State College, two from industrial firms, and one each from the state fish and game commissions. This committee was to identify potential waste treatment technologies, develop feasible financing approaches, and commence "an extensive educational program." Its goal was "to maintain the two-fold program of continuing recreational attractions and encouraging industries to locate on its river banks."[28]

Tasked with a specific charge and supported by a growing number of people throughout the state, the committee that formed at the Commonwealth Conference immediately set about its work. Water quality data for the entire watershed was limited to the City of Portland's survey. As evidenced during the Commonwealth Conference, there were different interpretations of the data and divergent conclusions about the river's health and future prospects. Of primary importance was more robust sampling to generate a larger dataset from which could be developed functional technologies, actionable policies, and feasible financing mechanisms. The committee's work initiated an unprecedented collection of quantifiable data on Willamette River quality and sources of pollution.

QUANTIFYING WATER POLLUTION
TRANSLATING SENSORY EXPERIENCES INTO EMPIRICAL DATA

New approaches to pollution abatement required new partnerships and new kinds of expertise. Advocates turned to a local resource, the Engineering Experiment Station at Oregon State College in the central Willamette Valley city of Corvallis. The College's Board of Regents established the experiment station in May 1927 for the purpose of "developing the research spirit in faculty and students" by investigating timely subjects and publishing results that promised "the greatest benefit to the people of Oregon, and particularly to the state's industries, utilities, and professional engineers."[29] In response to the abatement constituency that coalesced in the late 1920s and the pressing need for quantitative data, Professor Clair V. Langton and Engineering School Dean H. S. Rogers authored the experiment station's first bulletin in 1929, a preliminary report on water pollution control.[30] It outlined the research questions that helped frame a sanitary survey of the river above Portland in July–August 1929, which was the experiment station's second published bulletin.[31] Two additional bulletins supplemented these within a few years. The Portland City Council called upon the experiment station to conduct a survey of the river in and below Portland, results of which were published in September 1934. Another bulletin, initiated at the request of Governor Julius Meier in July 1933 and published in May 1936,[32] identified the locations and types of industrial and domestic wastes throughout the valley. The cumulative effect

Fig. 2.2. Downtown Portland (Southwest 3rd Avenue between Washington and Burnside Streets) during the Willamette River flood of June 1894 (Oregon Historical Society Research Library, bb015679).

of these publications (and a few other journal articles, academic theses, and reports) was that within a few years there was significantly more data available about watershed hydrology and sources of pollution. The river was no longer simply a mysterious, beautiful enigma, as Samuel Simpson characterized, or a filthy, abused, and possibly irredeemable working river, as newspaper editors lamented. The river was now something that could be characterized in measurable units.

The river and valley were integral products of a series of complex geological, hydrological, and climatological processes that had transpired over millions of years. Specialists in the first decades of the twentieth century may not have understood these processes with the same level of sophistication as their colleagues in subsequent decades, but many people who lived along or near the Willamette River were well aware the river was not simply a canal that effortlessly conveyed its waters and wastes to the sea. People knew from lived experience the river was sluggish below Willamette Falls because Pacific Ocean tidal fluctuations made their way one hundred miles up the Columbia River and a further twenty-six miles up the Willamette. At times, this caused water in and below Portland Harbor to be still as a pond, and at other times water even flowed upriver.[33] Compounding this, seasonal variations could be extreme. If spring snowmelt from the Cascades coincided with heavy rains, the river could rapidly bring so much water into the streets of downtown Portland residents needed boats to get around, as had occurred in 1894.[34] Conversely, during particularly dry summers, the river in Portland Harbor could fall so low as to expose sewer outfalls.[35] City and state civic and business leaders managed this complex natural system proactively and extensively to maintain the river as a key component of the region's transportation system and economy. Portland Harbor had been an important regional entrepôt since shortly after European Americans arrived in significant numbers in the 1850s. It was closer to the Pacific Ocean than Tacoma or Seattle, but the

Fig. 2.3. Derelict and dilapidated structures along Portland Harbor prior to construction of interceptor sewer and harbor wall, January 1927 (City of Portland Archives, A1999-004.70).

Fig. 2.4. Construction of interceptor sewer and harbor wall in Portland Harbor, August 1928 (City of Portland Archives, A1999-004.32).

meandering river carried a large sediment load. Since the lowest reach of the river flowed so slowly this heavier material settled to the bottom and made the river shallower. Because of this, the city began dredging in the mid-1860s to maintain a shipping channel from Portland Harbor north and west through the Columbia Bar to the Pacific Ocean; the US Army Corps of Engineers assumed this responsibility in the early 1870s.[36]

Thus, lived experience showed the river to be a complex, dynamic system. Influenced by tidal fluctuations with significant seasonal variations in water volumes, it was also a critical transportation corridor requiring extensive and regular maintenance. As the river was being called upon to serve many uses and fill many roles in an industrializing society, empirical data was needed to express watershed dynamics in quantifiable terms. These first water quality studies began to provide the data necessary to develop methods attempting to balance the river's "reasonable uses." Measurements showed the river in Portland Harbor could vary by as much as three feet during tidal stages, and with tidal fluctuations Columbia River water regularly made its way at least six miles up the Willamette.[37] As a result, it could take up to 7.6 days for water to flow from Sellwood, at the south end of Portland, northward to Multnomah Channel, a distance of only about seventeen miles.[38] Above Portland, the state placed a river gauge at Salem in October 1927 to record river flows. These readings showed minimum annual flows varied from about 3,000 to 5,000 cubic feet per second (cfs) (or about 37,500 gallons per second) during July, August, and September, while maximum flows exceeded this amount by up to four times, reaching almost 13,000 cfs (or about 97,500 gallons per second) during the winter and early spring.[39]

PORTLAND SEWERS DURING THE NEW DEAL

"We are the greatest sufferers and the greatest offenders."

The wildly fluctuating dynamics of the river and its inability to carry wastes downstream reliably were both motivating factors for the City of Portland to devise its downtown harbor wall and interceptor sewer project in the 1920s. In 1925, when City Engineer Laurgaard was finalizing these plans, he had faced calls from William Finley and others to integrate the interceptor into a more comprehensive sewer treatment infrastructure. Laurgaard disagreed with Finley's proposal. He understood that Portland's existing system of combined sewers would make treatment costly because of the large amount of mixed water and sewage in the system. However, building an entirely new system of strictly sanitary sewers would be prohibitively expensive. USPHS Sanitary Engineer Hommon weighed in on the debate by saying that since Portland's water quality survey had not yet commenced, there was a lack of data to determine if the city even needed to treat its wastes. Expressing sentiments he would later revisit in opposition to the A-SPL's 1927 legislative proposal, Mayor Baker did not support enhanced sewage infrastructure projects unless neighborhood residents petitioned for them, or they addressed clearly identified public health threats. His approach "consistently followed the policy of not burdening the property owners or taxpayers with unnecessary expenditures."[40]

Completed in 1929, the Laurgaard harbor wall and waterfront development plan did not include treatment facilities nor initiate a citywide approach to sewer infrastructure development.[41] Not long thereafter, in the midst of attempts to design and build a comprehensive sewer interceptor and waste treatment facility, city leaders found that a lack of quantifiable data interfered with their attempts to get much-needed federal funds as part of President Franklin D. Roosevelt's New Deal public works funding programs.

By early 1933, the city of Portland was home to more than three hundred thousand residents, all of whom were served by outfalls expelling raw sewage to Portland Harbor (forty-eight outfalls) or Columbia Slough (eleven outfalls). Except for the harbor wall interceptor project that consolidated

Fig. 2.5. Aerial of the Portland Harbor wall from above Burnside Bridge, January 1929 (City of Portland Archives, A1999-004.54).

twenty-eight individual outfalls into a single one, no other significant development for Portland's municipal waste system had occurred by the time President Roosevelt took office on March 4, 1933. This was not surprising, considering the onset of the Great Depression in late 1929, Mayor Baker's resistance, and the limited progress of Oregon's abatement advocates.

It was also not surprising given the response of President Roosevelt's predecessor to the economic collapse that occurred during the first years of the Depression. Herbert Hoover imbued his work as secretary of commerce with his own brand of conservationism. In relation to water pollution, he supported the Izaak Walton League's efforts and took some initiative in addressing pollution from oil spills. In spite of this, even well into the unprecedented economic emergency after the October 1929 stock market crash, Hoover remained wedded to politically conservative values and resisted most calls to lend federal assistance to the states. He sought a balance between what he viewed as the primary drivers of American democracy and economic health—state and local governments and private businesses—with *just enough* federal power to foster this balance.[42] Many historians have concluded that President Hoover's approach was inadequate to the emergency the United States faced, and that the much more proactive and interventionist approach of his successor was more inline with what the times demanded.[43] Evaluating the two approaches to water quality at the national level and within the Willamette watershed, there is no debate: the New Deal provided invaluable funding and resources to advance state and municipal efforts, thereby heralding a new era in pollution abatement throughout Oregon.

Roosevelt's first one hundred days in office have become noteworthy for the unequalled productivity of the president and legislators in passing laws that had a direct and lasting effect on the United States. The resulting varied programs expanded the federal government in unprecedented ways and introduced a slew of agencies and acronyms to the American lexicon: Civilian Conservation Corps (CCC), Works Progress Administration (WPA), National Recovery Act (NRA), Social Security Administration (SSA), Tennessee Valley Authority (TVA), Federal Deposit Insurance Corporation (FDIC), and others. Municipal leaders received benefits from these programs more quickly and more directly than did state officials. Among other Depression-relief initiatives, Roosevelt and cooperative legislators created the Civil Works Administration (CWA) to disburse funds for a variety of infrastructure projects—including sewage systems. The CWA quickly distributed its funds and the president subsequently replaced it with the Public Works Administration (PWA), with $3.3 billion to expend for the same kinds of projects.[44]

To be sure, some of Hoover's efforts influenced Roosevelt's public works programs and continued into the New Deal—Hoover authorized creation of the Reconstruction Finance Corporation (RFC), for example, in early 1932 to restore faith in the nation's financial system by making loans to banks and other institutions. As one historian has documented, Roosevelt's achievements exceeded Hoover's because Roosevelt, a Democrat with a more expansive view of the role of the federal government, was able to integrate the interests of state and local governments seeking infrastructure and other assistance with foundering private businesses to relieve unemployment using the mechanism of government-funded construction projects.[45] Roosevelt expanded the RFC's powers, for instance, and during his presidency the agency authorized loans for agriculture, housing, state government, and disaster relief, among others.[46]

The first few years of the Great Depression significantly affected state and local economies. In mid-1930 Portland depleted its emergency fund.

Three years later 13.3 percent of Portland's workforce—forty thousand people—were on relief and twenty-four thousand home owners had registered with Portland's Public Employment Bureau. Through the decade the city stagnated economically.[47] To address such dire conditions, state agencies and city governments were quick to apply for federal funds under the RFC, PWA, and other programs. By 1933, Governor Julius Meier had established a Civic Emergency Committee to administer the distribution of federal funds throughout the state for unemployment relief, and Portland commissioners had also created a committee with this same name to address municipal unemployment.[48] Early in 1933 Portland officials began working extensively with citizen groups organized for purposes only recently allied: to abate water pollution and alleviate unemployment. Primary among these groups were the Portland Anti-Pollution Council—not to be confused with the Anti-Stream Pollution League of the late 1920s—and the Civic Emergency Federation— not to be confused with similarly named committees organized at the behest of state and municipal leaders.[49]

Portland leaders and the advocacy groups desired to respond as quickly as possible to opportunities for federal funding of public works projects. In June 1933, City Engineer Olaf Laurgaard and his boss, Commissioner of Public Works A. L. Barbur, authorized a member of both the Anti-Pollution Council and Civic Emergency Federation to develop preliminary designs and budget estimates for an activated sludge treatment plant. This person, Walter E. Baer, was not a licensed engineer himself, but with colleagues in the Civic Emergency Federation he had already been working on preliminary plans for this and other urban infrastructure projects.[50] Further, the city had not been able to budget the funds necessary to develop plans and specifications for a comprehensive sewer interceptor and treatment system, so it lacked the internal resources to complete the job within the short time available.[51] Thus, Baer was in the right place at the right moment when city officials needed some kind of preliminary plans to respond to the offer of federal aid. Baer submitted his plans on June 8. City officials then hired twenty semiskilled laborers and two foremen to assist in mapping outfalls and other basic tasks while the city's sewer engineer developed more detailed plans and estimates, including "an outline map showing the location of the proposed conduits, pumping plant, and treating stations for a sewage disposal system."[52] The city also worked with a state of board of consulting engineers—recently established by Governor Meier—who reviewed the plans and estimates and submitted recommendations on July 19.[53]

Commissioners used these somewhat-refined plans as the basis for their federal loan application, but they also needed to show the federal government they had a viable loan repayment method. To do this they convened a special election on July 21, 1933, to ask Portland residents to approve a $6 million "self-liquidating" bond issue—the project would pay for itself through the sale of interest-bearing bonds.[54] Voters approved the plan by a wide margin—47,029 to 23,395, a result Barbur's replacement Ormond Bean characterized as "the emergency character of the times, as well as the intense campaign" of the Civic Emergency Federation, Anti-Pollution Council, and other groups.[55] In spite of work done refining Baer's preliminary plans and budget, and even with voter approval of the funding mechanism, in August 1933 the PWA deemed the plans inadequate and rejected the application.[56] Secretary of the Interior Harold Ickes and PWA administrators under his guidance disbursed billions in federal funds across the country, but proposals had to meet a core set of criteria. These included providing immediate unemployment relief while creating lasting and socially useful projects that adhered to sound engineering principles and technical guidelines. Worthy projects must also be proposed by financially stable applicants who would support the legal enforceability of securities used to fund the project.[57] Portland's application failed to meet the criteria of sound engineering and a viable repayment scheme.

Shortly after this setback, in September 1933 the city hired Boston engineer Harrison P. Eddy to consult. Commissioner Bean noted Eddy "has for many years been considered one of the great authorities on sewage disposal in the United States, having been engineer or consultant on over a billion dollars' worth of such projects." His authority was built over the course of forty years of experience that included cofounding a sanitary engineering firm in 1907 and consulting on sewage treatment design and construction for large municipalities throughout North America, including Chicago, Detroit, Buffalo, New Orleans, Milwaukee, and Pittsburgh. He had authored or coauthored technical papers and textbooks that established the foundation for sewage system design and engineering.[58] Of further importance to Bean and his fellow commissioners, Eddy was also on the PWA's Advisory Board, which, as Bean observed, "is the court of last resort in regard to engineering matters concerning PWA."[59] By bringing Eddy in to consult on Portland's project, city leaders were admitting their problems required both world-class expertise to address from a technical standpoint, and national-level influence to address from the financial perspective; conditions had become so severe that they were beyond the level of local expertise to resolve.

In spite of Eddy's input, however, in December 1933 the PWA rejected a $461,000 application for a section of sewer interceptor in southeast Portland.[60] Frustrations continued through 1934 for Portland city leaders, abatement advocates, and groups pressing for unemployment relief. In February, the PWA approved Portland's application for a $2 million grant for sewer-related work but could not allocate funds because they had already provided the state its full allotment.[61] In July, the PWA finally offered city leaders a $2.24 million grant contingent upon sale of $6 million in bonds.[62] Buyers for the bonds could not be found, however, because they still considered the plans inadequate. Though city staff and three consulting engineers enhanced Baer's original ideas and Eddy had consulted on the project, bond attorneys still found the plans to be "very indefinite . . . and so general that they meant practically nothing." Therefore, without a solid basis for a bond sale, Portland officials could not collect the PWA's grant money.[63] In light of these repeated failures, the Portland City Council submitted a charter amendment to voters in November to authorize $6 million in self-liquidating bonds. If buyers could not be found, the amendment stipulated the city would support the project out of general tax funds. Voters said "no" to the prospect of shouldering the entire financial burden while the federal government continued to offer grants and low-interest loans.[64]

Into early 1936, Portland city officials attempted to find ways to issue the sewer funding bonds legally, in spite of their string of setbacks. The Oregon Supreme Court halted these attempts in March with its decision that the $6 million funding method voters approved in July 1933 was tied to the original, incomplete plans and estimates Baer submitted—in other words, commissioners could not swap-out Baer's proposal with a more technically sound and feasible set of specifications and estimates. To resolve these legal and financial complications and start afresh, the city council prepared a new "pay-as-you-go" funding solution to replace the earlier method at the November 1936 election.[65] The pay-as-you-go approach proposed to build out Portland's sewer infrastructure incrementally as the city collected the requisite funds for each section through an additional service charge to city water rates. To help flesh out the details, Commissioner Bean convened a sewage disposal committee composed of former state game commissioner and current president

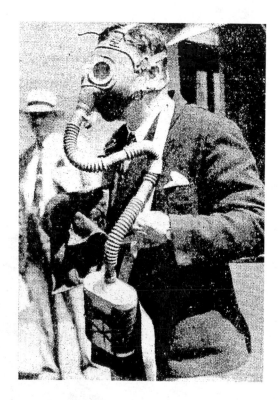

Fig. 2.6. Will R. Lewis, "foe of filth," dons a gas mask along the banks of the Willamette as a member of the City of Portland's sewage committee touring Portland Harbor in August 1936 (*Morning Oregonian*, Aug. 5, 1936, sec. 1, p. 3).

of the Oregon Wildlife Federation Edgar F. Averill and Portland attorney David Robinson—both of whom were active in the Portland Anti-Pollution Council in 1933—along with former state legislator and prominent businessman William F. Woodward, State Sanitary Engineer Carl E. Green, and Oregon State College Sanitary Engineer Fred Merryfield, among others.[66]

To experience conditions firsthand, committee members went on a cruise of Portland Harbor in early August 1936. Commissioner Bean commented that the river "certainly is black. I didn't realize how dirty it is," to which Averill replied, "Yes, and it keeps getting worse for the next couple of months." The *Oregonian* reporter who joined this trip continued:

Leaving from the foot of Stark Street, the boat nosed upstream on the west bank and passed three partially exposed sewers, all emptying into log booms, between the Morrison and Ross Island bridges. . . . William F. Woodward said . . . "There is almost criminal negligence of city officials to permit this . . ."[67]

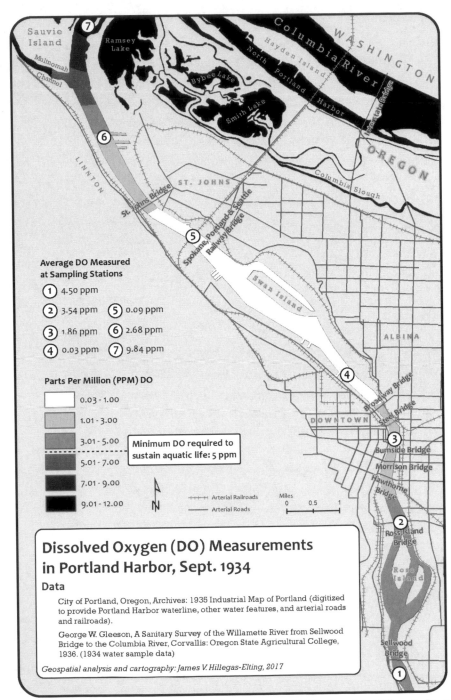

Fig. 2.7. Empirical data on dissolved oxygen collected in the mid-1930s showed the Willamette River through Portland Harbor to be well below the threshold to support salmonids and other native aquatic life-forms during annual low-flow periods (James V. Hillegas-Elting, 2017).

In spite of this direct evidence, at their next meeting committee member R. R. Bullivant, an attorney associated with the state's Dairy Cooperative Association, raised questions about "the degree of pollution of the river." He asserted the term pollution was "very indefinite," as was "the financial position of the city and the ability of the city to pay for any improvement." He "supposed there were foreign substances in the river which were not desirable," but asserted "there was evidence the public health was not in danger . . . and no adequate evidence that commercial fishing or the supply of fish is in danger." With the river visit at the forefront of their minds, some of his incredulous colleagues on the committee asked "whether or not Mr. Bullivant was a paid advocate," to which he responded he was not. William Woodward said "he regretted he had not touched a match to the mass of corruption they saw on their river trip." Another committee member suggested that "unless there were evidence of a reliable source or testimony and facts of experts who were willing to offer contrary evidence, it should be concluded that the river is polluted."[68]

A MATRYOSHKA DOLL OF ORGANIZATIONS AND ACRONYMS
NEW DEAL STATE AND REGIONAL PLANNING

As a member of Commissioner Bean's sewage disposal committee commented, the residents of Portland were "the greatest sufferers and the greatest offenders" among Willamette Valley cities. This was because wastes from all upriver communities passed through Portland and because the city contained, by far, the largest municipal population contributing the largest amount of raw sewage.[69] While city officials repeatedly failed to resolve the concurrent issues of sewer system engineering, financing, and public support, and while industries were not addressing their wastes, pollution levels increased year after year. One *Oregon Journal* reader commented upon the seasonal appearance of sewage sludge banks along the shores of the Willamette that "at times of unusually low water . . . [became] exposed and fester[ed] in the sun" and then attracted swarms of flies potentially carrying disease.[70]

Festering putridity was not healthy for fish caught either for recreation or for commerce, and the increasingly polluted Willamette eventually brought together recreational and commercial fishing interests. These groups had been at odds for decades, highlighted by their staunchly opposing views to the proposed closure of commercial fishing at Willamette Falls in the late 1910s and early 1920s. Representing recreational fishing interests, the

Multnomah Angler's Club was among sporting groups supporting the clo-sure, while the Columbia River Fisherman's Protective Union was at the fore-front of industry groups opposed. When the state decided to close the falls to commercial fishing, the industry was incensed.[71] A few years later, pollution had gotten severe enough to bring these former opponents together. In 1935, the Fisherman's Protective Union threatened to bring suit against polluters because of a large fish kill caused by a "lethal brown stain of pulp mill acid" that spread a "15-mile blanket of death" in Multnomah Channel downstream of Portland. Under authority of a 1927 state law empowering aggrieved parties to bring suit against polluters in any drainage west of the Cascade Mountains, the Fisherman's Protective Union charged municipalities and industries from Eugene to Astoria for damage to fisheries suffered as a result of the unsani-tary state of the river.[72] The suit faded from the newspapers after August 1936, but the alliance between sports and commercial fishing interests against their common foes persisted.

In their attempt to alleviate such conditions, Portland city officials had failed to secure RFC and PWA funding for the city's sewers. Oregon State officials had better luck getting funds from another New Deal agency, the Works Progress Administration (WPA). The WPA was another of Roosevelt's measures to relieve unemployment by infusing federal funds directly into state and local government projects. Established in 1935 and renamed the Works Projects Administration four years later, the WPA's funding supported a diverse array of public works projects and initiatives in the arts and media, among others.[73] In Oregon, the WPA funded many significant and lasting projects, including the Canby city hall, the State Library in Salem, an armory in Klamath Falls, and five spans across major Oregon estuaries. WPA monies also facilitated the reconstruction of Oregon's State Capitol building in Salem in 1938 after the original structure burned.[74] Oregonians' push to improve water quality benefited directly from the WPA in the employment of at least one staff member on the Oregon State Planning Board's Advisory Committee on Stream Purification.[75]

The Oregon State Planning Board (OSPB) created its Advisory Committee on Stream Purification in October 1935 in recognition of the importance of water quality as an integral facet of water resources planning.[76] Evolving from a predecessor body first organized under the auspices of President Hoover's RFC in 1932, the Oregon legislature created the OSPB in 1935 to provide fact-finding and advisory services used to promote conservation and sustained development of the state's natural resources.[77] Oregon's planning efforts were aligned with those of similar bodies in Washington, Idaho, and Montana within the Pacific Northwest Regional Planning Commission. Regional civic leaders formed this commission in January 1934 to coordinate Columbia Basin water resource management projects centering on transportation and business development.[78] Just as Oregon's planning board operated under the regional planning commission, a federal organization helped coordinate regional planning efforts. The federal government took this active role in June 1934 with creation of the National Resources Board (NRB). Conservatives assailed efforts to facilitate this kind of comprehensive national planning. When the US Supreme Court ruled the NRB unconstitutional, Roosevelt replaced it in June 1935 with the similarly named National Resources Committee (NRC). This replacement operated until 1939, when the president created the National Resources Planning Board (NRPB). In the context of progress on nationwide water quality issues, these changes are unimportant: between 1934 and the dissolution of the NRPB in 1943, the entity's purpose remained fundamentally unchanged.[79] In its day-to-day work, the NRB/NRC/NRPB provided funding, tools, and resources to regional and state planning organizations.

Within this Matryoshka doll of New Deal initiatives, Oregon's Advisory Committee on Stream Purification was a shorter-lived, fifteen-member body nested within a much larger state planning organization, which was itself nested in a much larger regional planning organization; this work, in turn, was part of a nationwide planning framework. The committee provided important contributions in addressing water quality statewide, and it took the lead on Willamette Valley water pollution research for the critical period of late 1935 until early 1937. Members collected, compiled, and disseminated information and recommended legislative solutions.[80] Enabling this work was unprecedented interest in and funding for water quality research. This is one tangible way in which New Deal support for state and local efforts differed markedly from what President Hoover and his predecessors had provided.

Advisory committee members were aided immensely by research at the local, state, and federal levels. Engineers and scientists at Oregon State College, in collaboration with others, had published the first substantive, quantitative reports about Willamette River water quality between 1929 and 1936. In July 1935, a team within the federal National Resources Committee provided the same kind of information for the entire nation. The NRC's advisory committee on water pollution—chaired by the nationally respected sanitary engineer Abel Wolman—published a survey concluding that pollution was outstripping efforts at abatement and urged cooperation between state agencies, municipalities, industries, and other interest groups to adopt treatment remedies to meet site- and use-specific water demands. The NRC supported maintaining responsibility for water pollution abatement at the state level, but recognized that interstate and international pollution issues would require federal regulation or interstate compacts. The NRC associated water pollution with decreased recreational opportunities, degraded aesthetics, and threats to aquatic life, but overall it found public interest in the matter lacking.[81] The NRC's water pollution report and creation of Oregon's Advisory Committee on Stream Purification linked local and national concerns about degraded water quality.

RESEARCH INTO ACTION
PRESENTING A STREAM PURIFICATION BILL TO THE GOVERNOR

The Advisory Committee on Stream Purification (ACSP) met for the first time in December 1935, a few months after the NRC's water quality report. The committee was a product of both the failures and achievements of ten-plus years of organized pollution abatement activities, as well as the unprecedented focus on water quality from state, regional, and local government officials. Its fifteen members were diverse and reflected interests that had been meeting on this topic for many years. These included a representative from the Pacific Coast Association of Pulp and Paper Manufacturers; vice president of the Inman-Poulsen Lumber Company; an executive with the lumber enterprise Crossett Western Company/Crossett-Watzek-Gates Industries; an Oregon State College civil engineering professor; and the mayor of Portland.[82] Fish and game advocates on the advisory committee included the state game supervisor and state fish commissioner—both of whom were also members of the Izaak Walton League, as were two additional members, Edgar Averill and David B. Charlton.[83] At least four members also belonged to the Portland City Club.[84]

Portland officials had failed to secure federal funds to upgrade its sewers, but in March 1936 state officials were able to get WPA monies to support attorney John C. Ronchetto's work for the advisory committee. He and other members produced four reports in less than a year.[85] The first report (August 1936) assessed water quality laws in Oregon and throughout the United States. The second (February 1937) outlined statewide pollution abatement efforts with a focus on the Willamette Valley. Another (June 1937) recommended principles for the formulation of future Oregon laws. Industrial pollution throughout Oregon was the subject of the final report (June 1937).

In early 1937, ACSP members applied their research and helped draft a bill for the state legislature to consider.[86] This proposal differed significantly from that of 1927 in that it did not specify the time frame within which cities and industries were required to build treatment plants. It also proposed to establish preservation of "the natural purity of navigable and other streams and of the lakes and coastal waters" for public health, recreation, and conservation as official state policy. To do this, the bill proposed creating a Sanitary Commission composed of the state's sanitary engineer, health officer, master fish warden, a member of the game commission, and a member of the general public. This body would approve sewage system plans, pursue individual, municipal, and corporate violators, and distribute state and federal funds. The commission would also have authority to conduct research, disseminate information, and foster cooperation among individuals, cities, and businesses.[87] Drafters had taken elements and inspiration from at least eight other states—Pennsylvania and Wisconsin as well as Washington, Ohio, and Rhode Island.[88] Democratic Senator Byron Carney of Milwaukie introduced the proposal as Senate Bill (SB) 392 on February 24, 1937.[89]

Carney served as an Oregon state senator from 1935 to 1939. Conservative Republican C. C. Chapman, editor of the *Oregon Voter*, said of Carney that he "was not a major factor in the senate" in spite of the respect he earned for his "booming voice, vigorous delivery," and the "fine scholarship" of his speeches. Carney was born in Illinois in 1875 and labored in various fields before he moved to Portland in 1918 to work in the shipyards. Among other things, he taught school, served as a Wyoming circuit court clerk, and was appointed as a First Lieutenant in the Ninth Illinois Volunteers during the Spanish–American War of 1898–1899 (the unit did not see active service). Additionally, he was a Methodist missionary in Illinois and Wyoming for seventeen years. While he served as an Oregon senator, he was listed as a "semi-retired carpenter."[90] Chapman labeled Carney a "left-wing" senator and characterized him as "the crusader to be expected as an ex-evangelist and early political follower of William Jennings Bryan." As such, Carney supported the grange and labor movements and critiqued some aspects of the American capitalist system in his advocacy for public ownership and a production-for-use (instead of profit) economy. Carney also supported Dr.

Francis Townsend's proposal for a nationalized system of support for senior citizens—"Townsendism"—a movement that served as the catalyst for the 1935 federal Social Security Act.[91] During his tenure as senator, Carney proposed legislation to reform the state's mental institutions, chaired the Senate Committee of Fish and Fisheries, and sponsored the senate bill repealing Oregon's criminal syndicalism law that restricted constitutional rights to assembly. He also supported public power initiatives, and in 1938 completed a survey on the use of power from Bonneville Dam at the request of Bonneville Power Administration head J. D. Ross.[92] These achievements led Oregon Commonwealth Federation Executive Secretary Monroe Sweetland to characterize Carney as "a leader of the liberal forces in outstanding support of every measure of social justice," and another Commonwealth Federation member to laud him as a "proven New Dealer."[93]

In February 1937, Carney's colleagues in the Senate requested relatively minor amendments to SB 392.[94] However, Carney withdrew it entirely and on March 3 submitted a revised version as SB 414.[95] The precise reason why Carney replaced SB 392 with SB 414 may never be known, but historian Floyd McKay's characterization of Salem politics in the late 1930s suggests a possible reason. According to McKay, the Oregon legislature was an "old boys' club" where work was often done "in smoky committee rooms whose doors were frequently closed when votes were taken."[96] Another clue is to be found in testimony F. H. (Frederic Harold) Young gave to the Oregon Senate committee considering Carney's bills. Young was a leading expert on taxation and finance who frequently spoke at events throughout the state and was no stranger to contentious political issues involving government administration and taxes. After graduating from the University of Oregon in 1914 and serving in World War I, he worked for a number of investment firms. For eleven years he was financial editor of C. C. Chapman's conservative *Oregon Voter*, and, among other affiliations, he was executive secretary of the Oregon division of the Security Owners' Association. He supported sales tax proposals in the 1930s but was staunchly opposed to public power initiatives and property tax measures.[97] Through the University of Oregon Alumni Association, Young had strong connections to influential men such as US District Court Judge Robert S. Bean, lawyer and legislator Homer D. Angell, and Portland attorney John C. Veatch.[98]

Speaking to the Oregon Senate's irrigation and drainage committee on March 1, 1937, Young said he opposed Senator Carney's bill for two primary reasons. One of his claims seems hyperbolic and echoes reactionary responses to pollution abatement that industries throughout the United States had made: Young asserted that were SB 392 to become law, it would "put virtually every paper manufacturing plant in Oregon out of business."[99] Such broad and oft-repeated claims are difficult to evaluate accurately, but they do reflect Young's previous stances supporting the transfer of tax burdens from businesses to individuals through a sales tax and opposing public power projects, a stance that would benefit private power companies such as the influential private utility Portland General Electric.

Young's second point of opposition was more substantive. He observed an array of federal and state water quality–related proposals that had yet to be resolved, and suggested legislators not enact Carney's bill while these were still in flux. At the federal level, Congress was debating the Vinson-Barkley bill (HR 2711), a proposal many municipal and state leaders opposed because it sought to centralize water pollution control administration at the federal level.[100] Within Oregon, the water pollution conundrum was among topics being studied by the state planning board, the state game commission, and a team working on a project to build multiple dams along Willamette River tributaries for storage, hydropower, irrigation, and water quality (the Willamette Valley Project). In light of these ongoing debates, Young requested "the state should delay any drastic action."[101] Whether by the influence of lobbyists in smoke-filled rooms, the entreaties of F. H. Young, or another compelling reason—or some combination of these—Carney's original bill was not long-lived.

Carney's replacement, SB 414, was much less robust than its predecessor. It provided for court injunction against polluting practices in much the same way as SB 392, but narrowly defined pollution as "sewage or any other noxious or deleterious substances by any *municipal corporation*" [italics mine]. Research had shown that pollution from Willamette Valley pulp and paper mill wastes *alone* was equivalent to about 601,000 persons—which was about twice as many inhabitants as Portland's 1930 population of 301,815; however, industrial and other wastes were excluded from SB 414's definition of pollution.[102] Also, while the bill retained language making it official state policy to ensure the "natural purity" of its waters, it did not centralize authority for maintaining water quality within a state body.[103] Centralization of such authority had been a prime recommendation of both the Advisory Committee on Stream Purification and the Anti-Stream Pollution League before this, in the late 1920s. Nevertheless, Carney's revised bill passed both houses on March

8.[104] The *Oregon Journal* lamented that SB 414 had "lost its teeth" by not creating "an official agency armed both with the authority and duty to set out actually to abate pollution as the public nuisance for which it is."[105] In spite of its perceived toothlessness, Governor Martin vetoed Carney's bill four days later.

Oregonians elected retired US Army General Charles Henry Martin to the governorship in 1934. After service in the Boxer Rebellion, Spanish–American War, Philippine–American War, and World War I, he and his wife Louise (Hughes) retired to Portland in 1927, where the Martins had real estate and other investments.[106] He ran for governor on the Democratic ticket after serving two terms as a Democratic representative in the US Congress, where he was first elected in November 1930.[107] Martin was avidly pro-business and perceived the proper role of government at all levels was to provide for individual business opportunities—and then get out of the way. Martin advocated for the use of state and federal taxpayer funds to support channelization and damming of the Columbia River, but primarily for the purpose of business development. Regarding the highly charged issue of hydropower development along the Columbia, Martin opposed public power, a stance that would directly benefit private power companies such as Portland General Electric, a company Martin's old friend—and one of his primary campaign contributors—Henry Cabell invested in.[108] Martin's "survival of the fittest" approach to economic development permeated his views on social and political issues. His views, at times, were quasi-fascist in that they advocated strong-arm tactics and intolerance of rivals. When faced with labor unrest along Portland's waterfront and elsewhere in summer 1935, Martin ordered law enforcement officers to "crack their damn heads!" Infamously, Martin held eugenicist beliefs, and publicly advocated for state execution of nine hundred elderly and enfeebled residents of the state's Fairview Home in Salem to help balance the budget.[109] Martin was a member of Congress during Roosevelt's first hundred days in office (March 4, 1933–June 16, 1933), during which the Democrats passed many landmark bills central to the president's "New Deal." As his biographer Gary Murrell documents, however, Martin was absent from the legislative chambers or did not vote on many of these proposals, including unemployment relief, industrial recovery, the Civilian Conservation Corps, and Public Works Administration, among others. Lacking explicit evidence for Martin's absence, Murrell posits that "despite his rhetorical support for Roosevelt's election and his declarations of loyalty, the general could not induce himself to ratify what he perceived as dangerously 'Socialistic' New Deal legislation."[110]

Martin vetoed Carney's SB 414 with the justification that it had been passed hurriedly, without proper consideration. He claimed the law was discriminatory because of the way it defined pollution. This definition would, in his opinion, place an undue financial burden on municipal taxpayers. Citing the projected $15 to $18 million cost of Portland's proposed sewage disposal infrastructure, Martin predicted taxpayers in Portland and other municipalities would "unquestionably attack the act for its unjust discrimination."[111] Fellow fiscal conservative F. H. Young criticized Martin's action by asserting Portland city officials influenced the governor's decision. He claimed the legislation "put Portland in the tender spot of doing something before the taste of a rejected bond issue for sewage disposal was out of its mouth," a reference to the city's repeated failures to pass sewer funding bond measures and secure federal public works funding since 1933.[112] Whatever the reasons behind the modification of Senator Carney's original bill and Governor Martin's disapproval of it, the veto set in motion a new, more urgent and more focused phase of Willamette River pollution abatement.

SUCCESS OUT OF FAILURE

While legislators, lobbyists, city officials, and abatement advocates had been debating in the halls of the Capitol Building at Salem and in the pages of the *Oregonian* and *Oregon Journal*, the Willamette River continued to putrefy. A notable example of this was the October 1935 "lethal brown stain of pulp mill acid" that had "spread a 15-mile blanket of death through the Willamette Slough" downstream from the St. Helens Pulp and Paper mill, resulting in dead fish "piled up in solid waves" along the banks. This event had spurred the Columbia River Fishermen's Protection Union to threaten to sue municipalities and industries from Eugene to Astoria for damage to the commercial fishing industry; with this egregious pollution, commercial- and sport-fishing interests had become allied in their advocacy for clean streams.[113] The following August, in 1936, state officials were called to investigate "the sudden appearance of a dark fluid substance" in Columbia Slough that had killed salmon and forced crawfish to crawl up on the shore.[114] That same year the Oregon State College Engineering Experiment Station published its bulletin on water quality in Portland Harbor, finding that its lower reaches were devoid of all dissolved oxygen during the low-flow months of July, August, and September.[115]

Responding to these conditions and lack of progress at the state or municipal levels, the US Bureau of Fisheries held a water pollution hearing in Portland in May 1937. Immediately thereafter, a new citizen group formed to carry the clean-streams banner: The Oregon Stream Purification League (OSPL). The state-sponsored (but temporary) Advisory Committee on Stream Purification had recently completed its work, and the Anti-Stream Pollution League had faded in the early 1930s. With failure to create a centralized state-level water quality agency, and Governor Martin's veto, the OSPL organized to fill the void.[116] It was a special committee under the auspices of Oregon Business & Investors, Inc., an organization similar to groups in other states bringing together business, government, and civic leaders concerned with issues of legislation and taxation; F. H. Young served as its president.[117] The OSPL itself was composed of influential leaders in conservation and state governance, such as State Treasurer Rufus C. Holman, Senator Frank Franciscovich, Senator Byron Carney, and naturalist William Finley.[118] Others actively involved in the league included members of women's groups and teacher's organizations, public health professionals, sanitary engineers—and a representative from a new ally to the cause, commercial fishing interests conveyed by the Columbia River Fishermen's Protective Union.[119]

As had its predecessor organizations, the OSPL understood its purpose as "centering public opinion" in the short term, and in the long term "enlighten[ing] the next session of the legislature." They also pledged to

Fig. 2.8. While Portland's busy harbor was vital to the regional economy, commercial and industrial activities contributed to river pollution. January 1937. *Oregon Journal* (Oregon Historical Society Research Library, bb006382).

include "as many women as there are men," to educate the public, and to "consolidate all of the diverse activities of the many existing but separate organizations" specifically on the issue of abating municipal sewage.[120] The OSPL initially focused its energy on abating sewage for at least four key reasons. First, sewage treatment technologies were much more advanced than industrial effluent treatment systems, making the former more practical to address. Second, municipalities, unlike industries, were eligible for many kinds of federal assistance—such as infrastructure funding from the PWA, planning aid from the NRB and its predecessors, and survey assistance from the public health service and bureau of fisheries. Third, the democratic process made city and state government officials more open to input and influence from advocacy groups, in contrast to the less-accessible hierarchies of private business firms. Finally, the influential F. H. Young, as head of the OSPL's parent organization, had recently shown staunch opposition to anything considered a burden on the state's industries. For these reasons the OSPL was much more likely to build consensus among its diverse membership and the general public, and thereby spur substantive state action if it focused on municipal rather than industrial wastes.

Senator Byron Carney chaired the league's committee drafting a citizen's initiative for the November 1938 ballot. Other committee members included Portland Commissioner of Public Works Ormond Bean and long-time abatement advocates Secretary of the State Board of Health Dr. Frederick Stricker and naturalist William L. Finley.[121] Filed with the state attorney general in December 1937, the proposal was similar to Carney's SB 392 and the 1927 proposal of the Anti-Stream Pollution League in that it established a state-level body in charge of water quality, the Oregon State Sanitary Authority (OSSA).[122] As with Carney's failed 1937 senate bills (but unlike the 1927 proposal), the OSPL proposal also established clean streams as Oregon public policy. The six members of the sanitary authority were to be appointed by the governor, and the state board of health would provide its state sanitary engineer to serve as secretary. The proposal would empower the sanitary authority to establish standards, conduct investigations, hold public hearings, and encourage voluntary cooperation with rules and regulations; if cooperation was not forthcoming, the sanitary authority was authorized to initiate proceedings against violators.[123] In all of these elements, the proposal was significantly stronger than any of the state's previous water quality laws. It also contained stronger enforcement provisions than the bill Governor Martin had vetoed.

Fig. 2.9. "Clean the Rivers!" campaign literature, fall 1938, which pollution abatement advocates prepared for Portland voters (City of Portland Archives, folder: Reports on Sewage Disposal Project 1951–1959, box: Early Sewer System Data Sewer Systems 1923–1972 (8890-02), A2011-024).

The initiative met with widespread public support. The politically divergent *Oregon Journal, Oregonian,* and *Oregon Voter* supported it for explicitly balancing interests of municipal officials, industry representatives, conservation groups, and the general public. The Portland City Club approved of it as well.[124] The OSPL, Izaak Walton League, Portland Chamber of Commerce, Oregon Grange, and many other groups helped educate the public and gather the requisite signatures to place the initiative on the November 1938 ballot.[125] Also on this ballot was an item addressed specifically to Portland voters. For the fourth time since 1933, residents were to be asked to approve a sewer system funding plan, this time a "pay-as-you-go" method. City commissioners who drafted this iteration—Ormond Bean and J. E. "Jake" Bennett—understood industries and other communities in the Willamette Valley were still looking to the state's second-leading polluter to take proactive abatement

Fig. 2.10. Portland Mayor Joseph Carson beating the drum for clean streams, October 1938 (Oregon Historical Society Research Library, ba018042).

steps. These men also saw the sewage infrastructure project as a way to allevi-ate unemployment. Commissioner Bean, at least, also supported it to avoid being forced into action from "panicky demand by the people because of health conditions" or by federal and state legislative mandate.[126] Throughout 1938, a wide range of advocates—representing public and private interests, organized alliances, and less formal groups—campaigned for passage of both the OSPL's state citizen's initiative and the City of Portland's funding plan. The proposals were intimately linked in addressing water quality within the state's most densely inhabited and industrialized watershed. Because of this, positive press for one cause benefited the other, evident in "Clean the Rivers!" campaign literature advocates targeted to Portland voters.

A favorite tactic of Oregon's abatement advocates was a hands-on, graphic tour of the befouled watercourse. As part of the campaign of Portland's unsuc-cessful attempt to pass a sewer funding measure in November 1936, *Oregonian* reporters accompanied advocates on a tour of Portland Harbor, where the newspaper's photographer snapped a picture of one man on the tour wearing a gas mask to bring attention to the river's unpleasant odor.[127] Campaigners in 1938 used similar tactics. In October they brought Miss Oregon down to the river to pose for cameras while she and others looked on at the slow suffocation of "finny guinea pigs"—hatchery fingerlings—that died within minutes of being immersed "in the poisonous waters."[128] Days before the election, Portland Mayor Joseph Carson even donned a deep-sea diving suit and immersed himself into thirty feet of "dun-colored water" that had "large quantities of sewer gas bubbling to the surface" and "varying amounts of floating sewage."[129]

Campaigners appealed to voters' intellects as much as they appealed to passionate rage and disgust. Dr. Richard B. Dillehunt, dean of the University of Oregon Medical School and chief surgeon at Shriner's Hospital, headed the campaign to pass the Portland measure. Dr. Abel Wolman also visited the state to campaign. As the nation's leading sanitary engineering expert and chair of the Water Resources Committee of the federal National Resources Committee, he was well aware both of the condition of the Willamette and the context of abatement efforts in Oregon relative to the rest of the United States.[130] Wolman's involvement echoed Harrison P. Eddy's contributions in 1933 by representing Portland city leaders' admission of the need to bring top-level expertise and influence to bear on local matters of pollution abatement; problems had become that acute. He told a Portland audience in August 1938, "Your river is pretty sloppy," but he saw the example of Portland's 1938 sewer funding proposal as proof that progress could be made without federal mandate.[131] Wolman perceived the proper federal role was providing research aid and financial assistance and in encouraging interstate cooperation, but "administration of pollution abatement is best performed by State and inter-state agencies."[132] This view echoed Portland Commissioner Bean and other city officials and reflected the conservative side of ongoing national debates regarding the proper relationship between state and federal agencies on water pollution and other issues.

Abatement advocates also appealed to voters' economic concerns that were rarely, if ever, far from the forefront—particularly during the Great Depression. Commissioners developed Portland's pay-as-you-go strategy expressly to provide a slow-but-steady stream of funds to construct the city's new sewage infrastructure one segment at a time. Staunch abatement advocate William Finley also presented economic arguments in favor of pollution abatement. In October 1938 he stated that "expenditure of $12,000,000 to clean up the pollution of the Willamette river will increase property values of the city by $50,000,000."[133] To speak to other voters' primary interests,

supporters of the citizen's initiative also harkened to Oregon's romantic pioneer past. The OSPL presented an argument in support of the initiative in
voters' pamphlets that urged Oregonians to recall that "the pure and sparkling
waters of Oregon's lakes and streams delighted the pioneers and early settlers."
They reminded voters Samuel Simpson had written a paean to the Beautiful
Willamette, but presented Simpson's elegy with a contemporary twist:

In its crystal depth inverted
Pours the sewage of the town
From the mills and the industries
Floods of waste come pouring down . . .[134]

As the November 8 votes were tallied, a proponent of pollution abatement might have reflected upon events of the past decade and expected a
victory. In 1927, sewage and industrial wastes choked Portland Harbor and
befouled the river up to Eugene, but no solid data yet existed to quantify the
pollution, and almost annually through the 1930s pollutants in the river killed
fish and forced crawfish out of the water. Responding to this in the 1920s was
a small, but increasingly more organized, pollution abatement constituency,
which Portland officials and industry representatives had given short shrift.
The work of the Advisory Committee on Stream Purification and involvement of F. H. Young's Oregon Business & Investors group had expanded this
constituency considerably. A 1927 state legislative proposal made it no further than the drawing board, and a 1937 bill had made it to the governor's
desk before being vetoed; with such momentum, it must have seemed likely
the 1938 proposals would be approved. This is what occurred: voters passed
the Oregon Stream Purification League's Water Purification and Prevention
of Pollution Bill by a landslide vote of 247,685 to 75,295, thereby establishing
the Oregon State Sanitary Authority.[135] The measure creating a pay-as-you-go
funding mechanism for Portland's sewer system also passed by a wide margin.[136] Oregon voters had overwhelmingly approved the strongest water quality measures in the history of the state. At the end of 1938, Oregonians looked
ahead to substantive changes in municipal and industrial waste disposal practices and hoped for a significantly cleaner Willamette River.

Chapter 3
Sewage and the City

Even without a soundtrack, the color video footage has a recognizable narrative structure and unambiguous message. The film begins in the pristine headwaters of the Willamette River, deep in Oregon's forested mountains. Clear water comes forth from the green earth to tumble over moss-covered rocks and boulders, moistening adjacent ferns and branches as gravity carries it downhill. Rivulets join the stream as it cascades until a strong torrent wends its way to the valley floor, where in places it slackens into pools, roils over boulders, and laps at shorelines. The first city the river meets as it flows north is Springfield, where a small pipe peaks from riverbank foliage, discharging some kind of liquid. Next is Eugene, where a sewer outfall of about four feet in diameter drains directly into a slack pond. From it issues cloudy, debris-laden, red liquid—perhaps from a meat-processing facility—and a distinct mixing zone forms in the river where red pollutants join clear waters. Downriver from Eugene the film records a patch of slimy red algal growth clinging to underwater rocks and twigs.

Somewhere between Eugene and Corvallis, footage shows another iteration of the "finny guinea pig" experiment clean stream advocates used during the 1938 citizens' initiative campaign. In a large glass jar of clear water, a few hatchery fingerlings are shown, perfectly lively. A technician then pours the jar into a screened box that sits in the Willamette's red water. Above the box is a timer. The fingerlings swim around casually for a short while before their movements become erratic: the fish move along the box's edges, frantically seeking to escape containment. They then begin poking their noses out of the water, seeking oxygen. After a short while, their movements become frenetic, spasmodic, as they twist in tortured circles seeking escape. A few seconds later, their movements begin to slow until they stop, turn on their sides, and float to the surface. Dead. The technician scoops the fish into a jar of red river water and exhibits the deceased fish to the camera, like so many inert pickles.

On down the river the film crew goes: more thick, red-colored effluents in Corvallis; more fingerling experiments near Crabtree on the Santiam River downstream of the pulp and paper mill. Nearly in the shadow of the State Capitol building in Salem can be seen an outfall pouring tan-colored foam and froth into the river, and the dun cloud creeps north along the riverbank for hundreds of feet.

The film provides a view of Willamette Falls and the two pulp and paper mills there before moving on to Portland Harbor, where truly spectacular pollution is evident. The same waters providing buoyancy to the stately retired battleship USS *Oregon* is saturated with filth and detritus of such variety and concentration that all previous examples from upriver communities seem quaint. In the harbor are thick mats of woody debris, human excrement, meat processing wastes, and dead fish and rats floating entwined in sludge. Large outfalls convey uninterrupted waterfalls of sadly colored muck and slime directly into the harbor. The images are so clear and unambiguous that one can almost smell the feces, rotting meat, and festering vegetation. Shockingly, these conditions fail to deter two hardy fishermen who cast their lines within the splash zone of an outfall. As if to mock Portlanders, the camera looks up from the river to see Olaf Laurgaard's imposing Ankeny Street sewer pump station along the city's western harbor wall, built not to divert wastes from the harbor but to keep them from backing-up into downtown basements during floods. The pump station merely conveys its raw sewage a few blocks farther downstream. The viewer witnesses other examples of seriously degraded water quality during the film's thirty-nine minutes. When footage ends at the confluence of the Willamette and Columbia, the message is clear: "reasonable use" has become unreasonable.

Frustration motivated conservationist William Joyce Smith to create this film in late summer 1940.[1] Smith was state manager of the National Life Insurance Company. His recreational pursuits and interest in the outdoors aligned in his service as president of the Oregon Wildlife Federation and membership in the Izaak Walton League.[2] He was also a director-at-large of the National Wildlife Federation in the early 1940s, at the same time that William L. Finley was one of the vice presidents of this organization.[3] The Oregon State Sanitary Authority (OSSA) helped Smith create his film by providing staff and equipment—it was OSSA technicians who subjected fingerlings to the lethal Willamette. Sanitary authority members viewed Smith's film with approval at their December 13, 1940, meeting and authorized funds to reproduce and distribute it.[4] Smith had been showing the film at community meetings for leaders of the Izaak Walton League, Salem Hunters' and Anglers' Club, Oregon Wildlife Federation, Troutdale Rod and Gun Club, and others as part of a campaign to pressure Portland officials into moving forward with substantive sewage abatement plans.[5] The color film was an effective form of communication because it graphically illustrated how pervasive pollution had become and how detrimental to fish life the formerly beautiful Willamette now was.

Smith was not the only Oregonian frustrated by a perceived lack of progress. Nearly two years after voters overwhelmingly passed laws to create the sanitary authority and approve a funding mechanism for Portland's sewer system upgrades, water quality continued to wane and city officials continued to resist moving forward with the sewer project. By late 1940 abatement advocates understood that even after creating a centralized state water quality agency, improving the health of Oregon waters would continue to require active citizen involvement. During winter 1938, advocates had reason to feel positive about the future of their Beautiful Willamette. Less than two years later, however, they were organizing once again to push for clean streams.

FUNDING THE UNFUNDED MANDATE AND DEFINING THE WORK TO BE DONE

The OSSA differed markedly from any of the state's previous approaches to water quality because it consolidated forty years of incremental and widely disbursed regulations into a single organization created expressly for this purpose. Administratively, the authority existed within the state's board of health. It had its own staff—quite small at first—but relied upon the board of health for additional staff and laboratory facilities. When voters created the OSSA in November 1938, Oregon became the eighth of fifty-seven states, districts, or Canadian provinces to centralize water quality oversight within a single body. Pennsylvania, Wisconsin, Illinois, and New Jersey were among the heavily industrialized US states to have done so previously, beginning in the early 1920s. Oregon was unique among other North American examples, however. It was the only state or province in which a citizen's initiative centralized water quality oversight. At least through the 1940s, Oregon was also the only state or province that established clean streams as an explicit governmental policy.

Finally, the 1938 citizen's initiative was rare for the time in providing for enforcement of water quality regulations. Many state legislatures that created water quality boards through the 1940s authorized them to provide research assistance and monitor progress, but not to enforce regulations.[6] In contrast, the OSSA had the authority to establish regulations, conduct investigations, and hold public hearings. If efforts at securing voluntary compliance failed, the OSSA could work with the state attorney general to enforce its regulations and decisions. This collaboration could include clarifying state laws, holding citation hearings at which the attorney general would be present, and initiating court proceedings against offenders.[7] Pioneering in these ways, upon its creation the OSSA lacked one critical element: funding.

The initiative creating the OSSA came with much fanfare, but no funding mechanism. Governor Charles H. Martin set in motion the citizens' campaign to create the sanitary authority with his March 1937 veto, but he did not even make it to the 1938 general election to get a chance to run for a second term. Martin had angered many in his party at the state and federal levels for his reactionary statements and increasingly active resistance to President Roosevelt's New Deal. The Oregon Commonwealth Federation was at the forefront of an alliance of Democratic Party factions comprised of military veterans, public power advocates, groups seeking unemployment relief, and others who collectively defeated Martin in the May 1938 gubernatorial primary.[8] Historian Jason Scott Smith found that Charles Martin was not alone in motivating resistance within his state's Democratic Party: Democrats in Georgia, South Carolina, Maryland, and New York also ousted conservative, anti–New Deal party members in 1938. Among this group, however, Charles Martin of Oregon joined Clarence D. Martin of Washington at the forefront of those who "most thoroughly disgusted liberals."[9] Martin's last day in office was January 9, 1939.[10]

One of Governor Martin's final and most significant tasks in late 1938 was to submit a budget proposal to state legislators for the 1939–1940 biennium. Because the sanitary authority had not been voted upon when Martin submitted his budget, he requested no funding for it.[11] When legislators came together in early 1939, they faced both an unbalanced budget and the need to appropriate funds to the OSSA. The authority itself requested a biennial appropriation of $50,000. Fiscally conservative C. C. Chapman, who had been elected as a Republican state representative in November 1938, warned providing the full funding request would increase the state's deficit nearly 400 percent, to $63,116.[12] Chapman's former financial editor and head of

Oregon Business & Investors, F. H. Young, supported the OSSA's budget request. Young was not alone, as representatives of the pulp and paper industry, woolen mills, fish and game interests, and others were among the supporters.[13] In spite of this, in early March 1939, the legislature's joint ways and means committee proposed an appropriation of only $7,500. Fish commission chair and OSSA member John C. Veatch testified that $23,000 would be the absolute minimum for the authority to operate, saying "it would be better to refuse any appropriation than to allow [only] $7,500." Incoming governor Charles Sprague responded that, in his view, $15,000 would be adequate.[14] In the end, legislators authorized only $7,500, 15 percent of the OSSA's original request and .035 percent of a total biennial state general fund appropriation of $21,290,778.[15] Legislators authorized an additional $2,500 for the period January 1 through June 30, 1941, when the state shifted to a July 1 through June 30 biennial budget period; $10,000 total was still just 20 percent of the authority's original request.[16] Over the next few years authority members would continue to lobby for more funds, but in the midst of the Depression this would be all that the legislature would approve. When the OSSA finally received the significantly higher appropriation of $23,574 for the 1941–1943 biennial budget, members found it difficult to fill staffing needs due to increased employment of sanitary engineers by the US Army, Navy, and Public Health Service.[17] This difficulty only increased through World War II and was equally true for the City of Portland with regard to the design of the city's interceptors and sewage plant.[18]

In the midst of debates over its funding, the OSSA held its first meeting on February 25, 1939. As stipulated in the initiative, members included State Health Officer Dr. Frederick D. Stricker of Portland; State Engineer Charles E. Stricklin of Salem; State Fish Commission Chair John C. Veatch of Portland; and State Sanitary Engineer Carl E. Green. Other members were Blaine Hallock of Baker, Albert Burch of Medford, and Harold F. Wendel of Portland.[19] At their first meeting, members elected Wendel to serve as Chair. Wendel was born in Piqua, Ohio, in 1892 and came to the Pacific Northwest for the first time in 1917 as First Lieutenant in the US Army Ordnance Corps assigned to Fort Lewis, Washington. After the war, he joined the Lipman Wolfe & Co. department store company in Portland, then headed by his uncle, Adolphe Wolfe. He soon rose to the executive level with this firm and would remain with it the rest of his life.[20] He was a member of Temple Beth Israel and served on the Reed College Board of Regents and the boards of the

Fig. 3.1. Harold F. Wendel, Chair of the Oregon State Sanitary Authority 1939-1967. Gladys Gilbert photographer, 1941. *Oregon Journal* (Oregon Historical Society Research Library, bb015681).

Portland Symphony Society, Portland Chamber of Commerce, the Portland Council of the Boy Scouts of America, and other civic organizations. Some years later, he would earn the American Veterans Committee Freedom Award for his positive example of racial integration in hiring practices at Lipman Wolfe & Co.[21] Upon assuming the role of OSSA chair, Wendel expressed the pragmatic, cooperative approach he and the authority would take: "Being extremely practical," he said, the authority was well aware of the "many problems that exist in the effort of communities to provide sewage disposal facilities" and in the challenges industries faced in addressing their pollution, and he and the authority would favor any substantive abatement steps.[22] With this foundation, the OSSA immediately set about its work.

One of the authority's first tasks was to "expedite the Portland sewage disposal project." To do this, members instructed their secretary to "prepare a letter to be addressed to the city of Portland, in which the interests of the authority relative to the progress of the city's sewage treatment project would be made known." Another task was to gather baseline data from industries and municipalities. Reports from the Oregon State College Engineering Experiment Station earlier in the decade provided essential background material, but by 1939 this information was becoming dated with the continued growth of Oregon's population and economy. To ascertain "basic information for the intelligent planning of a stream pollution reduction program," the authority solicited voluntary survey responses from cities and industries. They understood that quantitative data would be the only way to identify pollution sources, understand effects, and generate practical solutions—and, if necessary, support a case at law against intransigent polluters.[23]

A FOCUS ON WASTES THAT ARE KNOWN TO CAUSE WATER POLLUTION

With some amount of funding secured, an organizational structure in place, and fundamental principles articulated, the OSSA next turned its attention to establishing a metric that both it and polluters could rely upon to gauge whether or not state water quality standards were being achieved. One way to approach this was to agree upon a definition of "pollution." Authority members discussed the difficulty of arriving at a definitive understanding of the kinds and quantities of substances that constituted "pollution," but did not come to a conclusion. In the Willamette Valley, "pollution" referred to raw sewage, both sulfite and sulfate waste liquors from pulp and paper mills, woody debris from lumber mills, meat processing effluents, flax retting wastes, and a host of other materials that, in sufficiently high concentrations, overwhelmed the river's capacity to assimilate. In other Oregon watersheds silt from mining or logging operations was a significant pollution source. With such diverse wastes to consider, Authority members concluded that they should not attempt to define precisely what constituted "pollution" without further research and consultation with the state attorney general.[24]

The attorney general advised the OSSA "not to attempt to define 'pollution' at this time, but, instead, to devote its time, primarily, to the control of sewage and industrial wastes which are known to cause water pollution."[25] Though at first this bit of circular logic seems confusing, it conveyed practical advice: rather than get lost in the impossible task of arriving at a universal definition of pollution, the authority should, instead, take practical action by focusing on those cases where the harmful effects of effluents was most egregious. The international community of scientists and sanitary engineering experts had not yet developed the tools and methods needed to understand pollution in all its complexities, so it would not have been realistic to expect Oregon's water quality agency to have done so.

The attorney general's advice was also consistent with state law and practices since the late nineteenth century. Though Oregonians had created the sanitary authority and authorized the City of Portland to build expensive sewage treatment infrastructure, in 1939 the very term "pollution" remained undefined and contentious. Specifying precisely what constituted pollution was complex—as complex as changing state laws, developing actionable engineering specifications, and determining effective ways to fund large-scale infrastructure projects. Oregon statutes since the late nineteenth century had, in various ways, prohibited the disposal of solid or liquid wastes that would despoil waters intended for domestic, commercial, or recreational uses. The majority of these statutes, however, relied upon a definition of "pollution" that was more *qualitative* than *quantitative*: if someone dumped putrefying animal wastes into a river, and a downstream water user suffered from it by getting sick, this was a relatively straightforward example of water pollution. Such direct causal relationships between substances deposited in waterways and public health did not rely on precise measurements of organic compounds, the effect of decaying matter on fish and benthic life, or the counting of bacteria harmful to public health. Complicating the qualitative determinants of water pollution were the compounding factors of population growth, industrial expansion, and increasingly complex and concentrated industrial wastes.

Tools to measure these variables evolved along with the ability of public health and sanitary engineering professionals to interpret the resulting data. As scientific knowledge increased and with rapid urbanization and industrialization in the United States and Western Europe, the term "pollution" itself evolved. Historian Adam Rome finds that until the Civil War Americans used "pollution" to mean the degradation of an individual's moral integrity. In mainstream Christian America, this meant acting in contradiction to community norms and dictates of the church. It was not until after the Civil War that Americans began to apply "pollution" to environmental degradation. By 1890, some observers were regularly using the phrase "river pollution," but even as late as 1915 Americans did not have a single word describing contamination of land, air, and water.[26] By the 1930s, Oregonians and other Americans generally equated "pollution" with a change in expected environmental conditions sufficient to require a modification of accustomed activities. People still characterized this kind of pollution in moral terms (and many still do so well into the twenty-first century). However, a concept that applied almost exclusively to moral perversion and corruption up to the middle of the nineteenth century had, two generations later, become closely associated with the negative environmental and public health consequences of modern life.[27]

Organizations such as the Lawrence Experiment Station of the Massachusetts Board of Health (beginning in 1886) and the US Public Health Services' Water Quality Laboratory (beginning in 1911) helped establish the earliest quantifiable definitions of the amorphous concept of "pollution." In 1910–1911, Earle B. Phelps of the Cincinnati laboratory worked with US Army Corps of Engineers Colonel William M. Black to conduct oxidation tests of various wastes in New York Harbor. The US Public Health Service extrapolated their findings into a general theory and model of stream purification.[28] Black and Phelps's study determined that as different kinds of organic wastes decompose they consume varied amounts of oxygen dissolved in the water. Generally, the stronger the wastes, the more oxygen is consumed during the decomposition process. This is called *biochemical oxygen demand* (BOD). Aquatic life requires *dissolved oxygen* (DO) for survival. Different life forms tolerate different amounts of DO, but as DO decreases there is a point at which water conditions will not support higher fish species such as salmon and trout. Eventually, water depleted of DO will no longer sustain photosynthetic life, and at critically low levels, the aquatic environment turns fetid and lifeless. Since BOD is a measurement of the oxygen requirements of a particular waste type during decomposition over a given period of time, Phelps and his sanitary engineering colleague H. W. Streeter developed standard BOD measurements at between five to twenty days. These measurements are expressed in units such as "BOD_5" or "BOD_{20}": a given quantity of effluent consumes X amount of dissolved oxygen over N days as it breaks down. Higher BOD measurements generally correlate with lower dissolved oxygen in the receiving water. Water temperature was also a significant factor in evaluating BOD, because higher-temperature water contains less dissolved oxygen than cooler water.[29] By analyzing these factors together, Phelps and Streeter created the "oxygen sag curve" water quality assessment model. Though it did not measure all pollutants—public health experts observed it had nothing to say about bacteria—the Streeter-Phelps approach became the first empirical water quality model.[30] With subsequent refinement, it remains an important metric well into the twenty-first century.

Given the complexity of this evolving science, rather than specifying a strict definition of "water pollution," in March 1940 OSSA members adopted a three-tiered water quality classification system that replaced qualitative, subjective interpretations with a metric that incorporated quantitative, empirical data within a flexible framework. It broadly reflected the approach of the Massachusetts Board of Health's Lawrence Experiment Station, first developed in the 1880s, as well as the approach Pennsylvania adopted in 1923 and Secretary of Commerce Herbert Hoover supported in his 1927 speech to the Izaak Walton League—but with an important difference. Oregon's system established three stream classes. Class A streams were to be maintained at the highest level of quality. They were to be suitable for swimming, recreation, sustaining aquatic life, and use as municipal water supplies, and could not receive untreated wastes. Class B waters were to be relatively clean but might require a higher level of treatment for domestic use than class A. Class C waters could receive temporary discharge of untreated wastes if the OSSA determined that such discharges would not preclude other reasonable uses.[31] Where this system differed from the Massachusetts and Pennsylvania examples was in the OSSA's rejection of giving over one class of streams entirely to untreated municipal and industrial wastes. They also tried to create flexible classification categories that fostered pollution mitigation while recognizing longstanding realities that had been evolving since white settlers began arriving en masse in the 1840s.

Oregonians were not alone in struggling with water pollution. In early 1939, the federal National Resources Committee contextualized Oregon's situation by publishing the nation's first comprehensive water quality report. Leading the group of experts behind this work was sanitary engineer Abel Wolman, who had visited Oregon a few months previously to campaign on behalf of Portland's successful sewer funding proposal. One map in this report exemplified the problem as starkly as would William J. Smith's film the next year. It showed that pollution in the Willamette and Lower Columbia river watersheds was on par with the more heavily industrialized northeast and Great Lakes. As such, the basin joined the western Puget Sound as the most polluted watershed in the Pacific Northwest and, along with a stretch of southwestern California, one of the four most polluted watersheds in the western United States.[32] In comparison, it is no surprise that the Northeast and Great Lakes were experiencing severe pollution. As of 1930, these regions made up 14 percent of the nation's land area but were home to 65 percent of its urban population. Population density in the Northeast and Great Lakes of 40.5 people per square mile was significantly higher than the 14.1 per-square-mile average for the rest of the country. Compounding this higher population concentration, these regions were also home to a large percentage of the country's industrial facilities.[33] That the Willamette was

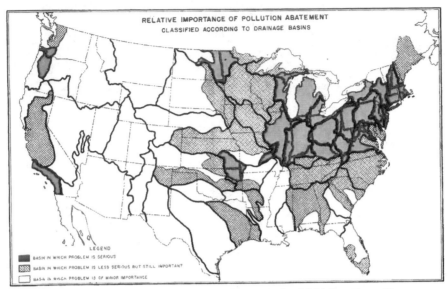

Fig. 3.2. A 1939 map from the National Resources Committee Special Advisory Committee on Water Pollution showing pollution in the Willamette and Lower Columbia watersheds (upper left) to be on par with pollution in the heavily industrialized Great Lakes and Northeast regions. *Report on Water Pollution in the United States* (Washington, DC, US Government Printing Office, 1939).

as polluted as watersheds near Chicago, Detroit, Pittsburgh, Boston, and New York, and rivers such as the Illinois, Cuyahoga, Potomac, and Hudson, clearly indicated its severity. This severity had drawn Wolman's attention as it had sanitary engineer Harrison P. Eddy's earlier in the 1930s.

PORTLAND'S MUNICIPAL WASTES

"The most difficult problem in this project is the financial problem."

During its first year of existence, the OSSA had addressed the fundamental issues of funding, staffing, and administration. While opting not to establish a universal definition of "pollution," it did set up a stream classification system and began collecting baseline data from cities and industries. By early 1940 authority members were prepared to fulfill their mandate. They held their first meeting focused on industrial pollution in May 1940, attended by representatives from Oregon's pulp and paper industry as well as meat packing, textile, and fruit- and vegetable-processing firms.[34] Also in 1940, the OSSA collaborated with the Washington State Pollution Commission to conduct the first survey of water pollution in the lower Columbia River.[35] These were important parts of the OSSA's efforts, and would remain so for decades to come, but for the first years of its existence the authority's primary focus was the City of Portland's sewage pollution problem.

Portland was a laggard within and beyond Oregon. Communities outside the Willamette Valley led the state in building sewage treatment systems. The city of Wasco in north-central Oregon built the state's first sewage treatment system of any kind—a septic tank—in 1904 to serve its three hundred residents. La Grande, in northeast Oregon, built the state's first trickling filter treatment plant in 1925 in response to lawsuits against the city for fouling irrigation ditches with raw sewage. In the 1930s, federal funds available as part of the Works Progress Administration and other New Deal programs helped finance treatment plants in at least eleven Oregon cities, including Banks, Gresham, and Hillsboro within the Willamette watershed.[36] At the national level, measured as percentage of urban population receiving at least primary sewage treatment, as of early 1940 the top five cities were Milwaukee (85 percent), Cleveland (75 percent), Columbus (75 percent), Indianapolis (75 percent), and Chicago (70 percent). Portland, with a population of about 300,000, joined Pittsburgh (750,000), Cincinnati (500,000), Duluth (115,000), Kansas City (450,000), and St. Louis (950,000) among the nation's larger cities with no treatment.[37] Portland's neighbor to the north, Seattle, with a 1930 population of about 365,000, was only slightly further along in treating its sewage; though New Deal funds had aided in construction of intercepting sewer lines to route raw sewage away from Lake Washington, with the exception of sewage given minimal treatment by three low-efficiency settling tanks, most municipal wastes were routed directly to Puget Sound.[38] The City of Portland had failed to secure New Deal funds in the 1930s while continuing to expel raw sewage from forty-eight outfalls to Portland Harbor and eleven to the Columbia Slough.[39] Recognizing the city's lack of progress in developing engineering plans and establishing a funding mechanism, the authority passed a resolution during its first meeting to remind officials "to expedite the Portland sewage disposal project."[40]

Sanitary Engineer Abel Wolman aided in the campaign to pass Portland's 1938 sewer funding proposal and was lead author of the National Resources Committee's water quality survey. He would contribute to Oregon's pollution abatement efforts again when Portland commissioners called upon his expertise to analyze options, prepare estimates, and recommend a course of action for the city's sewer interceptors and treatment facility. Just days after the November 1938 vote, commissioners authorized $5,000 to finance Wolman's

Fig. 3.3. Untreated sewer outfall under the Burnside Bridge, Portland, 1930s. *Oregon Journal* (Oregon Historical Society Research Library, bb004452).

work, a move the *Oregonian* announced with the headline "Sewage Project Gets Under Way in City."[41] When Wolman and his team delivered their "Report on the Collection and Disposal of Sewage" in August 1939—or, as it became known, the "Wolman Report"—it reflected the best in sewage treatment science, engineering, and technology.[42] The report characterized the Willamette River as "unsafe for bathing, unsuitable for recreation purposes and impossible for sustained fish and aquatic life," particularly during low flow periods. Sharing the view that Commissioner Bean and others expressed earlier in the decade, Wolman considered the Columbia Slough to be in worse condition than the Willamette and recommended cessation of sewage discharge into the slough as soon as possible.[43] In spite of work done to improve Walter Baer's preliminary plans of 1933, the Wolman Report concluded that "local understanding of the engineering plan and of the methods of financing is not yet completely clarified." These deficiencies stymied Portland officials' attempts to secure New Deal funding and Wolman discovered these shortcomings still plagued the project. The report also observed that cost estimates for the treatment system had risen significantly since 1933, from $6 to $9 million.[44]

The Wolman Report called attention to another critical consideration: the unique hydrology of the Willamette watershed, which was important because of the central role that "assimilative capacity" played in waste disposal. Quantifying the ability of a given body of water to receive wastes without becoming degraded was fundamental to water quality science and engineering, and water quality was but one element in a conservationist approach to natural resource management.[45] Based on this logic, the Wolman Report observed that the network of dams proposed as part of the Willamette Valley Project would be an important part of any overall sewage treatment plan. Without augmented seasonal flow from tributary waters retained in reservoirs, the river during low-flow periods was an "unusually sluggish body of water with delayed carrying off of discharged materials." Because of this, it would be difficult to maintain the minimum 5 parts per million (ppm) DO required for fish survival."[46] City engineers were well aware of the unique hydrological conditions that made their sewer project difficult. One engineer involved in designing Portland's sewer interceptor system in the late 1930s noted that he had found "no city of similar size in the United States with such conflicting problems."[47] The Wolman Report concurred with this and contrasted the Willamette's sluggishness with the much higher flow in the Columbia River. Echoing Eddy's findings from 1933, Wolman and his team concluded the Columbia would be able to assimilate *all* of Portland's *untreated* sewage "with no deleterious effect in oxygen balance upon any of the present or potential uses of the Columbia River." However, the authors recommended that if the city decided to pipe effluent into the Columbia that wastes first be given primary treatment to avoid "any conceivable evidence of sewage discharge." In their view, there was no scientific reason for doing so but the public would likely protest the idea of dumping raw sewage into the Columbia, as they had earlier in the decade in response to one of Eddy's options.[48]

While Wolman and his colleagues were preparing their suggestions about how the system might be built, city commissioners were trying to figure out how to pay for it. Their mandate from the 1938 vote was to charge *up to* an additional 33.3 percent of monthly water service bills to fund sewer infrastructure. Shortly after the election the city council took up the matter of

determining the service charge and faced "various strenuous remonstrances .
. . especially by citizens of Portland who were engaged in business enterprises
of various kinds," as Oregon Attorney General Lyman Latourette would later
report. Because of this push-back, Latourette recalled, "the council concluded
that it would be necessary to have a Board of Equalization, in order to study
more definitively the local conditions with reference to industries, business
houses[,] and residences." In early March 1939 the board's three members
began their work, but this seemingly mundane administrative process sparked
controversy, confirming Latourette's view that "the most difficult problem in
this project is the financial problem."[49] Contention centered on whether or not
the board contained a representative cross-section of key constituencies that
would be affected by what might be a significant hike in monthly water service
charges. By 1939 Oregonians had experienced more than a decade of economic recession and depression, and though New Deal programs had brought
some relief, unemployment remained high and prospects for a full recovery
were elusive. City commissioners were aware that many Portland residents
would be burdened by a significant spike in water service rates and established
the equalization board to help balance these concerns with the need to raise
funds. Commissioners J. E. Bennett and Ralph C. Clyde represented the views
of the Oregon Commonwealth Federation and others in claiming that the
board could only achieve this if it was expanded to include a wage earner and
a housewife. As initially constituted, the board contained representatives with
backgrounds in engineering, finance, and city administration. Liberal Monroe
Sweetland, representing the Commonwealth Federation, stated that he had
been opposed to the sewage project from the outset but would be more amenable to the plan if the board was expanded to include representation from the
working class. Edgar Averill opposed such an adjustment on the grounds that
the proposal was from a "small pressure group" that had been opposed to the
project for years and sought to delay action further by including opponents on
the board.[50] This was an oblique reference to the contention that surrounded
the Portland sewage system topic in late 1933, when various groups organized
to advocate for the project as an unemployment relief measure. One faction
within the Portland Anti-Pollution League had strongly opposed consulting
engineer Harrison P. Eddy's conclusion that the Columbia River could assimilate all of Portland's untreated sewage.

Sweetland and the Commonwealth Federation were not opposed to
building sewage treatment infrastructure—they were opposed to financing

schemes that threatened to place an undue financial burden on working-class taxpayers. As head of the Commonwealth Federation, neither he nor his organization took strong positions advocating for water pollution abatement. He had, however, lauded the work of Byron Carney, his colleague among the Commonwealth Federation's leadership, for Carney's key role in writing the 1938 citizen's initiative to create the sanitary authority. Though closely associated during the 1938 election campaign, Carney's work and efforts to fund Portland's sewers were distinct. The root of Sweetland's opposition is apparent in his preferences for participation on the equalization board: advocating for working-class participation reflected the Commonwealth Federation's support for other liberal causes such as public power, civil rights, consumer protection, collective bargaining, and free medical care for those in need.[51] After considering these arguments, in late March 1939 the Portland City Council rejected the proposal to expand board membership on a vote of two to three.[52]

In July 1939, the three-member equalization board recommended city leaders not levy the full 33.3 percent authorized by voters but only about half that amount, a rate sufficient to realize $275,000 annually. Commissioners did not follow this recommendation, instead establishing an even lower rate sufficient to generate only $130,000 per year.[53] Sanitary Authority Chair Wendel estimated Portland lost $35,000 every day the City Council did not levy the full charge.[54] The city's decision had the support of some influential constituents—for example F. H. Young of Oregon Business and Investors, Inc., agreed that increasing water service charges was not justified. This was not what the sanitary authority and other advocates had been hoping for, however, because the City Council's decision translated into a drastically slower pay-as-you-go program of sewer construction. Speaking on behalf of the Izaak Walton League, Portland Anti-Pollution Council, and Oregon Stream Purification League in May 1940, Edgar Averill urged the authority to order the City of Portland to proceed with the sewage treatment project.[55] Advocates took the position that *in spite* of expense and engineering complexities, building sewer infrastructure justified levying the fully authorized charge because it would help clean up the Willamette and relieve unemployment.[56] Voter approval of the two water quality–related items on the November 1938 ballot indicated a strong majority supported this approach. The sanitary authority agreed with abatement advocates on these points and asked Averill and his network to help formulate a legal brief in preparation for a citation hearing the following month, at which city officials were to appear before the authority "and show cause, if any they have, why the nuisance resulting from the disposal of city sewage shall not be abated."[57]

After receiving voter approval for their pay-as-you-go funding plan in 1938, city commissioners promptly solicited Abel Wolman's expertise to develop design recommendations. Their March 1939 sewer funding decision and actions over the next few years, however, show that a majority among them did not interpret voter approval as a mandate to carry through with design and construction. Testimony at the June 1940 citation hearing brought to the foreground their rationale, centered on the challenge of financing. Upwards of $150,000 was necessary to complete required surveying and engineering work before construction could begin. City residents and businesses would not support the financial burden of the full water service charge unless they were confident actionable engineering plans and specifications were in place. Even with plans and specifications, there was no feasible way for the city to collect the $7 million to $10 million required for timely construction. Without federal or state financial assistance, city officials argued they were proceeding as expeditiously as possible and, therefore, the sanitary authority should not take legal enforcement action against them. In response, authority member John Veatch replied: "The reason we issued the citation was because the city of Portland was doing nothing. We want to know what has been done and what is going to be done." He continued: "I do not think the Authority has any desire to tell the city how to pursue its problem. We only want to know if steps are being taken."[58]

City officials made little headway in addressing these issues over the next few years. To justify inaction, they continually referred to the same barriers put forth in mid-1940. Authority members also opined that some city council members lacked political will for fear of being ousted at the next election.[59] After December 7, 1941, Commissioner Bowes added another reason for the city's inactivity: with the United States' entry into World War II, competent engineers were being enlisted into military service. The sanitary authority's response when city officials cited this reason in late 1942 was that its staff was also thinned, and for the same reason, but it was legally required to push city officials to levy the full amount voters authorized. If it was not possible to develop detailed plans because of a lack of engineering expertise, the authority determined that the city still should set aside as much money as possible to prepare for postwar sewer projects, even if this was the best they could do during the war emergency.[60]

In reaction to the City of Portland's intransigence, the OSSA played a central role in organizing a public campaign to increase pressure on city leaders and educate voters on the need for sewage treatment funding. An early and important element of this campaign was William J. Smith's color film of late summer 1940, which showed OSSA technicians at work sampling water conditions. The film provided a portable and compelling piece of visual evidence showcasing the Willamette River's degraded state. The OSSA hired the services of a public relations counsel in early 1943 to plan and implement a formal media campaign including "letters, editorials, cartoons and other publicity materials," a speakers' manual, and presentations on the radio and "before certain influential groups." The objective was to "create public interest which in turn would convince the City Council that the people of Portland are in favor of an increased sewer-user service charge, levied to finance construction of sewage collection and disposal works."[61] The campaign launched during summer 1943. The *Oregonian* editorial board wrote in July that "while it may have been desirable five years ago to keep the [water service charge] levy at a minimum, now it is even more desirable to invoke it at the maximum." The editorial continued:

> The position of this city, in its attitude toward the cleansing of the [Willamette] river by adequate sewage disposal, is embarrassingly unenviable. Instead of leading the program, as our voters instructed the city council, we have procrastinated for five years—and for several times five years before the vote. The metropolis of the state not only has set a bad example to its sister cities on one of the most beautiful rivers in the world, but it has retarded, by its seeming indifference, their enthusiasm for reform.[62]

In addition to groups active on this issue since the 1920s and 1930s—such as the Izaak Walton League, Oregon Wildlife Federation, and Columbia River Fisherman's Protective Union—the Portland Chamber of Commerce, Central Labor Council, and Federation of Women's Clubs also actively supported the OSSA's push against the City of Portland.[63] The degraded water quality conditions these groups were organizing to address had been captured in Smith's film, which he continued to show at public forums. He reported to the sanitary authority in fall 1943 that, upon viewing the film, "as yet there has not been a single group who have not offered their support to a program to clean up this river."[64] In early October 1943, five men who had recently completed a raft journey from Eugene to Portland to promote war bond sales provided a complementary soundtrack to Smith's silent film. During a presentation to Portland clean stream advocates, they described "filth" and "slime" encountered on the river "that gave off an odor that was so strong it was offensive a block away."[65] The *Salem Capitol Journal* questioned the value of this trip in war bond sales but supported it for highlighting horrid conditions: the raft trip "widely advertised the fact that the Willamette is utilized as an open sewer and placed the stigma where it belongs, on Portland's municipal authorities, who have refused to obey the mandate of the state anti-pollution law and of the people of Portland who long ago voted for the construction of a sewage disposal plant to free the river from filth."[66] Such observations echoed those from earlier accounts, such as an *Oregon Journal* article finding that not only salmon and trout but crawfish, carp, and other pollution-tolerant species were dying in large numbers, and residents between Salem and Oregon City had to "sweep hordes of dead and dying crawfish" from ferry slips and other surfaces.[67] A letter to the *Oregonian* from a Portland resident stated, "Many mornings at dawn I have seen thousands of filthy rats feeding in the scum at the water's edge. The rats and their filth return to shore, where goods of commerce are stored. The chance of pestilence and epidemic is ever present under present conditions."[68]

Another key supporter in this growing alliance was Oregon's Postwar Readjustment and Development Commission.[69] Commissioner Bowes appointed the forty-seven-member commission in February 1943, at the behest of Mayor Earl Riley. Riley's motivation was to formulate a comprehensive public works plan to develop regional transportation and economic infrastructure, provide jobs for returning servicemen, and fill the economic gap expected between the end of the war and full resumption of peacetime production.[70] The sanitary authority was successful in garnering the postwar commission's support because, as graphic recent testimony illustrated, the river was filthy and growing increasingly so. The pollution abatement cause also had a strong voice on the commission. Among its members were many experts who had long been involved in clean stream efforts, including Oregon State College engineer George W. Gleeson; attorney, fish commission chair, and OSSA member John C. Veatch; bacteriologist and member of the 1930s Advisory Committee on Stream Purification David B. Charlton; and OSSA member sanitary engineer Kenneth H. Spies.[71] Though not a member himself, industrialist Edgar Kaiser also had strong influence on the commission. His shipyards producing vessels for the Allied war effort in Portland and across the Columbia River in Vancouver, Washington, were significant contributors both to the overall war effort and the local economy, and this made him highly influential in Portland city politics during World War II. By August 1943, Kaiser felt the commission was moving too slowly, so he invited renowned New York City urban planner Robert Moses to help speed progress.[72] He had taken similar unilateral action when he created a new city in Oregon to house African American and other shipyard workers—called Vanport because it was located between Portland and Vancouver.[73] In November Moses and his team published their report, *Portland Improvement*. It provided a blueprint for large-scale postwar infrastructure development projects involving transportation, public buildings, airport improvements, and parks. Out of the $24 million total package, the single-largest amount—$12 million—would fund the city's sewage interceptor and treatment system. As with its recommendations for a freeway loop around the city, *Portland Improvement* did not put forth a unique sewage infrastructure plan; rather, the report relied upon previous work. In the case of the sewer system, *Portland Improvement* supported Abel Wolman's 1939 recommendations.[74]

PORTLAND'S MUNICIPAL WASTES

"Before we pretty up our city we should first clean it up."

By the time the Sanitary Authority convened another citation hearing in October 1943, significant momentum had been built from the public relations campaign and inclusion of Portland's sewer infrastructure within Moses's *Portland Improvement*. The authority called city officials to "show cause, if any exists, why [the City of Portland] should not now take the steps available leading to compliance with the laws of Oregon relating to abatement of stream pollution, and why appropriate proceedings should not be instituted against the said city should it fail to do so."[75] Since its first meetings, the OSSA wanted Portland officials—representing by far the largest urbanized area in Oregon—to take leadership in a basin-wide sewage treatment program. "It might be the salvation of the entire country if we do this," Wendel had said. He coupled his call for clean streams with the claim that setting aside funds to employ servicemen returning from World War II would help ensure the state's postwar economic health.[76] State Senator Lew Wallace, another critic of Portland's inaction, found it ironic that officials approved funds to build a new highway along the western bank of the Willamette River that would carry tourists within close proximity of a shamefully polluted waterway. Wallace's observations in a 1941 *Oregon Journal* editorial were still relevant in late 1943. He characterized the city's approach as "deplorably inconsistent," and found it "interesting to observe that although years have elapsed since the votes on sewage disposal and nothing has been done," it took just a little more than a year to begin construction of the highway after approval of the bond issue. Wallace excoriated city officials for building "an expensive boulevard, to be traveled by tourists from all parts of the United States, side-by-side with an open sewer such as the Willamette river is at the present time."[77] Sanitary Authority Chair Harold Wendel continued this theme during the hearing:

> The city does not seem to believe in the present philosophy of planning and saving for the future; the financial program which the city advances is greatly inadequate; and in the minds of the Authority it is considered doubtful that any great progress will be made in the consummation of the plans. It is the opinion of the Authority that the state is entitled to the maximum compliance that can be made, being done lawfully and legally under the circumstances, toward making progress toward compliance. . . . We feel it is regrettable that the maximum charge has not been levied during this entire period. If it had been we should now have over two millions of dollars in the treasury. . . . The Board of Equalization, by their own recommendation, would have produced more than twice the revenue contemplated by the city.[78]

In response, former commissioner J. E. Bennett and current commissioner William A. Bowes provided seemingly conflicting interpretations of the city's actions—Bennett as a coauthor of the 1938 funding plan, and Bowes from the perspective of the current commissioner of public works in charge of implementing the plan.

Over the course of a political career spanning 1925 to 1970, J. E. "Jake" Bennett was a contentious figure, known for his willingness to take unpopular stands and ruffle feathers. He would be remembered as a "curmudgeon's curmudgeon" who "had little fear of indulging in verbal over-kill," but also for his skill in connecting with voters by providing his fellow politicians "a living example of what a lot of people they didn't know were thinking about, what they wanted from government and life, [and] what they mistrusted and disliked in 'the establishment.'"[79] He served as state representative (1925–1927, 1947–1949, and 1963–1969), state senator (1929–1931), and Portland city commissioner (1932–1941 and 1950–1952). Initially a Republican, in 1956 he switched to the Democratic Party. All the while he opposed the sales tax and fought to lower taxes for homeowners and renters. As a city commissioner in the early 1930s, Bennett was a vocal (and often minority) supporter of advocacy groups pressing the city to be more active in unemployment relief projects such as Walter Baer's sewage treatment proposal. After coauthoring Portland's 1938 sewer funding mechanism, he was a member of the sewer funding equalization board.[80]

Bennett testified that he had been "a member of the City Council that hastily considered the [sewage] project to produce jobs," but that "it was left pretty much up to Commissioner [Ormond] Bean to prepare a program that was workable"—the 1938 pay-as-you-go plan. Commissioners soon realized that the program was not workable, Bennett said, but did not put a stop to it: "We discovered that the scheme the council used in order to push the issue was a mistake." This conclusion reflects testimony Mayor Joseph Carson conveyed to the authority three years previously, when he observed that even if the city levied the full water service charge it would take up to twenty-five years to design and build Portland's sewage treatment system on a pay-as-you-go basis.[81] Bennett continued: "It was my position and is now that the minute we discovered that we had presented a scheme that was unworkable, we should have stopped collection of money." For these reasons, Bennett concluded, the City of Portland was wrong to continue to collect *any* pay-as-you-go amount.[82]

Commissioner Bowes disagreed. The Portland City Council selected him to replace Ormond Bean in June 1939, after Bean departed to serve as head of the Oregon Public Utility Commission.[83] Bowes oversaw the city's public works department during his long tenure as a city commissioner (1939 to 1969); before entering politics as a Republican, he worked as a printer and was secretary of the Multnomah Typographical Union.[84] He would come to be a controversial member of the city council who was not afraid to stand alone on issues and hold fast to his position. He was also trusted by Portland voters, who elected him to an unprecedented eight terms on the city council.[85] He was fiscally conservative, as evidenced by his opposition to the development of public housing during World War II and creation of a public utilities district. At the same time, he supported a range of urban public works projects to enhance infrastructure and provide jobs. Within months of the United States' entry into World War II, Bowes championed job creation and expansion of the city's automobile infrastructure as an influx of workers came to labor in the shipyards and other war-related industries.[86] Like Bennett, Bowes was often "one of the most controversial members the council" who was not afraid to stand alone and was "embroiled in an endless series of political brawls."[87]

In response to Bennett's claims that Portland's funding plan was "unworkable," Bowes vouched for the plan but lamented that even though Abel Wolman had provided basic engineering diagrams and calculations, the city could not produce detailed specifications because "with the war, every competent engineer was taken in either one branch of the service or another." Bowes told the authority that "as far as my department is concerned, we are honestly trying to develop a sewer system that will satisfactorily clean up the Willamette River," and that his department "will not submit anything to the people of Portland until we definitively know what type is needed for a treatment plant that we know will do the job."[88]

Both perspectives contained valid points. Both men had direct knowledge of the issue, but with Bennett no longer a city commissioner in 1943 he may have felt freer than Bowes to admit its flaws. City officials admitted by late 1939 that by levying only a fraction of the authorized amount, funds would be insufficient to develop detailed engineering plans and estimates necessary before construction was possible.[89] In 1938, however, with the groundswell of support for creation of the sanitary authority, three failures over five years to secure voter approval for Portland's sewer system, and the city's inability to get New Deal funding, the Bean-Bennett pay-as-you-go plan was the best alternative to doing nothing at all. The city council's collective fiscal conservatism on this topic echoes what historian Carl Abbott has found on other issues before the council such as housing, transportation, and urban planning; Abbott makes particular note of Commissioner Bowes's stances that included strong support of business interests and resistance to federal involvement in local affairs.[90] Regardless of Bennett's or Bowes' positions, or the increasingly contentious dynamic between polluters and abatement advocates, the Willamette River was not made cleaner by the City of Portland's inability to raise funds.

In December 1943, sanitary authority members sought State Attorney General George Neuner's legal counsel. Harold Wendel asked Neuner if Oregon courts would sustain an OSSA mandate that the city "act in the matter by increasing the sewer service charges to the maximum authorized by the voters, and by requiring the completion of engineering plans and specifications?"[91] Neuner's reply referred to an inquiry the authority made in 1940 on the same fundamental question—could the OSSA force the City of Portland to take specific abatement measures? Neuner concluded the OSSA could "legally go no further than to *recommend* to the city the adoption of an ordinance" [italics mine]. To do otherwise would "constitute an attempted invasion of the legislative power of the city council," something even the state legislature was not legally empowered to do. Neuner also concluded, however, that the city's record of noncompliance, coupled with the collection of empirical data clearly identifying the negative effects of pollution from the city's untreated wastes, was strong evidence of the lack of a good-faith effort. This evidence would likely be sufficient to convince the court to rule in favor of an OSSA-initiated injunction against Portland officials to require the city to act.[92] What this meant was that the sanitary authority could not dictate to Portland officials *how* to cease polluting, but they were likely to win a legal battle requiring city officials to do *something* as soon as possible. These findings from the attorney general laid a foundation for the sanitary authority to take its first enforcement action, should it decide to do so.

Confronting a vocal alliance and faced with the threat of legal action, in late January 1944 three Portland city commissioners—Dorothy McCullough Lee, Fred Peterson, and Kenneth Cooper—began crafting a measure for Portland voters for the upcoming May election. Their proposal would replace the 1938 pay-as-you-go system with a $12 million loan supported by a voter-approved bond. This funding plan followed Robert Moses's recommendations—and, by extension, Abel Wolman's.[93] Commissioner Lee called this "the No. 1 project of the Moses report, because before we pretty up our city we should first clean it up."[94]

Two influential constituencies opposed the Lee-Peterson-Cooper $12 million bond proposal. One was Portland's commissioner of public works himself, William Bowes. His arguments remained consistent with reservations he had been presenting to the sanitary authority since 1940, a critical component of which he summarized before the city council in February 1944:

Never since the beginning of the sewage disposal question in Portland [in the mid-1920s] has there been complete and detailed plans and specifications drawn; never has the problem been explored to the

point where a contractor would bid on a contract for the interceptor
and the treatment plant.

Bowes put forth an alternative to the Lee-Peterson-Cooper proposal: he
asked for $200,000 to hire engineers to develop plans and specifications.[95]
Though he supported the Moses plan in general, he was wary of moving
ahead with funding schemes if the specific details and financial requirements
were not known.[96] Joining Bowes in opposition were two men with statewide
influence in tax policy and real estate development, F. H. Young and Frank
H. Hilton. Both influential Republicans, in the 1930s they had variously
been allies and opponents during previous debates on taxation and finance.
In the matter of Portland's sewage in the 1940s, their interests overlapped
significantly.

During Portland's sewer system funding debates, Young represented
Oregon Business & Investors, Inc., and Hilton the Apartment House Owners
Association.[97] Young and Oregon Business & Investors, Inc., were critical in
establishing the Oregon Stream Purification League, which, in turn, led the
successful 1938 citizen's initiative creating the sanitary authority. In February
1942, Young became secretary of the Citizens' Committee for Wartime Tax
Saving. The group "intended to take up cudgels against non-essential public
spending" in light of "all-out war in the Pacific." Hilton was also part of this
group.[98]

Hilton had been a member of the Apartment House Owners Association
and Portland Realty Board since the early 1930s, affiliations that spoke directly
to his involvement in sewer debates centered on increasing water service
charges.[99] His interests and influence in local and state politics were broad and
included service as a Republican state senator in the early 1930s, state representative in the late 1930s, and, beginning in June 1937, briefly serving as acting municipal judge for Multnomah County; he would be a state senator again
in the mid-1940s.[100] Hilton was one of six candidates who filed for the 1940
Portland mayoral race and put forth a platform emphasizing "industrial expansion and a closed throttle on tax spending." Hilton lost a close race to Earl
Riley but asserted Portland was "more tax conscious" as a result of his efforts.[101]
Oregon politics in the first half of the twentieth century were not as partisan
as they would be by the end of the century, and Hilton's career reflected this.
In 1930 he supported the successful gubernatorial campaign of Independent
candidate Julius Meier.[102] In early 1934, Hilton and Young debated opposite

sides of a proposed state sales tax, with Hilton in opposition.[103] Concurrently, Hilton supported liberal causes such as the Townsendite plan to provide federal senior citizen assistance, and while in the state legislature in the late 1930s, Hilton cosponsored a bill providing for old-age assistance.[104] In March 1938, Hilton was one of the honorary pallbearers at the funeral of Portland attorney Harry L. Gross. This association is relevant to the history of water quality in Oregon because Gross was a Republican who had long been active in liberal political movements, including involvement in the Oregon Commonwealth Federation and representing labor groups affiliated with the left-leaning Congress of Industrial Organizations (CIO). Other honorary pallbearers at Gross's funeral included the author of the 1938 citizen's initiative, State Senator Byron J. Carney; then–Portland city commissioner J. E. Bennett; Oregon Commonwealth Federation Chair Monroe Sweetland; and liberal future US district court justice Gus J. Solomon.[105]

Both Frank Hilton (on behalf of Oregon Apartment House Association) and F. H. Young (on behalf of Oregon Business & Investors and the Citizens' Committee for Wartime Tax Saving) joined Bowes in opposing the full water service charge, on the familiar grounds that the city lacked detailed plans, specifications, and budget. Hilton stated during the October 1943 citation hearing that his conversations in the community led him to conclude that if city commissioners agreed to levy the full service charge for the purpose of building a pool of funds for postwar construction,

> the Commissioners would have been recalled or would have been
> defeated at the election. At the present time, I definitely and positively
> know that the people feel that rather than an increase in the tax, it
> should be eliminated entirely until the war is over and comprehensive
> planning can be done. We admit by the statements of all here that
> we have not yet a plan that is at all satisfactory. The law permitting
> the present tax is a law enacted by the people under a positive
> misapprehension—money taken under false pretenses. People didn't
> anticipate that the sewer tax would just be a drop in the bucket,
> accomplishing nothing; they thought it meant bringing about a self-
> liquidating sewer system.[106]

With such resistance, in early February 1944 the Portland City Council voted to postpone consideration of the Lee-Peterson-Cooper proposal. In response, on February 21 the Portland Chamber of Commerce organized approximately two hundred "representative citizens" into an "anti-sewage group."[107] The group's sponsors and nominating committee reflected a diverse range of interests and included Mrs. C. D. Cummins, president of the Portland Council of Parents and Teachers; Dr. Blair Holcomb, president of the Multnomah County Medical Society; Palmer Hoyt, publisher of the *Oregonian*; Louise Palmer Weber, "prominent clubwoman"; and representatives from the Portland Central Labor Council and the Oregon State Industrial Union. Women were conspicuous among the group's executive committee and elected officers, illustrating the broad-based membership within the abatement movement.[108] *Oregon Journal* editor Marshall N. Dana chaired the group's first meeting and was an exemplar of another category of Oregonians becoming increasingly active in addressing Willamette River pollution.[109] Unlike Edgar Averill and OSSA Chair Harold Wendel, it was not necessarily part of Dana's daily work activities to be directly involved in water quality issues; additionally, unlike Oregon State College engineers or bacteriologist David Charlton, he did not possess the technical skills and specialized training to quantify water pollution. Like William Joyce Smith and many others who increasingly joined the cause, Dana was interested in natural resource conservation in the broadest sense: using them wisely so they would be sustained for present needs *and* future generations.

Though the collective participation of all members of the Portland Chamber of Commerce's advocacy group would be needed, Marshall Dana was a more prominent and influential participant than others. He served forty-five years in the news business with both the *Oregon Journal* and *Oregonian* and was highly active in conservation issues. Along with William Finley, Dana was among the founding members of the Portland Chapter of the Izaak Walton League in 1922.[110] In 1932 he was selected as the first president of the National Reclamation Association, and shortly thereafter helped establish and then chaired the Pacific Northwest Regional Planning Commission.[111] Dana's interest in regional planning within the multistate Columbia basin included preserving the scenic beauty of the Columbia River Gorge, conserving both commercial and sport fisheries, establishing port facilities, and building dams to improve irrigation and navigation.[112] He believed a healthy democracy required active civic participation and was a strong proponent of solutions developed at the state level.[113] At critical times in Oregon's pollution abatement narrative—such as 1926, 1938, and, now early 1944—people such as Dana joined with those who had a day-to-day interest in the matter to help propel change.

Reacting immediately to this pressure, the Portland City Council voted on February 24 to spend $200,000 to develop sewer system engineering plans for postwar implementation and scheduled a meeting for the council to decide whether to submit the $12 million bond to voters in May.[114] The persistence of the OSSA and broad-based membership of the Portland Chamber of Commerce's anti-sewage group spurred this change, as did worsening river conditions. As part of the rejuvenated abatement campaign, shingle mill owners and workers attended a city council meeting in early March to provide firsthand accounts of the situation along the Columbia Slough. These men urged immediate action because of "filthy conditions that prevail[led] . . . which for years have been a health menace." Severe pollution from sewage and meat-processing firms decreased shingle production by making workers ill and forming "shoals and bars" of accumulated detritus that interfered with operations.[115] The next day, the Portland City Council unanimously agreed to submit the $12 million sewage disposal bond issue to voters. Although he voted in favor, Commissioner Bowes warned that the $12 million was only an estimate because detailed plans and specifications had not yet been completed. Other commissioners argued that this action was better than nothing.[116] Voters approved the measure on May 19 as part of a $24 million postwar funding package for schools, roads, docks, and sewers recommended in the Moses Report.[117]

Commissioner Bowes's worries were not unfounded. The project still lacked the specificity required to establish clear design details and cost estimates, and in its breadth and complexity opened the possibility to significant overruns. Manifesting Wolman's sewer recommendations would require much labor and materials. Since there was to be a single treatment plant on the north side of the city, sewer interceptor and trunk lines would have to be drilled *under* the Willamette River, and large pumping stations at Swan Island and other strategic locations would be required to convey wastes uphill.[118] At any point in the excavation or construction, engineers could find that an aspect of the original plan was not feasible and that a more expensive alternative was the only solution; one or two instances like this and $12 million would be insufficient. Two additional economic considerations were also important. First, with World War II still raging in both Europe and the Pacific, it might be years before construction commenced, which raised the prospect of price inflation for materials and labor. Second, $12 million in 1944 was no small amount, both in relative and real terms. Twelve million was roughly equivalent to $135 million in 2018. Twelve million spent on the sewer system of just one Oregon municipality in 1944 was equivalent to what the state government authorized to fund most of its operations—its board of health, highway department, prison system, state hospitals, relief programs, and many others—for just one year.[119] Additionally, while about half of the voter-approved $24 million for projects identified in the Moses plan would go to visible public amenities such as schools, roads, and parks, most of the $12 million spent on sewers would be buried in the ground. To voters and municipal leaders in 1944 who were not yet out of the Great Depression and still embroiled in a world war, it was a significant commitment to invest such a large amount for the city's sewer treatment infrastructure.

The successful May 1944 vote was not only a victory in the cause of a cleaner Willamette River, but also marked a distinct shift in Portlanders' values. City residents had taken

Fig. 3.4. Brochure from the Citizens Sewage Disposal Committee urging Portland voters to support the May 1944 bond issue (City of Portland Archives, A2001-007).

what might seem to have been a similar vote in 1933 when they approved a $6 million bond for a sewage system following the "Baer plan"; this vote was markedly different, however, in that it was tied to millions of dollars in federal grants. When federal dollars were not to be had and commissioners sought different ways to fund the program in 1934 and 1936, voters twice said "no." The successful 1938 vote benefited from momentum behind the proposal to create the sanitary authority, but it was also presented in the more palatable form of an incremental, pay-as-you-go plan. In May 1944, voters

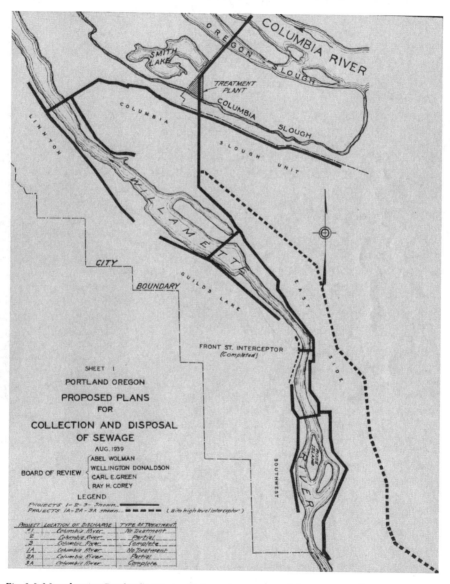

Fig. 3.5. Map showing Portland's proposed interceptor sewers and the treatment plant as Abel Wolman recommended and as voters approved in November 1938 (City of Portland Archives, folder: Reports on Sewage Disposal Project 1951–1959, box: Early Sewage System Data Sewer Systems 1923–1972 (8890-02), A2011-024).

said they were willing to pay for the complex sewer project in its entirety in the interest of their local environment, without the promise of federal contributions. Influencing this decision was unavoidable evidence of significantly worsening pollution, which increasingly affected more people by curtailing their accustomed activities. Statewide pulp and paper industry production had more than doubled between 1920 and 1940, from 150,000 to 360,000 tons. This production was concentrated within the Willamette Valley, where practically 100 percent of the industry's wastes made their way directly into rivers and streams.[120] Urban growth within the Willamette Valley also contributed more pollution. With an increase in war-related employment—most notably Kaiser's shipbuilding ventures—the Portland area had been growing steadily even before the United States' entry into World War II. By the end of the conflict, Portland and its suburbs, including Vancouver, Washington, had gained 250,000 residents.[121] In the Portland area, all of these people's untreated wastes went directly into the Willamette River, Columbia Slough, or tributaries.

To address the most egregious source of this latter kind of pollution, at least, Portland city officials now appeared to have a workable solution. They moved swiftly and in July 1944 hired A. M. Rawn, general manager of the Los Angeles County Sanitation District, to conduct a detailed feasibility study following recommendations of the 1939 Wolman Report.[122] Wolman's plan consisted of interceptors along the east and west banks of Portland's Willamette River waterfront and along the south bank of Columbia Slough. These large pipes intercepted smaller lines and directed sewage by way of gravity and pumping stations to a primary treatment facility along Columbia Boulevard in North Portland, which would sanitize the effluent and remove about one-third of its biochemical oxygen demand (BOD). The treated effluent would be pumped to a Columbia River outfall on the north side of Hayden Island.[123] In early September, Rawn and Portland city engineers determined the forty-acre site for the sewage treatment plant and the city began purchasing lots, an action the Oregonian noted was the city's "first definite step taken toward sewage disposal."[124] Complementing these milestones, in summer 1944 Governor Earl Snell approved the Postwar Readjustment and Development Commission's recommendation to provide state funding for further comprehensive Willamette River water quality studies.[125]

By early 1945, some secondary aspects of Portland's sewage project were still being debated—such as the possibility of making a profit by using

Work Starts on Sewage Disposal Plant

Fig. 3.6. *Oregonian* editorial cartoon by Quincy Scott observing the start of construction of Portland's sewage treatment plant (*Oregonian*, July 18, 1947, sec. 1, p. 8).

sewage sludge as an agricultural fertilizer. Making money from processed and dried sewage or methane gas was an integral part of the 1933 Baer plan, and had periodically come up in subsequent debates. Rawn concluded in March 1945 that up to 300,000 cubic feet of methane gas could be produced per day and used to generate electricity, but doing so would not be financially viable because cheap regional hydropower was already available. He asserted profit might be made from the sale of sewage sludge as fertilizer, but only if the sludge could be dried without cost, as was possible in sunny Los Angeles. Neither of these options would be economically feasible for Portland, however.[126] Notwithstanding such debates, many observers likely shared *Oregonian* reporter Virgil Smith's opinion: "Oregon Waters: Crystal Ball Shows Them Clean Again." Virgil Smith lauded the work of the OSSA but also recognized the long struggle to compel Portland officials to treat municipal wastes involved sustained efforts from dedicated citizens and groups. Smith also identified "a new weapon" in the water pollution abatement struggle: "the need for jobs after the war." Although Willamette River water quality had been deteriorating for decades, clean stream advocates found it difficult to spur public action in the form of approving funding solutions and pressuring city commissioners to act. As the example of the Willamette River had thus far shown, unless health, recreational, or economic effects were experienced directly, "the actual connection between pollution and values [was] difficult to make." Jobs, however, were "something that everyone understands."[127] In surveying the progress of a number of Oregon cities since 1940, Virgil Smith concluded that the abundance of sewage infrastructure projects being planned showed that "the biggest battle has been won. The people, the voters, not only in Portland but elsewhere, have authorized work and consented to being taxed to pay for it."[128] He observed that most pulp, paper, vegetable-processing, meat-packing, flax-retting, and other industries along the Willamette had been "inclined to hide behind the skirts of the municipalities in which they operate." For this reason, "it will be much easier to get them in line once the cities do their part." Articulating an economic argument for abating pollution, he concluded that the OSSA pursued not only aesthetics and public health but also saw the issue as being "sound business which will bring good financial returns" in commercial fishing and tourism.[129]

Construction commenced on Portland's long-awaited sewer project on July 17, 1947, with a groundbreaking ceremony for the east unit of the

Fig. 3.7. Unidentified workers excavating a section of Portland's sewer tunnels, ca. 1950. City of Portland Archives, A2005-005.59.122.

Columbia Slough interceptor in the far north of the city. Focus on this part of the project reflected the priorities of city officials and Abel Wolman's recommendations.[130] The *Oregonian* reported that many who had long been involved in the struggle watched "with satisfaction and no little emotion" as the ceremonial first earth was removed. Taking the opportunity to remind Oregonians that the water pollution fight was far from over in spite of the momentous occasion, Harold Wendel said that other Willamette Valley communities could "no longer . . . point their finger to Portland as the greatest violator in stream pollution." Commissioner Bowes estimated the entire project would take approximately three and one half years to complete. Portland Mayor Earl Riley and pollution abatement advocate William Joyce Smith also delivered comments.[131] By late October 1947, OSSA Sanitary Engineer Curtiss M. Everts was able to report that fifteen Oregon cities, including Portland, had sewage treatment plants under construction or recently completed, and eighty-six Oregon communities were considering such plans.[132] Although the City of Portland and other Willamette Valley communities were finally building sewage treatment systems, pollution from industrial pulp and paper manufacturing processes was only increasing—and becoming increasingly more detrimental.[133]

Chapter 4
Postwar Effluents

Twenty years of pressure upon Portland officials had finally resulted in the construction of the city's sewage treatment plant, beginning in 1947. In 1948, twenty years of national debate bore results in the first substantive federal water quality program. Observers may have considered the late 1940s as heralding a new era during which the Willamette River would be cleaned up, once and for all. Environmental degradation continued, however. On September 18, 1949, the *Oregon Journal* presented its readers with a disturbing, yet all too familiar, report. Thousands of crawfish had expired in the most recent die-off to afflict the lower Willamette. In oxygen-depleted waters downstream of Portland, sections of riverbank were thick with thousands of dead orange-shelled crustaceans, while hundreds more futilely clambered over these carcasses to get out of the deadly water. Biologists called this particular die-off—by then a nearly annual event—the worse they had yet seen: in addition to crawfish, trout, and salmon, "hardy bullhead" were also expiring in large numbers.[1] Water quality seemed only to have gotten worse ten years into the sanitary authority's work.

David Charlton, representing the Oregon Division of the Izaak Walton League, continued to press the issue. While other members of the Oregon IWLA were becoming increasingly involved in challenging the construction of dams on the Deschutes and other rivers, Charlton remained committed to water quality. The committee he chaired published a report in December 1948 berating the sanitary authority's ineffectiveness. Referencing the authority's own studies and meeting minutes as evidence of gross pollution, industry stalling tactics, and the inadequate technical qualifications of most authority members, the report concluded that "it appears the Authority as presently constituted and as it operates is not qualified to do the job." It called for outright abolishment unless significant changes were made. "If the Authority is to be retained," the report stressed, "there should be set up a well qualified director of pollution control activities who will devote full time to the job, who will be informed on the subject, have an adequate staff, and who will have some authority to take action."[2]

The committee's report also sharply criticized the pulp and paper industry for continuing to discharge untreated wastes into the Willamette and lower Columbia rivers. Charlton concluded that the industry's recent presentations in front of the OSSA were not merely incomplete, but willful misrepresentations— in his view "a disgrace." "The recorded remarks of the top executives of the companies were very brief in certain instances, and in others amazing ignorance was revealed," he wrote. Being an experienced chemist, he observed that "the discussion which took place following the formal presentation . . . was very amateur." Industry representatives claimed there were no known methods of treating sulfite wastes from the pulping process, but Charlton possessed research findings from successful treatment processes then being applied at mills in Europe, Canada, and Washington State. Rather than challenge industry statements that ignored these options, the lone technical expert among the OSSA, the state sanitary engineer, "did not comment

Fig. 4.1. December 1950 inspection of sewer line by Portland City Engineer L. G. Apperson and others (City of Portland Archives, A2000-031).

on the erroneous and incomplete statements that were presented." The pulp and paper industry's single expert at these hearings "only answered a few brief questions and there was no indication that he had made preparation for the hearing or that he had taken part in any studies bearing upon the subject."[3]

Charlton's report illustrates the shift in Oregon's water quality debate in the late 1940s. With construction moving ahead on Portland's sewage treatment system, the focus of clean stream advocates turned to industrial wastes— particularly those of the pulp and paper industry. For its part, industry representatives were focused solely on achieving bottom-line financial results for owners and stakeholders. They resisted changes to the status quo involving complex engineering questions, substantial technical challenges, and unclear economic returns. Staunch abatement advocates such as David Charlton and the Izaak Walton League pushed for effective, comprehensive, and immediate action founded on quantitative evidence, sound science, and advanced technologies. These advocates were much less concerned with cost to industry, and felt their position was justified by the water quality goals Oregon citizens approved in 1938. The sanitary authority occupied a position between these two extremes. Aware that hardline enforcement would be costly, time-consuming, and potentially counterproductive, the authority continued its approach of working with industry to find solutions. It also justified this approach in referring to its foundational statute, the 1938 citizens' initiative, noting that the OSSA's ultimate charge was to find balance among the often-conflicting considerations of economy, public health, and the environment.

CHARLTON CRITIQUES BOTH INDUSTRY AND THE SANITARY AUTHORITY

Just as it had been necessary to pressure Portland commissioners in the early 1940s to enact real changes to the city's sanitation infrastructure, so was it necessary to pressure both industry and the OSSA in the late 1940s. Charlton was at the forefront of advocates doing so. In conducting research for the Oregon IWLA's 1948 report, he attended a sanitary authority meeting at which he registered dissatisfaction with the industry's lack of progress. He reminded authority members of industry officials' statements as early as 1937 that they would be willing to try feasible abatement methods. Since the industry as a whole had enacted no substantive abatement measures in more than a decade, Charlton was convinced that industry statements were deliberate stalling tactics. Based on his own direct involvement in Willamette River water quality issues since the 1920s and evidence from OSSA reports, he asserted that mills were not making as much progress as their publicity indicated.[4]

David Berry Charlton was born on January 26, 1904, in Vancouver, British Columbia, and earned a bachelor's degree in chemistry from the University of British Columbia in 1925. He served as a laboratory technician with the Portland Bureau of Health from 1926 to 1927, during which he first became involved in Willamette River water quality. After a tenure as a bacteriology instructor at Oregon State College from 1929 to 1931, he earned his PhD from Iowa State University. In 1933 he returned to Portland and founded Charlton Laboratories, providing chemical and bacteriological research on materials, products, and commodities for manufacturers, food companies, logistics firms, government agencies, and the US military.[5] One of his company's first contracts came from the City of Portland to survey water pollution in the Willamette River from the Sellwood Bridge downriver to the Columbia.[6] From 1935 to 1937, Charlton served on the Oregon State Planning Board's Advisory Committee on Stream Purification.

Charlton was also a long-time member of the Izaak Walton League. He was one of William Finley's recruits to the Portland Chapter in the early 1930s; of this he later wrote:

> My becoming a Waltonian in 1934 resulted from my visit to the
> Portland chapter meeting when there was much uncertainty on how to
> get Portland to do something on its sewage disposal problem. Perhaps
> my greatest inspiration and awareness of wildlife and conservation
> problems come from contact with Bill and Irene Finley.[7]

In addition to his professional experience and his long involvement in the IWLA, Charlton was a member of the Portland Chamber of Commerce and the Portland City Club. In the mid-1940s he served on the Postwar Readjustment and Development Commission, contributing to plans for Portland's sewage treatment system that were a component of Robert Moses's *Portland Improvement*. He would later receive a Beaver Award from the Oregon IWLA for "outstanding work in conservation" (1956) and an Oscar Award from the Oregon Museum of Science and Industry (1960) for his help in founding the institution.[8, 9]

Charlton's critiques, therefore, were based on solid evidence from his own water quality studies and his experience gathering information from

Fig. 4.2. Dr. David B. Charlton (standing right) is replaced as president of the Oregon Museum of Science and Industry (OMSI) by L.C. Binford (seated left) while State Senator S. Eugene Allen looks on, November 18, 1953 (Oregon Historical Society Research Library, bb016030).

state officials, private citizens, and engineering firms throughout North America to learn more about current technologies, policies, and practices. In September 1948 he wrote to Clarence W. Klassen, a nationally respected sanitation engineer and head of the Illinois Sanitary Water Board, observing that "our streams are more polluted than they ever were, though there are plans and some construction with respect to municipal wastes."[10] He also corresponded with state officials in Pennsylvania, representatives of Wisconsin's Sulphite Pulp Manufacturers' Research League, and the Vickers-Vulcan Process Engineering Company of Montreal, which was then developing pollution abatement technologies.[11]

Charlton also actively tracked efforts to abate industrial pollution along Wisconsin's Fox River, where the state's Waltonians had been lobbying for changes in pulp and paper waste discharge practices since the 1920s.[12] Charlton wrote an overview of Oregon's problem with pulp and paper wastes for the *Oregon Voter* periodical, and after reading it, Virgil J. Meunch, Wisconsin IWLA president, observed in a December 1949 letter to Charlton how similar the situations were in the two states:

I am amazed at how closely it parallels the same problem we have in Wisconsin. I am writing you . . . in the hope that we can be of mutual assistance, thru [sic] the League, to a final solution of the sulphite pollution problem. If it can be solved in one area there is no reason why it can't be solved elsewhere.[13]

Charlton and Meunch helped convince national IWLA leadership to establish the organization's first permanent industrial pollution abatement committee, and both men served as core members. Reflecting the most pressing industrial pollutant in Oregon and Wisconsin, this committee initially focused on sulfite wastes.[14]

While he strongly criticized the sanitary authority, Charlton was particularly incensed at what he perceived as stalling by the pulp and paper industry. In correspondence with Klassen in Illinois, he wrote of the industry's recently created research arm, the National Council for Stream Improvement (NCSI): "Our pulp and paper mills have done nothing so far," he wrote, "except contribute to their industry-sponsored research program." As the industry's own testimony at a January 1948 OSSA hearing had concluded, this research program had failed to produce any substantive abatement practices or technologies.[15]

As they pressed for pollution abatement, Charlton and other Oregonians did not call for the river to be pure. They understood that some level of pollution was to be expected in a modern industrialized society.[16] Waltonians as a group were motivated by considerations of economics, recreation, and morals, and applied technical expertise and scientific evidence to support their position.[17] They argued for a "common sense medium" between the extremes of "asthetic [sic] theorists preaching conservation for conservation's sake" and sportsmen interested only in "taking more and bigger fish."[18] As Charlton expressed it, "'The Beautiful Willamette' of poetry may not return in our present era of civilization," but with the continued push of clean stream advocates, the river would "look better, smell better, support fish life and be less offensive."[19] Thus, while he and his allies pushed for immediate action, they did not pretend pollution might ever be solved completely. Even so, this perspective may not have considered some significant confounding factors the industry faced. These included the water-intensive process of the making of pulp and paper, complex chemical and biological elements within the resultant wastes, and ever-changing market conditions within which the industry operated. As the Pacific Northwest pulp and paper industry expanded significantly during the first half of the twentieth century, there were more mills using more water and discharging more untreated wastes into the Willamette and lower Columbia rivers.

MAKING PULP AND PAPER AND THE MAKING OF OREGON'S PULP AND PAPER INDUSTRY

Clean water is a critical component of pulp and paper making. Creating wood pulp requires separating wood fibers (cellulose) from sugars (lignins) and other noncellulosic materials. To do so, chemicals are combined with vast quantities of water in a high-temperature process to produce "cooking liquors." There are three primary chemical processes used to separate cellulose, with the chemicals differing based on the type of wood being pulped.[20]

In the *sulfite* process, calcium bisulfate breaks down long-fibered, nonresinous softwoods (such as white fir, hemlock, spruce, and balsam) to produce light-colored pulp that is easily bleached. German researchers developed this process in the 1840s. Such pulp can be used in a wide array of unbleached and bleached products. At the lower end of quality, this includes newsprint, catalogue and book papers, tissue and wrapping papers, and boxboard. At the higher-quality end of the spectrum, this process produces writing and parchment paper, and, when highly purified, produces transparent cellulose sheeting (cellophane) and rayon.[21] With one exception, the first generation of Willamette Valley mills (those built through 1927) used this process: Oregon City, West Linn, Lebanon, Newberg, and Salem.

Later in the nineteenth century, German researchers developed the *sulfate* (or *kraft*, German for "strong") process. This method cooks wood pulp in a solution of sodium hydroxide and sodium sulfide. Douglas fir, hemlock, spruce, southern pine, jack pine, and other resinous woods are commonly pulped using this process. Unbleached kraft pulp is useful for a variety of products requiring strength but where the dark color does not matter, such as wrapping and bag papers and container board. Bleached varieties of sulfate pulp are useful for bag paper, container board, and more highly refined products such as writing, book, tissue, and waxing papers.[22] Willamette Valley mills built from 1949 used this process: Springfield, Albany, and Halsey. The St. Helens mill, built in 1926 on the Oregon side of the lower Columbia River, also used this process.

The third primary way to make wood pulp was, as of 1950, the least important for the Willamette River pollution story. The *soda* process uses a sodium hydroxide solution, and is most effective on short-fibered hardwoods including poplar, cottonwood gum, beech, birch, and maple. Soda pulp, when bleached, is very white. When combined with bleached long-fiber pulp, it is often used as a filler in the production of high-grade papers.[23]

All three processes shared two important traits: they required large amounts of clean water, and they produced a great deal of liquid wastes. Economic historian John A. Guthrie writes that "next to wood, water is the most crucial natural resource needed for pulp and paper manufacture." Water was used to transport logs to the mills, remove bark with power-spraying devices, and generate steam to run mill equipment. It was a critical component in bleaching processes and to convey pulp through the paper-making machines.[24] The industry's greatest use of water, by far, was its most polluting: mixed with chemicals and heated to create the sulfite, sulfate, (kraft), and soda cooking liquors. After separating out wood fibers needed to make different kinds of paper and paperboard products, with very few exceptions mills considered the 55 percent or so of suspended solids remaining in the cooking liquors as wastes to be disposed of in the nearest body of water.[25] These lignins, fibers, and other solid materials contributed significant amounts of biochemical oxygen demand (BOD) to receiving streams.

In the absence of comprehensive water consumption data for the industry through the 1940s, an analysis of average industry water use from a slightly later period can provide some insights applicable to the Willamette River water quality story. In 1959, an average mill used 57,000 gallons of water to produce one ton of pulp. Driven by the need to be more cost-competitive, growing citizen activism, and actions by some state governments, by 1969 the average US mill had reduced water use by about 8 percent to 53,000 gallons.[26]

Average industry water use data can be misleading, of course, because the amount of water to produce one ton of pulp at any given mill is highly variable. Critical variables include the equipment used, type of pulp made, and quality of paper manufactured. For example, as of about 1970, groundwood pulping required up to 1,000 gallons of water for every ton of pulp produced, whereas the sulfite process used up to 30,000 gallons per ton. Bleaching groundwood pulp used an additional 1,000 gallons per ton of pulp, while the kraft and sulfite processes used up to 100,000 gallons. Guthrie determined that national averages of water use among pulp and paper mills in 1970 were 62,700 gallons per ton of sulfite pulp; 34,500 gallons per ton of unbleached kraft; and 53,500 gallons per ton of bleached kraft—though extremes of water used in each category varied widely from these averages. Additionally, Guthrie's data showed about 10 percent of water used was "'consumed' in the sense that it is unavailable for further uses." Therefore, 90 percent of water used during production was delivered back to a receiving stream, laden with solids and chemicals.[27]

Given the qualifications noted above, three assumptions about industry water use in the 1920s, 1930s, and 1940s seem reasonable to apply to the five Willamette Valley pulp and paper mills. First, firms probably used it more efficiently over time. Second, it should be safe to assume that each mill used an average of at least 57,000 gallons of water (probably much more) for every ton of pulp produced. Finally, 90 percent of this water—or about 51,300 gallons per mill per ton of pulp per day—made its way back into the Willamette as untreated wastes with high BOD levels.

Making pulp and paper required water-intensive, highly polluting processes. Pollution from a single mill within a watershed would be compounded by construction of more mills, and as production capacity increased at each site, pollution would also increase. This occurred within the Willamette and Lower Columbia watersheds and across the country during the first half of the twentieth century. National annual pulp production increased fourfold between 1914 and 1948, from 2,893,000 tons to 12,872,000 tons. On the West Coast, production increased even more rapidly: Oregon and Washington (with some small contributions from California) provided 6 percent of total national production in 1919, 16 percent in 1929, 21 percent in 1934, and 17 percent in 1947.[28] Pulp production more than doubled in Oregon between 1920 and 1940, from 150,000 to 360,000 tons.[29] By 1950, Washington was the leading pulp producer in the nation, and Oregon and Washington together produced about 59 percent of all domestic pulp sold in the United States.[30]

The period 1900 to 1940 saw significant pulp and paper industry migration from the Northeast and Great Lakes regions into the Pacific Northwest.[31] Four primary developments spurred this migration into Oregon. By the turn of the century, the timber industry had logged-over the Northeast and Great Lakes regions and needed to find new forests to consume. Concurrently, in the Pacific Northwest, the timber industry was extending logging railroads into the hinterlands at an unprecedented rate, while rapid population growth increased regional demand for paper products. Researchers discovered that the sulfite pulping process worked well for western hemlock, and the relatively new sulfate pulping process was effective for Douglas fir, both abundant species in Oregon.[32] Finally, pulp and paper production increases in Oregon and Washington between 1920 and 1940 was also related to a downturn in the region's lumber industry and a concurrent push to find new products and uses for materials previously considered wastes.[33]

Such significant national and regional shifts increased the industry's economic importance in Oregon. As of early 1939, the state's paper industry manufactured products valued at an estimated $18,240,000 annually. This was only 2.2 percent of the nationwide paper industry total but about 17.4 percent of the total for all products manufactured in Oregon.[34] From 1947 to 1957, industry employment in Oregon and Washington combined increased 35 percent, while the dollar value of industry payrolls increased 120 percent. In 1957, Oregon and Washington produced 17.1 percent of the wood pulp in the United States, an amount that exceeded combined production in the Soviet Union and Finland.[35] That year, Oregon had nine pulp and paper mills (seven in the Willamette Valley), and employed 4,197 workers at an annual payroll of $22,070,595. In 1956 Oregon mills shipped $126,000,000 worth of products.[36] As the *Oregonian* editorialized in 1957, the growth of the pulp and paper industry in Oregon was "one really bright spot" in the state's economy.[37] While the pulp and paper industry was one of the most important employers in a state heavily reliant upon extractive industries, it was concurrently polluting the Willamette River far in excess of any other industry or municipality.

THE SCIENCE OF WATER POLLUTION FROM PULP AND PAPER MANUFACTURING

By the 1940s, scientists and engineers were well aware that the sulfite pulping process produced much more water pollution per ton of pulp than the sulfate (kraft) process. Chemicals could be recovered and reused in the latter process, but years of research had shown that chemical recovery was not feasible with the sulfite process, nor could useful byproducts be created in an economically viable way.[38] In the matter of water pollution, it was not merely the *amount* of pulp and paper produced that was critical; it was also the *kind* of pulp created.

Most Willamette Valley mills specialized in the production of sulfite pulp—the more polluting variety—because this process permitted use of a common species in the coastal Pacific Northwest, the western hemlock.[39] Between 1914 and 1948, the relatively less-polluting kraft pulp increased from 2 to 47 percent of total national production, while more-polluting sulfite production decreased from 40 to 22 percent. At West Coast mills, kraft production also increased, but, rather than follow the national declining trend, sulfite production also expanded.[40] The result in the Willamette Valley was that pulp and paper industry production increased significantly, and of a kind that was much more polluting.

Well into the 1970s, specialists characterized water pollution from the pulp and paper, food processing, textile, and some other industries primarily in terms of its biochemical oxygen demand (BOD) and interrelated effect upon the dissolved oxygen (DO) levels in receiving waters. With these metrics, pollution from these kinds of industrial wastes were calculated in the same way as pollution from raw sewage; unlike sewage, however, these industrial wastes did not carry fecal coliform bacteria. Based on the standard metrics of BOD and DO, sanitary engineers evaluated relative levels of pollution using the concept of "population equivalent." This measurement is as straightforward as it sounds: an engineering measurement of the pollution load on a body of water equivalent to the amount of sewage produced by a given number of people. For example, untreated sewage from one hundred people would have a population equivalent of one hundred; standard primary treatment (also referred to as "mechanical treatment") involves screening and sedimentation to reduce sewage strength approximately 33 percent, bringing the population equivalent to seventy-seven. Secondary treatment (through the activated sludge and other processes) reduces the strength approximately 90 percent, bringing the population equivalent to ten. Tertiary treatment reduces sewage strength beyond 90 percent to near the level suitable for potable water supply but requires markedly higher investments in process design, equipment, chemical additives, and staff training than primary and secondary treatment.[41]

Using these metrics, Willamette watershed data compiled since the 1930s had clearly shown pulp and paper mill wastes to be significant pollutants. Wastes from the valley's five pulp and paper mills contributed a population equivalent BOD of about 1,500,000 in 1948 and 2,243,600 just five years later, an increase of 49.6 percent. For comparison, in 1950, Portland numbered 373,628 residents, and Multnomah County 471,537.[42] Willamette Valley pulp and paper mills, therefore, produced about five times more oxygen-consuming wastes than the city of Portland and four times more than all of Multnomah County.

Prior to the refinement of these fundamental water quality metrics, citizen activists and government agencies found it difficult to develop actionable pollution abatement policies and methods. The pulp and paper industry also found it easy to avoid taking responsibility. In 1926, industry executives in Oregon claimed their wastes did not negatively affect streams or life-forms therein, and the absence of contrary evidence served as de facto support of their inaction.[43] These claims could stand because no scientific evidence had found a direct connection between untreated mill effluents and degraded water quality—but empirical data proving or disproving such a connection did not yet exist. Scientific evidence collected in the 1930s and 1940s in Oregon and elsewhere, however, enabled abatement advocates to counter industry executives' assertions. A key study in the case of the Willamette River was George Gleeson and Fred Merryfield's 1936 report. Their findings relied upon BOD and DO measurements to conclude that "the pulp and paper pollution problem is neither new nor local," and that "the disposal of the wastes from pulp and paper mills, particularly the sulphite liquor, has become a great problem wherever sulphite plants are located."[44]

There were, of course, many industrial water pollutants that could not be measured in relation to BOD, DO, or population equivalent. Specialists in the 1940s had already become aware of some of these, and they would identify others in ensuing decades. For example, wastes from coal mining operations had been recognized as a significant polluter since the early 1900s, and this knowledge spurred the creation of Pennsylvania's sanitary water board in 1923. About fifty years later, scientists finally understood that dioxins were persistent and highly carcinogenic chemicals, but it was not until the 1980s that researchers definitively linked one source of dioxin pollution to pulp and paper mill wastes. However, when Charlton and the OSSA turned their attention to the pulp and paper industry in the late 1940s, their goal was to find solutions to the extreme BOD loadings contributed from mill wastes that decreased DO levels more significantly than all other industrial wastes and municipal sewage outlets *combined*.

Water quality in and near Portland continued to deteriorate, and excessive pollution levels continued to be more apparent during the annual low-water periods of late summer through early fall. As citizen activism increased and as quantifiable scientific data began to disprove industry claims, pulp and paper companies had little choice but to transition from a strategy of denial into one of action—albeit limited.

POLLUTION ABATEMENT RESEARCH UNDERTAKEN BY THE PULP AND PAPER INDUSTRY

Assertions from David Charlton and the Izaak Walton League that the pulp and paper industry had not been doing enough to stop pollution came about fifteen years after some firms in the industry grudgingly recognized their waste disposal practices needed to change.[45] This recognition was not driven by corporate altruism, but by the concerted efforts of abatement advocates and government actions in key pulp- and paper-producing states.

Mills in Wisconsin and Maine were at the forefront of change because citizens in these states took the lead nationally in demanding the industry do something about pollution. Wisconsin IWLA members and other advocates successfully lobbied for the creation of the state's Committee on Water Pollution in 1927. A primary focus of this group was to address the extreme degradation of Wisconsin's thirty-five-mile-long Fox River, polluted from twenty-one pulp and paper mills. Not long thereafter, Wisconsin mills formed the Sulphite Pulp Manufacturers Committee on Waste Disposal.[46] In the Northeastern US, in 1942 the Maine Supreme Court limited the amount of sulfite wastes that each mill could put into the Androscoggin River. In the lead-up to that decision, the three pulp and paper companies in the watershed formed the Androscoggin River Technical Committee to conduct research and "find ways to eliminate nuisance conditions."[47]

During these years pulp and paper companies were not alone among polluting extractive industries to begin to change in response to increasing social pressure. The Federated American Engineering Societies suggested in the early 1920s that industry ought to begin to look for productive uses of its wastes as a way to conserve natural resources and decrease pollution.[48] The American Petroleum Institute funded a US Bureau of Mines water pollution study in 1922 and conducted its own water pollution survey in 1927. In the mid-1930s, during contentious congressional debates on a federal water pollution control bill, the Manufacturing Chemists Association established a stream pollution abatement committee to present a unified industry position at hearings.[49]

Shortly before the end of 1941, representatives of various pulp and paper companies took a path similar to the Manufacturing Chemists Association and began forming a national industry research group. The Executive Committee of the Technical Association of the Pulp and Paper Industry and the Board of Governors of the American Pulp and Paper Association appointed a stream improvement committee to look into the matter.[50] Based on this committee's recommendations, industry representatives formed the National Council for Stream Improvement of the Pulp, Paper, and Paperboard Industries, Inc. (NCSI) in 1943.[51]

The NCSI's policy was to "cooperate with state and local authorities and to investigate problems largely by university research." Eighty-five percent of US pulp, paper, and paperboard manufacturers belonged to the NCSI by early 1947; the NCSI's 1946 budget was $140,000, mostly for research. By the end of the 1940s, the NCSI was funding research at more than ten institutions across the United States, including Oregon State College, University of Washington, the Institute of Paper Chemistry in Wisconsin, the Mellon Institute in Pennsylvania, and universities in Maine, Louisiana, Alabama, and Texas.[52] A sanitary engineering professor at Purdue University lauded the NCSI's work just four years after its creation when he wrote that the organization provided evidence that "some manufacturers have taken the industrial waste problem seriously." He asserted that the NCSI was "not formed to fight legal battles" but to serve as a research group "bent on finding solutions to the waste disposal and recovery problems." In this regard, he opined that the NCSI "is probably the largest group to undertake coordinated research and investigation for a whole industry."[53]

Industry representatives often touted the work of the NCSI in Oregon's newspapers and at sanitary authority hearings. Research centered on systems and technologies that would either result in marketable byproducts derived from wastes or enable mills to decrease measurable pollution while continuing to use streams as waste sinks.[54] Research topics included trickling filtration systems similar to a proven sewage treatment technique; storing wastes in large lagoons for release during periods of increased river flow; and the artificial aeration of entire river sections to help break down organic materials. Studies also involved producing linoleum cement, artificial vanilla flavor (vanillin), cattle fodder, insecticides, fungicides, and other products. As of the late 1940s, however, most of this work had proven ineffective or economically impractical.[55]

Some critics—David Charlton among them—charged that the NCSI was little more than an excuse to stall. It is the case that the NCSI did not contribute as much money as abatement advocates desired, nor did the group achieve pollution abatement results as quickly as advocates hoped. However, from the outset, the NCSI funded substantive hydrological, biological, and technological research at universities across the nation. A notable component of their research provided quantifiable data applied in the service of enabling the industry to abate measurable pollution while continuing to use streams as waste sinks.

One tangible measure of the industry's commitment to abating water pollution is the development of effective technologies and saleable byproducts through the work of NCSI-funded researchers. By this measure its record was less than stellar. Another measure is the NCSI's level of funding and proportion of this funding to industry profits. One might reason that if the industry was committed to finding solutions then its intention could be measured in relative dollar amounts invested in research toward that end. This latter metric can be evaluated in comparison to Oregon State funding for pollution abatement work through the sanitary authority. Such a comparison raises a number of challenges. First, the NCSI and OSSA were not tasked with the same kinds of work. The NCSI used a portion of is funds for administrative expenses and most of the rest for research on developing marketable byproducts from wastes and understanding watershed hydrology. OSSA monies supported administration, laboratory analyses, and enforcement, but little funding for research on hydrology, engineering, and technologies.

A second, more fundamental, challenge arises in getting the data necessary to compare NCSI and OSSA budgets directly and in relation to funding sources. The NCSI received its funds from yearly dues payments of individual members. Data about how much funding the group received is available. It is difficult, however, to determine a reliable set of figures comparable over time that would help one understand how this funding compared to profits of individual mills or even the industry as a whole. The industry did not regularly make financial data public. Further, not only did the industry rate of return change annually, it also changed annually for each mill. For example, one analysis of mill profitability between 1900 and 1940 shows variation according to the type of pulping process employed; whether or not a given mill was a pulp mill, a paper mill, or an integrated pulp and paper mill; the age of the mill's manufacturing technologies; and the kinds of paper produced.[56] Tracking state budgets and OSSA funding, on the other hand, is much easier because this information is part of the public record.

Considering the limitations of data from the pulp and paper industry, comparing what is available with Oregon State data nonetheless makes for an intriguing analysis into the relative level of importance the industry and the state placed on water pollution abatement. The first step in doing so is to understand changes in the respective funding sources for the OSSA and NCSI during the 1940s and early 1950s.

The Oregon legislature allocated funds to the OSSA through the state general fund. Figure 4.3 shows biennial general fund appropriations through the 1939–1941 and 1953–1955 biennia. The general fund grew significantly: between 1939 and 1954, it increased by a factor of 9.1, from $21,290,778 to $192,776,039.[57]

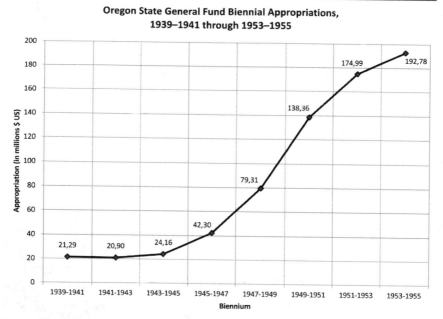

Oregon State General Fund Biennial Appropriations, 1939–1941 through 1953–1955

Fig. 4.3. Oregon State General Fund biennial appropriations, 1939–1941 through 1953–1955 (Data: See biennial budget documents submitted by Oregon governors and published by the Budget Division of the Executive Department, Salem).

Figure 4.4 provides broadly comparative data on the pulp and paper industry during the same period. This US Census of Manufactures data shows that national industry profits grew steadily from $862,866,000 in 1939 to $4,580,944,000 in 1954—a factor of 5.2.[58] Comparing these figures shows that while both amounts increased year-over-year, industry profits were orders of magnitude higher than state general fund appropriations. With this foundation, the next step is to compare OSSA and NCSI budgets in relation to these funding sources.

Figure 4.5 compares OSSA and NCSI funding between 1943 and 1949, the years in which the most reliable data from the pulp and paper industry has been found. In the midst of rising general fund appropriations and industry profits, the two amounts again differed in orders of magnitude, but this time the roles are reversed: OSSA funding was consistently about ten times more than NCSI funding, relative to respective funding sources. This figure also shows that the OSSA received more than three times as much of the state's general fund as the NCSI received from its funding source between 1943 and 1949: .0522 compared to .0172 percent. This information is also represented in Table 4.1.

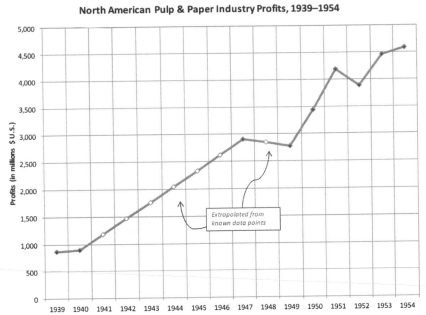

Fig. 4.4. North American pulp and paper industry profits, 1939–1954 (Data: US Dept. of Commerce Bureau of the Census, *US Census of Manufactures*, Washington, DC: US Government Printing Office, 1954).

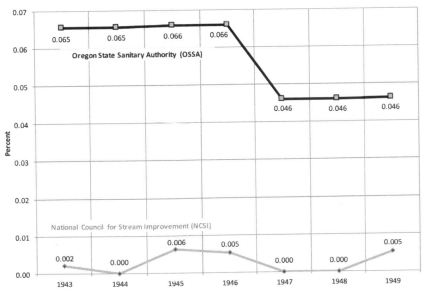

Fig. 4.5. Percent OSSA and NCSI funding relative to parent organization overall budget, 1943–1949 (Data: Biennial budget documents submitted by Oregon governors and published by the Budget Division of the Executive Department, Salem; US Dept. of Commerce Bureau of the Census, *US Census of Manufactures*, Washington, DC: US Government Printing Office, 1954).

**Percent OSSA and NCSI Funding Relative to
Parent Organization Overall Budget, 1943–1949**

		NCSI	OSSA
Parent Organization Funds (Pulp & Paper industry profits; Oregon State general fund)	1943	$1,756,141,571	$12,082,353
	1949	$2,777,231,000	$69,182,380
Aggregate Organization Funds, 1943–1949		*$17,295,440,426*	*$214,657,871*
Percent Change, 1943–1949		*+158.1*	*+572.6*
Institute Funding (per year)	1943	$38,000	$7,900
	1949	$150,000	$32,014
Aggregate Institute Budget, 1943–1949		*$478,000*	*$112,165*
Percent Change, 1943–1949		*+394.7*	*+405.2*
Percent of Aggregate Institution Budget In Relation to Aggregate Organization Funds, 1943–1949		.0172	.0522

Table 4.1. Percent OSSA and NCSI funding relative to parent organization overall budget, 1943–1949 (Data: Biennial budget documents submitted by Oregon governors and published by the Budget Division of the Executive Department, Salem; US Dept. of Commerce Bureau of the Census, *US Census of Manufactures*, Washington, DC: US Government Printing Office, 1954).

It is important to keep in mind that NCSI funding is measured in national terms, whereas OSSA funding pertains only to one of forty-eight states. Translating industry-level data into units per mill and per state can aid in comparisons between the NCSI and OSSA. The 1947 US Census of Manufactures lists 226 pulp mills and 665 paper and board mills (891 mills total) in 24 of 48 US states. At the per-mill unit of analysis, $480,000 in cumulative NCSI funding over seven years equates to about $538 per mill, or about $77 per mill per year. Per state, $480,000 equates to about $20,000 over seven years for each of the 24 states, or about $2,857 per state each year. Funding for the Oregon State Sanitary Authority, by contrast, never fell below $3,750 per year, and averaged about $11,372 per year between 1939 and 1949.[59]

The pulp and paper industry expanded rapidly between 1900 and 1950 to become economically dominant in the Willamette Valley and throughout the state. As it did so the amount of untreated wastes released into the watershed grew drastically. Collectively, the five pulp and paper mills located along the Willamette River and its tributaries as of 1948 discharged hundreds of thousands of gallons per day of waste cooking liquors and suspended wood fibers. These wastes contributed more pollution to the watershed than the entire valley population.[60] Oregon mills helped fund the NCSI, and creation of this research funding organization was a notable departure from the previous era in which the industry ignored or downplayed pollution, but by the late 1940s the NCSI's tangible contributions toward finding abatement solutions were limited, at best. While industry profits continued to grow, contributions to the NCSI were minimal in relation both to industry profits and increased OSSA appropriations. David Charlton and the Oregon Izaak Walton League characterized as a stalling tactic the industry's inability to develop satisfactory methods to reuse or reduce wastes while continuing to defer to NCSI research. Whether or not the industry was intentionally trying to stall, the claim has some merit. NCSI-funded research illustrated the difficulty of finding ways to transform or treat mill wastes, but relative to the high levels of waste the industry produced, it had not delivered results that would alleviate pollution. Companies externalized pollution costs and effects by using rivers and streams as waste sinks, disregarding all other actual and potential uses.

While it can be useful in some instances to view Oregon-based pulp and paper companies as a monolithic entity, however, in fact each mill and parent company approached the water pollution abatement issue somewhat differently. For example, from the late 1940s, the Crown Zellerbach mills at West Linn and Lebanon and the Weyerhaeuser mill at Longview, Washington, cooperated appreciably more than others with state water quality agencies by providing regular progress reports and conducting research on sulfite waste utilization and treatment options.[61] Such differences stemmed as much from unique corporate and facility cultures to the kinds of production technologies in use, products created, relative profit margins, and mill locations. The science, engineering, and physics involved in water pollution also made things difficult: sulfite wastes offered a considerably more complex treatment conundrum than municipal wastes.[62] Confounding the matter further, whereas federal funding and research assistance had been available for municipalities and state governments since the 1910s, there were no similar programs for the private sector. All of these factors influenced the openness of a given mill or firm to implement abatement practices and systems.

THE OSSA EDUCATES ITSELF ABOUT PULP AND PAPER INDUSTRY POLLUTION

As a first step in trying to understand the complexities inherent in pulp and paper industry pollution, sanitary authority members turned their attention to the topic not long after they established their administrative structure and operating processes. Shifting the focus of their limited resources from municipal sewage treatment—particularly within the City of Portland—the authority held its first hearing on industrial pollution in May 1940. Pulp and paper officials appeared with representatives from the meat-packing, textile, and fruit- and vegetable-processing industries. Authority Chair Harold Wendel opened the discussion by referring to a recent *Oregon Journal* article "which stated that stream pollution in the state was increasing and that fish are facing a losing fight for existence." Reflecting a common position among industry officials—and even some sanitary engineers—since the 1920s, Cassius Peck of the Crown Zellerbach Company asserted that sulfite waste liquors were not harmful to fish if given adequate dilution: "It is a debatable question as to whether or not sulphite waste liquor does any harm to streams or fish life . . . fishermen should be convinced that these wastes are not harming fish life." He urged the OSSA to keep in mind the relative value of fishing in comparison to the pulp and paper industry, insinuating that the latter was more economically important and should be given priority. To allow his industry to continue dumping untreated wastes, he suggested the OSSA adopt the Pennsylvania Sanitary Water Board's approach to stream classification, where some streams were functionally given over to waste. OSSA member John C. Veatch dismissed this request outright. "The Willamette river can't very well be put in just one class," Veatch stated, "for the reason that there is considerable amount of industries depending on that one stream . . . we have the paper and fishing industries, both of which are very valuable to the state, which are dependent upon one particular stream, and it would be pretty hard to classify the stream for the interest of one industry and the detriment of the other." The hearing concluded with all industry representatives in agreement that the OSSA should not compel them to abate pollution without the cities moving first. [63]

Following this initial hearing, progress addressing industrial water pollution in Oregon was slow throughout World War II. The war emergency took precedence generally, and facing staffing and budgetary limitations the authority focused much of its attention on Portland's sewers. In March 1945, the Crown Zellerbach Corporation was the first pulp and paper company to approach the OSSA with a plan to alleviate its sulfite waste problem. By modifying pulping equipment at its West Linn mill, the company claimed it would be able to evaporate waste effluents to recover valuable chemicals and ignite wood sugars to produce thermal energy for onsite applications. Such a system would result in a 70 percent reduction of pollution discharged into the Willamette from this mill, even while production expanded. [64] Crown Zellerbach's project became unfeasible a few months later, however, when the US Army Corps of

Fig 4.6. Aerial view of Willamette Falls looking north, ca. 1940s–1950s. Crown Zellerbach's West Linn mill is on the left, with the Willamette Falls Locks immediately to the left of the mill. Publishers' Pulp & Paper Company mill at Oregon City is on the right (Oregon Historical Society Research Library, ba014383).

Engineers announced an upgrade to the Willamette Falls Locks that would alter waste impoundment infrastructure in the mill's immediate vicinity.[65]

With the change in Crown Zellerbach's plans and a consistent lack of progress from the four other Willamette Valley mills, the OSSA called company representatives to a special meeting in January 1948. Authority Chair Wendel opened this session stating that

> the main force of the Sanitary Authority has been directed against municipalities; and many municipalities are moving rapidly towards consummation of abatement of pollution in the waters of the state of Oregon, and the public demand is being felt on the industrial side of the problem. The public would like to know from the pulp and paper industry what their plans are and what their intentions are in this regard.

Among the topics pulp and paper representatives were called on to address was the claim that "the industry as a whole and Oregon plants in particular are spending large sums in trying to find a solution" to stream pollution, and what the results of these investments meant in practical terms.[66] In response, industry officials referenced numerous examples of NCSI-funded research at Oregon State College, the University of Washington, and elsewhere on methods to treat or use waste sulfite liquors. In addition to trickling filtration, lagooning, and stream aeration, studies also involved producing linoleum cement, artificial vanilla flavor (vanillin), ingredients for insecticides and fungicides, and other products. These officials admitted that most of this work had thus far proven ineffective or economically impractical and that there was as yet "no known way of cleaning up sulphite liquor wastes."[67]

These admissions spurred a heated discussion. Wendel stated that "when the cities have completed these [sewer] building projects and the river is still polluted, the public feeling is very much in favor of having industry clean up their share of the pollution." Authority member John C. Veatch concurred, finding all indications pointing toward increased public interest in pollution abatement. "The movement for cleaning up the [Willamette] river is going to gain momentum instead of quieting down," he said, and continued,

> I further believe it is up to the industry to bear that in mind. It is not only in the state of Oregon, but the same movement is spreading in eastern states, and the only hope is that someone finds the proper method for elimination of the pollution caused by this industry. The fishermen on the Columbia River are preparing to bring action against the plant at Camas for causing such pollution. That is just one indication of what you are going to be up against. I do believe that most of the plants are facing a very serious situation—a bigger problem perhaps than that of any other industry.

Upon hearing further equivocal testimony from industry officials, Oregon's Assistant Attorney General Rex Kimmel concluded that the "present status of the program looking toward pollution abatement seems to be that the paper industry has nothing to contribute to solve the problem."[68]

Thus, later in 1948 when David Charlton submitted the Oregon Izaak Walton League's report criticizing both the OSSA and pulp and paper companies, the recent context was that authority members were becoming increasingly frustrated with the industry's lack of progress, while industry officials felt they were doing their best by contributing to NCSI research. Charlton's expertise provided yet more quantitative evidence against the industry, and the Izaak Walton League's position represented the voice of increasingly more Oregonians opposed to business-as-usual. The report concluded that "the Willamette River is more grossly polluted than ever," primarily from the pulp and paper and food-processing industries, yet "studies upon which an accurate estimation could be made have not yet been carried out." Ten years after passage of the citizen initiative creating the OSSA, "the accomplishments of these industries in waste disposal through their own efforts as a result of encouragement by the Sanitary Authority have been almost negligible." The report challenged pulp industry representatives who asserted there were no economical ways to treat or dispose of waste sulfite liquors by referencing two specific methods. One fermented wood sugars to produce alcohol, and the other involved an alternate sulfite pulping process being tested at the Weyerhaeuser plant in Longview, Washington. This process substituted magnesium oxide for calcium bisulfate as a pulping base, from which solids were more readily evaporated for burning (and generating energy for onsite use) and a higher percentage of chemicals recycled. At OSSA meetings and in public statements, industry officials implied that Willamette Valley pulp mills would adopt this and other technologies if proven both technologically and economically feasible.[69]

In reply to examples of turning wastes into saleable byproducts, Wendel observed such methods were not always the panacea Charlton and the IWLA imagined. If all mills started producing alcohol, Wendel argued, much more would be produced than could be sold, which would drive down prices and make further production unprofitable. Regarding the alternate pulp-making process, Weyerhaeuser officials were still far from knowing definitively whether or not it would be effective, scalable, and economical.[70]

Notwithstanding these technical and economic arguments, there was some validity to Charlton's claim that the industry was not making adequate progress in spite of NCSI research—and using collective NCSI contributions to avoid individual mill-by-mill responsibility. Evidence of this can be seen in information State Sanitary Engineer and OSSA Secretary Carl Green shared in December 1948. Over the course of nearly a decade he had repeatedly requested effluent data from the five Willamette Valley mills. Some of the mills had not yet even begun to measure the amount and strength of their own effluents—or, if they were measuring it, they had chosen not to report this information. Green's conclusion was that "these industries had no idea as to the extent of the wastes which were discharged into the waters of the state."[71] The authority relied a great deal upon voluntary cooperation, and Willamette Valley mills were not cooperating. In spite of industry's claims to the contrary and its repeated references to NCSI-funded research, it had failed to comply even with this straightforward task.

Reacting to increased criticism as well as industry intransigence, the authority actively sought public participation at its February 1949 meeting. Even though the OSSA's quarterly meetings had always been open to the public, Wendel observed that "very few persons have ever attended." He wanted to dispel claims from Charlton and others that the authority "concealed facts" and was confident that public observation would convince skeptics that stream pollution was complex and "cannot be corrected without public co-operation and support."[72] At this meeting conservationists, industry representatives, and state and federal officials heard about progress with the alternate pulping process at Weyerhaeuser's Longview mill and on research in Wisconsin. Experiments in that state had shown that holding waste liquors in large lagoons often led to groundwater contamination as concentrated wastes seeped out, and that reducing the wastes' BOD using trickling filtration technology was prohibitively expensive. The most cost-effective methods for dealing with sulfite wastes in Wisconsin appeared to be using these wastes as a road binder or growing yeast for cattle fodder. As the meeting concluded, a representative of the Portland Chamber of Commerce was convinced that industrial pollution abatement was "being handled in an intelligent manner by the Authority," faced as it was with a lack of adequate technologies to treat or use sulfite wastes.[73]

While progress abating industrial pollution was slow and contentious, in the late 1940s authority members reflected on significant recent achievements among Oregon municipalities. Portland's sewage treatment project was still on schedule to be completed in 1952, and the OSSA interpreted this achievement as indicating most of the state's sewage pollution problem would be solved.[74] Pollution from pulp and paper mill effluent was still a growing problem, however. In spring 1949, sanitary engineers found that each of the five Willamette Valley pulp and paper mills discharged the equivalent pollution load of a city of between 400,000 and 500,000 residents, as measured by BOD.[75] Combined, these discharges equated to a population equivalent of 2,851,000, or 90 to 95 percent of the river's total pollution load.[76] For comparison, the combined BOD from all other industries above Portland was just 450,000, and the sewage population equivalent for communities above

Portland was only 203,750; in 1950 Portland numbered 373,628 residents and Multnomah County 471,538.[77] This was further evidence that the pulp and paper industry was by far the largest point source polluter in the Willamette Valley, as it had been since at least the 1930s.

THE OSSA'S ULTIMATUM TO INDUSTRY: COMPLY OR SHUT DOWN

The OSSA tried to work cooperatively with the industry, understanding that technologically feasible and cost-effective abatement solutions were lacking. Authority members remained open to constructive solutions, however temporary.[78] Nevertheless, firms had not yet been able to develop technological or engineering solutions that allowed them to find economical ways to utilize their wastes. Industry representatives used this narrow focus on technology and economics to support assertions that pollution abatement was not feasible. Corporate resistance to pollution abatement based on such arguments was not unique to the pulp and paper industry or to the situation in Oregon. For example, historian John T. Cumbler finds industrial polluters presenting this same argument in Connecticut in the early twentieth century: "The search for a scientific and technological transcendence could easily become a slippery slope to acceptance of environmental loss until a solution appeared."[79] Such arguments stalled pollution abatement efforts throughout the United States.

By July 1949, sanitary authority members concurred that the mills would not make any real progress unless forced to do so. Armed with data from their 1949 survey, the authority exercised its enforcement powers and gave the five Willamette Valley pulp and paper mills until December 31, 1951, to cease polluting or shut down. They set the compliance date to coincide with the expected completion of Portland's treatment plant. Wendel foresaw that "as the Portland sewerage system was completed, the public would be very much concerned if private industry continued to pollute the waters of the state."[80] The Washington State Pollution Control Commission had recently issued its own ultimatum to mills in the Everett area, ordering abatement by September 1951.[81] Historian Terence Kehoe describes a "cooperative pragmatic" approach to addressing industrial pollution that existed between state water quality agencies and polluters in the Great Lakes region prior to the 1960s, "based on the principles of voluntarism and informal cooperation, administrative expertise, and localism." In the Pacific Northwest, at least, such cooperative pragmatism started to erode by the late 1940s as state officials and the pulp and paper industry began to diverge in their understanding of what constituted "cooperation."[82]

Pressure increased for the industry to make changes, and debates continued within the OSSA's hearing room, in the state's newspapers, and at public events. The Multnomah Anglers and Hunters Club sponsored one such event, hosting industry representatives and David Charlton. To Charlton's assertions that industry was not making adequate progress, a Crown Zellerbach official referenced hundreds of thousands of dollars the industry had contributed to NCSI research. Charlton countered by claiming that "research frequently is used as a delaying mechanism." He bemoaned what he perceived as the OSSA's lack of progress and found that, by default, leadership had been left to sportsmen's groups.[83] Another of Charlton's critiques against the OSSA was that "news accounts and even biennial reports from the Authority often stressed progress made without putting equal emphasis on the lack of progress, thus the public might get a false impression that all is well."[84]

Notwithstanding growing public pressure and Charlton's quantitative evidence, pulp and paper officials were not prepared to meet the sanitary authority's mandate to cease polluting by the end of December 1951, so in November 1949 they petitioned for a five-year extension. The attorney for Publishers' Pulp & Paper Company told authority members that "as far as setting the deadline for abating [sulfite] pollution by December 31, 1951, the date might just as well be set for any day after tomorrow, because there is no practical known solution for the problem." Harold Wendel reminded them that it was the OSSA's duty to carry out the will of the people, and this meant continuing the call for industry to change its waste management approach. When industry officials requested the OSSA's help doing so, Chairman Wendel explained that "it was not the duty of the Authority to solve the financial problems of industry or to conduct their research for them."[85]

Unable to sway the OSSA, officials from the Publishers' Pulp & Paper Company lodged a formal objection and planned to file an appeal with the state attorney general. Their central argument was that pollution abatement imposed an undue economic burden. OSSA member Barney A. McPhillips drew attention to the many municipalities the authority had pushed into building sewer treatment works in spite of financial difficulty. "Are we to understand," McPhillips queried,

> that the Publishers' Paper Company is asking the Authority to desist from trying to enforce the pollution law, in spite of the fact that all municipalities representing the greater portion of the people of the

state of Oregon are being forced to comply, and that other industries also are being required to abate pollution?

Publishers' attorney countered that an important difference was a solution for treating sewage existed, but there was no solution for sulfite wastes. In reply, OSSA members expressed hope that the intervening two and one half years before the deadline would be sufficient for the industry to come up with solutions.[86] Newspaper editors applauded this toughening stance. The *Oregonian* recognized the industry faced complex, costly issues, but drew upon increasing public sentiment viewing such arguments as delaying tactics.[87] The *Oregon Journal* supported more forceful action against industrial polluters, and stated that compelling compliance would "be a form of expiation by our civilization for some of its sins against nature."[88] Influential newspaper editors were joining with leading abatement advocates, the OSSA, and growing public opinion supporting enhanced abatement measures for both cities and industries.

As tensions grew, Crown Zellerbach officials were compelled to clarify statements made during the February hearing. They asserted that, in spite of what the company's attorney was overheard telling the OSSA, it was *not* threatening to close its West Linn and Lebanon mills if the sanitary authority maintained its strict deadline. The company claimed it was "zealously attempting to solve the many problems involved in sulphite waste disposal"; however, if it had not developed an abatement system by December 31, 1951, and if the authority was not willing to compromise, then it "would have no alternative but to obey the order of the lawful authority" and close its mills. "Any closure," therefore, "would result only from Crown's inability to comply with orders made by lawful authority."[89] With this statement the company was responding to the tide of public sentiment that was becoming less tolerant of the Willamette River's continued use as a waste sink.[90] These officials were also trying to draw a nuanced distinction while implying that blame for possible mill closures would be on the OSSA and Oregon laws, *not* on Crown Zellerbach. Wendel and other authority members had long recognized there were no simple solutions to treating or reusing sulfite wastes, but they had consistently stated an openness to *any* plausible abatement option. Rather than respond to the sanitary authority's requests to provide basic data on their effluents, industry officials had instead consistently avoided action, requested extensions, and deferred to the NCSI—and, when finally ordered to change their practices, an industry critical to the state's economy implicitly threatened to cease operations.

The issue continued to escalate. In February 1950, the OSSA convened a hearing for representatives to explain to Oregon Attorney General George Neuner why the sanitary authority should not follow through with its decision to compel the industry to abate pollution or cease operations.[91] The OSSA did not record minutes from this hearing, but later documented Attorney General Neuner's findings of fact that summarized the most recent OSSA water quality data and made a determination on the contentious deadline. Quantitative facts included evidence that the Willamette River below Salem and the Santiam River below Lebanon were both devoid of sufficient DO during the low-flow months of July through October. Although the five Willamette Valley mills contributed 90 to 95 percent of oxygen-depleting discharges, these mills were not implementing any known abatement options.[92] Approximately 50 percent of municipalities along the Willamette River had built primary sewage treatment facilities by 1950 but, with the continued discharge of pulp mill effluent, this would not be enough to elevate Willamette River DO levels to support fish. Therefore, the attorney general found that pulp and paper industry pollution violated Oregon laws and OSSA edicts and ordered the mills to cease discharging effluents from July 1 to October 31 and at any other low-water period. He conceded to industry requests in some part by giving the mills an extra five months to comply, but made it clear that industry officials were not to rely on completion of the thirteen-dam Willamette Valley Project as a partial remedy, though releases from reservoirs would increase summer and fall water flows. Within these limitations, his extension of the compliance date to May 1, 1952, was a pragmatic approach reflecting the difficulties involved in developing abatement technologies and the importance of the industry to the state.[93]

As these regulatory negotiations continued, water quality declined further. In late September 1950, another large-scale fish kill occurred in Portland Harbor. Sanitary Authority Engineer Kenneth Spies was surprised the dead fish he saw along the banks had been able to make it that far. He noted recent OSSA tests showing "hardly any oxygen left in solution in the Willamette from Cottage Grove through Portland to the Columbia," now an annual occurrence during low-flow months.[94] Two research efforts in 1950 echoed Spies's findings. The NCSI conducted its own survey (published in 1951) and found the river above Oregon City in worse condition than at any time in the past, as gauged by DO content.[95] A few months later, President Truman's National Water Pollution Control Advisory Board visited Portland during its national tour of major river basins. Truman established this advisory board in early 1950 to help realize the goals of the Federal Water Pollution Control Act of 1948.[96] After hearing comments on abatement progress, board members found the region to be "behind the times" relative to other areas of the country in terms of state water quality administration and industry compliance with abatement measures.[97]

In spite of pulp and paper industry threats and continued seasonal die-offs, by the middle of 1951 state officials perceived progress toward sulfite waste abatement. In June, the OSSA and Oregon Health Board complimented the pulp and paper industry for finally building waste storage lagoons. The Crown Zellerbach and Publishers' companies had been researching this practice in partnership with the Oregon State College Engineering Experiment Station since November 1950. This technique involved excavating reservoirs in which mill wastes could be stored when river levels were low; when water levels increased sufficiently, wastes would be metered into the river and carried downstream. It was not a method for treating or repurposing wastes, but it did keep effluents out of the river when water levels were too low to assimilate them. Construction of other lagoons at Crown Zellerbach's West Linn mill, Spaulding's Newberg mill, and Oregon Pulp & Paper's Salem mill continued throughout 1952.[98] State officials expressed guarded confidence these lagoons would keep the river free of mill wastes during low-flow months as early as 1952.[99]

Publishers' Pulp & Paper Company officials were involved in two other notable abatement projects. Clackamas County staff and the company's Oregon City mill collaborated on experiments in summer 1951 to use sulfite wastes as binder for gravel roads. These trials followed successful experiments in Washington, Wisconsin, New Jersey, Maine, and Scandinavia. County officials found the practice more cost effective than using waste oil, but the solution would not be appropriate for communities incurring the costs of shipping wastes far from centers of production.[100] As experiments continued, however, Publishers' officials soon realized that Clackamas County would not be able to use a sufficient volume of the company's waste liquors to make any real difference. To address this issue, Publishers' gained approval from the OSSA in early 1953 for another option: barge wastes downstream in 120,000-gallon batches to be dumped directly into the lower Columbia River.[101] In late July

Fig. 4.7. Barge filled with sulfite wastes from the Publishers' Paper Company mill in Oregon City for transport to the Lower Columbia River, 1953 (National Council for Air and Stream Improvement, Corvallis, Oreg.).

1953, Publishers' launched its first barge from Oregon City and released it "harmlessly" into the Columbia.[102] Given some limitations, the OSSA found initial results from these measures positive.[103] At the very least, these methods decreased waste discharges into the Willamette River during low-flow periods.[104]

Echoing the complexity of the issue, in the early 1950s *Oregonian* editors expressed contradictory opinions on the progress of abating Willamette River pollution. Early in 1952, the *Oregonian* found abatement efforts to be "a slow and discouraging fight, in which the state authorities seem to have fallen behind public opinion."[105] However, a few months later the newspaper congratulated state residents and foresaw that water pollution was nearing an end. Influencing this positive interpretation was the soon-to-be-completed Portland sewer system and an "increasingly healthy attitude of pulp and paper companies along the Willamette" that illustrated "the wisdom of authority policy."[106] As the 1950s progressed, however, solutions that state and federal agencies pursued did not keep pace with environmental pressures put on the Willamette watershed by expanding population, increased industrialization, and the compounding effects of other forms of pollution.

Chapter 5
Polluted Paradise

In early 1954, the *Oregonian* asserted that "no other state has a better post-war record than Oregon in reducing pollution of its streams and lakes," an achievement for which "official and unofficial advocates of clean rivers who have led the crusade untiringly" may reflect upon "with great satisfaction."[1] One clear example of this achievement was that since the graphic, large-scale die-offs in 1949 and 1950, fish kills were not nearly as common nor as large as they had been since the 1910s.

Facts on the ground, however, challenged the congratulatory tone of such editorials by demonstrating that pollution abatement was not a perpetually ascending line of progress. Years after beginning his long career helping to ensure water quality throughout the state, retired OSSA biologist Glen D. Carter recalled his own observations of Portland Harbor river conditions in the mid-1950s:

> When the city held its annual Rose Festival in June, US Navy ships traditionally sailed into the harbor with great fanfare and tied up along the harbor seawall. During the 1950s, when harbor waters appeared to be the greasiest ever, the ships would arrive spotlessly clean in accordance with Navy regulations. After a few days in the harbor, however, every ship and small boat accumulated a two-foot-wide belt of heavy tar, grease, and oil at waterline. When the Navy's visit ended, the mighty ships would steam downriver with sailors hanging over the sides in boatswain's chairs, each grasping a bucket of solvent and a stiff brush to remove the rings of Oregon tar and grease.[2]

Through the 1950s, the same newspaper editors who praised state officials and industry representatives in one column would, only days or weeks later, fill their columns with castigation and lamentation. This reflects not schizophrenic editors but, rather, the complex and often frustrating facts of water pollution. Water quality had improved by some qualitative and quantitative measures, at least—but setbacks often followed improvements. When excoriating polluters, for example, editors characterized the river as "Cleaner, But Not Clean," because efforts had thus far failed "to get ahead of the pollution treadmill."[3] During summer 1956 the river in and near Portland was still below the bare minimum four parts per million (ppm) dissolved oxygen requirement, and the sanitary authority continued to discover high coliform bacteria counts from untreated sewage.[4]

Carter's recollections and newspaper editorials illustrate how reductions in the most egregious and visible point source pollution from the river brought to the foreground more complex and varied waste sources, undermining previous achievements and requiring new approaches. Three specific examples from the 1950s into the early 1960s help illustrate this: the *Sphaerotilus* ("slime") problem in the lower Columbia River; wastes from the meat-processing industry along Columbia Slough at Portland's northern boundary; and the continued inadequacy of Portland's sewage treatment infrastructure.

Before highlighting these three examples, it is important to understand some critical regional and national context interweaving them. First was continued population growth and densification, combined with increased industrialization, which amplified negative environmental effects, particularly in the tri-county Portland metropolitan area. To help quantify some of these, both the OSSA and the pulp and paper industry embarked upon extensive watershed-wide research projects to gather quantitative data. A third important bit of context was the increasingly more active role federal agencies played in regional water quality issues. Finally, the character of the Willamette River itself changed drastically with the completion of the largest Willamette Valley Project dams; these reservoirs offered planners the ability to regulate river flows, and by the early 1960s this ability became integral to pollution abatement practice and policy.

WILLAMETTE VALLEY DEMOGRAPHICS AND
INDUSTRIALIZATION

The most easily identifiable cause of continued water pollution in the Willamette River during the 1950s was significant population growth and industrial expansion throughout the valley. Nowhere was this more apparent than in the Portland metropolitan region. Portland and its suburbs (including Vancouver, Washington) had gained 250,000 residents between 1939 and 1945 as part of war-related employment.[5] After World War II, Oregon's population continued to expand significantly: during the 1947–1948 biennium, Oregon was first among the forty-eight states in population growth at 49.3 percent.[6] This accelerated growth and development continued to be felt most strongly within the Willamette Valley, and particularly in the state's largest urban area, Portland. In 1950, Portland contained 373,628 and Multnomah County 471,537 residents. The next three most populous cities and counties were also within the Willamette River watershed. Between 1940 and 1950, the city of Portland grew by 23.3 percent, and Multnomah County grew by 32.8 percent. The following decade, Portland actually lost 952 residents to arrive at a 1960 population of 372,676, but Multnomah County as a whole grew 18.3 percent to 522,813. Both adjacent counties (Washington and Clackamas) grew significantly as well: 50.5 percent and 30.4 percent, respectively.[7]

To varying degrees, communities across the United States experienced increased population and industrial expansion immediately after World War II. Such rapid changes taxed existing transportation and sanitation infrastructures.[8] In Oregon, this dynamic further degraded the Willamette River, lower Columbia River, and Columbia Slough, in spite of the fact that residents had been hard at work for over a decade trying to address water pollution. "'Progress' has cancelled out 'progress,'" *Oregonian* editor Herbert Lundy wrote in August 1957, because "population growth and industrial expansion have kept pace with gains in pollution control."[9] Quoting David Charlton, Lundy continued, "The progress made to date is commendable but not startling," because

> in spite of construction of sewage treatment plants, water releases from two large reservoirs (Detroit and Lookout Point) and good water years, the water condition is at best marginal . . . [also] there are no large industrial plants actually treating wastes or making by-products, as is being done in many states. The time will soon come when

industries will have to take such steps. Likewise, domestic sewage plants in the larger cities will have to be equipped for secondary treatment.[10]

While more than 90 percent of Oregon municipalities had built primary sewage treatment systems by summer 1956, these systems would be over-loaded as populations increased, with the result that insufficiently treated sewage would once again enter the river. Reflecting on this, the *Oregonian* editorialized in August 1956 that "we have not licked the problem of stream pollution in Oregon. Rather, we are just holding our own in the battle to keep it from getting worse."[11]

QUANTIFYING WATER QUALITY

Abatement advocates understood that any attempt to address environ-mental degradation caused by population growth, industrial development, and other factors would require quantitative data. Specialists at the City of Portland, sanitary authority, and even within the pulp and paper industry each contributed to this work.

An integral aspect of Portland's sewage treatment project had been the construction of a water quality–testing laboratory adjacent to the Columbia Boulevard Wastewater Treatment Plant. Engineers began sampling Willamette River water at this facility in summer 1949 to provide a baseline and continuing data on the effects of the improved sewer infrastructure.[12] Portland's chief chemist reported in July 1952, nine months after the treat-ment plant had come online, that the Willamette River showed marked improvement as measured by both a decrease in bacteria counts and an increase in dissolved oxygen. This was a direct result of a ten-million-gallon-per-day decrease in untreated sewage—from fifty million gallons—dumped into the river.[13]

Beginning in 1949, the sanitary authority received sufficient funding and had increased staffing to a level that it was able to produce regular water quality reports for the entire main stem of the Willamette River. Its first comprehensive study in early 1949 showed definitively that pulp and paper mills were by far the largest point source polluters in the watershed, producing 90 to 95 percent of the river's total pollution load.[14] These find-ings updated and enhanced those of preliminary studies undertaken in the 1930s.

In response to increasing pressure from government agencies and citizen groups within and beyond Oregon, the pulp and paper industry's National Council for Stream Improvement (NCSI) also assessed Willamette River water quality. In May 1951, the NSCI published its comprehensive survey of the lower ninety miles of the Willamette, and ten years later the NCSI published an update.[15] Sanitary Engineer Clarence J. Velz authored both studies. He chaired the Department of Environmental Health at the University of Michigan's School of Public Health and published extensively in professional journals from the 1940s into the 1980s. This research shows his primary interest in developing, applying, and refining models and techniques to establish watershed-specific plans that would enable industries to continue to use rivers, lakes, and estuaries as integral elements of waste infrastructure systems. Much of his work spanning five decades was funded wholly or in part by the NCSI.[16] He sought to understand the unique characteristics of the Willamette so as to "forecast stream conditions under various drought probabilities and pollution loadings" that would then "serve as a guide in arriving at a rational solution . . . without waste or abuse of the natural purification resources of the river."[17] With these unambiguous and utilitarian motivations, Velz and the NCSI produced the most thorough analysis to date of the river's complex hydrology and waste loading.

Velz's research came in the midst of increasing pressure upon the pulp and paper industry from the OSSA, David Charlton, and other Waltonians throughout the United States, and other clean stream advocates. In mid-1949, sanitary authority members had concluded that Willamette Valley mills would not make any real progress unless forced. Armed with data from their 1949 survey and backed by growing public sentiment, the authority had exercised its enforcement powers and levied an ultimatum: the five Willamette Valley mills were to cease polluting the river by December 31, 1951, or shut down.[18] Velz's reports made use of—and confirmed—water quality surveys conducted since the late 1920s by the Oregon State College Engineering Experiment Station, sanitary authority, and others. His research methods and conclusions also reflected the dominant viewpoints of sanitary engineers and public health officials throughout North America and Europe.

American sanitary and civil engineers and public health professionals approached the water pollution issue somewhat differently in the late 1940s than had their Progressive Era colleagues at the turn of the twentieth century. This perspective retained some elements of the "wise use" conservation approach—in particular, engineers' desire to develop policies and technologies that would enable rivers to be used for multiple purposes. Nonetheless, mid-twentieth-century engineering was comparatively more inclusive in that it considered watersheds as systems and pollution as more than an issue of nuisances and threats solely to public health.[19] Conceiving of watersheds in such a manner can broadly be understood as a precursor to the holistic, ecological, scientific view that would gain prominence in the late 1960s.[20] However, this interpretation only goes so far, because Velz and his colleagues applied their efforts explicitly to avoid "the wastefulness of over-correction in some areas and the ineffectiveness of under-correction in other areas."[21] Their motivations, then, were not to improve water quality for the sake of the river itself, or for the nonhuman life it supported, but to minimize industry expenditures on abatement technologies and lessen the potential for litigation.

Velz reflected contemporary understanding of rivers as systems and of the need to find a middle ground that would still allow these systems to be used as an integral part of the overall waste treatment network. He wrote in 1947: "Streams cannot be generalized. Each has its particular characteristics." His belief was that "a sound economical solution . . . rests . . . fundamentally on evaluation of self-purification capacity for each particular stream."[22] This "sound economical solution" would come as a result of applying robust sampling techniques within each watershed to quantify two of the three primary water quality metrics: biochemical oxygen demand and dissolved oxygen. With this information, waste effluents could then be modulated to the degree necessary to continue using the receiving waters as waste sinks.[23]

Velz's research explained more thoroughly the many characteristics that made the Willamette unique. Since at least the 1930s, experts were aware of the complexities of the Willamette River system that made abating pollution difficult.[24] For example, one engineer involved in designing Portland's sewer interceptor system in the late 1930s noted that he had found "no city of similar size in the United States with such conflicting problems."[25] Velz's work enabled these characteristics to be quantified in much more detail than had previously been possible. He did so by identifying the river's "problems." One problem was the long stretches where velocity slowed considerably. Two of the most significant slack areas were the "Newberg Pool," about twenty-five river miles above Willamette Falls, and the twenty-five-mile stretch from its confluence with the Columbia through Portland Harbor to Willamette Falls. Another problem was that Pacific Ocean tides significantly influenced river velocity in this lower stretch, sometimes to the point that the river flowed *backward*. Tidal fluctuations also caused the periodic backflow of colder water from the Columbia River at least six miles up the Willamette. A third significant problem Velz discussed was that the Willamette experienced large swings in yearly high- and low-water levels. These directly affected the river's volume and rate of flow.[26] Since sanitary engineers understood that a river's capacity to neutralize— "assimilate"—organic wastes was directly correlated with the increased aeration potential of flowing water, any factors that slowed the velocity of a given unit of water, by definition, reduced this potential and increased pollution loadings.[27] Velz concluded his analysis with a model individual pulp and paper mills could use to generate the waste load reductions needed to achieve the minimum required five ppm dissolved oxygen content with any given level of water flow. It was not Velz's role to advocate for a specific abatement program, but to provide reliable data and modeling so that even if the complex hydrological realities of the river could not be remade completely, at least there would be a predictable system upon which the industry could base business plans.[28]

Since 1950 the sanitary authority had been publishing results of its annual river surveys. In 1957, the authority produced an interim report intended to consolidate these surveys and provide a "state of the river" overview (some of this data is presented in Table 5.1 on page 169).[29] Based on this report, media outlets concluded that the Willamette was "cleaner, but not clean" because river pollution control was "on [a] treadmill." The OSSA's 1957 data and the NCSI's 1951 conclusions supported this assessment.[30] Further, the sewered population in the basin had grown 20 percent and the untreated industrial waste load had increased 41 percent in just seven years.[31] The sanitary authority found these changes had negated expert assumptions about the adequate level of municipal and industrial waste treatment required. Because of this, the authority determined that its May 1950 order establishing maximum pulp and paper effluent discharges was too high. To generate this conclusion members inserted their own data into Velz's model.[32] Significant pollutant increases had brought about unpredictable effects within the complex Willamette River system, severely limiting—if not nullifying—the model Velz had so confidently developed only seven years previously.

FEDERAL REGULATIONS AND SYMPOSIA

"... make more objective the research on problems relating to water
pollution"

While the OSSA had been shifting its focus from Portland's sewage to pulp
and paper industry wastes, the federal government was gradually becoming more involved in water pollution abatement. The Izaak Walton League
and other groups had been pressuring Congress for a concerted federal
response to water pollution since the early 1930s. Later that decade the
National Resources Committee and some US senators and representatives
had proposed water quality legislation, but World War II diverted attention from this subject.[33] While the US Public Health Service maintained its
research laboratory in Cincinnati, by and large the federal government did
not conduct its own research. Instead, this work was largely left to the states,
private and public institutions of higher education, and private industry.[34]
Not long after the end of World War II, however, with renewed pressure
from the IWLA and other reformers, Congress passed the Water Pollution
Control Act (Public Law 845) on June 30, 1948. Senators Alben W. Barkley
(D-Kentucky) and Robert A. Taft (R-Ohio) sponsored this legislation after
unsuccessful bills had been introduced during the 1945 legislative session.
The Water Pollution Control Act did not fulfill all reformers' goals, but it was
the first comprehensive federal water pollution law enacted in the United
States.[35] The law can be considered "comprehensive" because it sought to
consolidate water quality research, infrastructure funding, and some level of
enforcement authority within a single agency, the US Public Health Service
(USPHS). In these broad terms it echoed at the federal level the role the
OSSA played within Oregon. The law increased federal technical and financial assistance to cities and states; in Oregon, the OSSA would be responsible
for distributing federal funds to municipalities and sewer districts. Congress
limited the USPHS enforcement authority to interstate waters and issues of
public health. In doing so, Congress made sure not to impinge upon a state's
right to administer and enforce water quality management programs, a ten-
sion evident in national water quality debates since the 1930s.[36]

Federal legislators were reacting to widespread changes across the
United States during and immediately after World War II when they passed
the Water Pollution Control Act. Intensifying industrialization and growing
urban populations throughout the West accompanied the end of the war.
Results were obvious in the late 1940s and included increased amounts,

concentrations, and diversity of municipal and industrial wastes. Existing federal laws were proving inadequate: the 1899 Rivers and Harbors Act pertained solely to the obstruction of navigable waterways, and the 1924 Oil Pollution Act was applicable only to petroleum pollution in tidal waters. While the need for a national approach to water pollution abatement was becoming increasingly apparent, the 1948 law fell short in many ways. It was only temporary: while Congress extended the law for three years in 1953, the previous year it stopped funding its state grants-in-aid program. Congress also never funded the law's municipal waste treatment loan program.[37]

During the three biennial budget periods after enactment of the 1948 Water Pollution Control Act, funding to the OSSA for its operations and for granting within the state fluctuated significantly. The OSSA received a little over $18,000 per biennium between 1949 and 1953, or about 20 percent of its 1951–1953 budget of $90,248. For the 1953–1955 period, however, it received just $1,244, or only about 1.5 percent of its $86,189 two-year budget.[38] Federal funds for OSSA operations and various statewide municipal sewage projects would expand greatly with the 1956 amendments to the 1948 law, as would federal support for research and interstate collaboration. The 1956 amendments also brought increased federal infrastructure funding and research support. Congress responded at the national level to the same kinds of complexities experienced in Oregon, as well as to continued pressure from grassroots organizations such as the IWLA and the League of Women Voters.[39] A few years earlier, David Charlton had written to the bill's sponsor, Representative John A. Blatnik (D–Minnesota), that "there is a growing demand by citizens and various organized groups that increasing pollution of our streams be stopped and that present pollution reduced to the minimum and kept under effective control by federal legislation, if the states are unable or unwilling to do so."[40]

Representative Blatnik and his allies had other motivations to strengthen national water pollution laws. They perceived water quality not as solely a matter of sportfishing, recreation, or aesthetics, but as an important resource management issue critical to urban and industrial growth. Thus, they saw the issue foremost as one of economic development and, because of this, worthy of federal dollars. Historian Paul C. Milazzo concludes that Blatnik and others perceived the allocation of pollution abatement funds as a means of "exploiting the venerable rhetoric that celebrated development and the institutional channels that delivered government resources, better known as 'pork.'"[41] Blatnik and his allies sought to protect water quality by linking this goal to federally funded infrastructure projects, and in this way garner support from fellow congressional leaders who may not have otherwise been interested in the clean stream issue. To do this, he joined with federal public health officials in arguing to strengthen the 1948 legislation by explicitly linking water quality with economic development. Such correlation was not new. The National Resources Council in the 1930s and President Truman's water pollution board in 1950 both made this connection.[42] Additionally, water pollution abatement advocates in California equated clean streams with the economic health of commercial fisheries, and in Oregon, Portland city officials garnered public support of the $12 million sewer bond measure in 1944 by characterizing the project as a postwar jobs creation measure.[43] Pollution abatement had long been directly connected with significant and expensive public works projects to spur employment and economic development. What made things different at the federal level in 1956 was that this would be a *federal law*, not merely a federal agency recommendation or a state policy. The 1948 Water Pollution Control Act might have been the first *comprehensive* federal water quality legislation, but, eight years later, what made the first *permanent* federal water pollution act significant was that it tied pollution abatement explicitly to substantial federal appropriations. It was, therefore, a measure for *proactive* steps rather than merely a *reaction to* degraded water conditions. Blatnik and the act's other supporters used federal public works enticements to gain the support of state and municipal officials who had previously been opposed to such legislation. Along with this came the compromise of enforcement mechanisms only minimally stronger than in 1948.[44]

Under the 1956 amendments, federal funds to the sanitary authority increased significantly. The authority received $21,785 during the 1955–1957 biennium (16.9 percent of its total biennial budget) $66,745 the following biennium (27.3 percent), and $71,032 (23.6 percent) for 1959–1961.[45] Under the original 1948 law and 1956 amendments, the OSSA disbursed federal funds to worthy municipal sewage infrastructure projects, and these dollars were tracked independently of funds allocated for the authority's day-to-day work. In 1957 alone the OSSA disbursed $647,125 to an array of projects throughout the state, and handed out increasing amounts to various cities and sewer districts over subsequent years.[46] Federal funds to help address water pollution in the Willamette Valley and throughout Oregon were important, and increasingly more so throughout the 1950s.

In addition to a substantial grants program to spur municipal sewer construction, the 1956 act also authorized the US Public Health Service to foster interstate cooperation and a broad-based, collaborative research program.[47] The public health service sponsored biannual water quality symposia in the Pacific Northwest beginning in late 1957. Earlier that year, Sanitary Engineer Edward F. Eldridge ended his nine-year tenure with the Washington State Pollution Control Commission to coordinate water quality research at the public health service's newly established Portland technical research station. In this position he embarked upon a "constant soap-box campaign" pushing for more research at regional universities. He found the issue to be increasingly more complex, particularly because of the wide variety of new chemicals and other substances being developed by industry.[48] Reflecting an ecological approach to water quality, Eldridge's scientific and technological challenges included "discover[ing] means of permitting man to progress in his environment without destroying the natural organisms that make that environment possible."[49]

In his role Eldridge also hosted the water pollution symposia. He opened the first of these in Portland by stating his main objectives: to "further and make more objective the research on problems relating to water pollution" and identify research gaps.[50] Subsequent symposia addressed financing water pollution research projects, the *Sphaerotilus* problem, siltation, radioactive wastes, and "the social and economic aspects of water-resources quality control."[51] Beyond facts and figures, another important contribution of these symposia was to provide a regular venue bringing together state and federal officials, industry representatives, engineers, academics, pollution abatement advocates, and other interested parties. The "slime" problem in the lower Columbia River would prove to be just one example of how water pollution was a joint concern for both Washington State and Oregon officials. Experience throughout the nation also showed that water pollution was a complex, multifaceted threat that did not respect political boundaries. Federal leadership in the form of serving as a central hub for research and best practices, and for convening stakeholders and constituents, offered the prospect of fostering more broad-minded and lasting water pollution abatement approaches.

THE WILLAMETTE VALLEY PROJECT

By the early 1960s, Congress had increased federal funding for pollution abatement measures and had taken steps to help fund and foster coordinated research into a wide array of water quality issues. Other entities within the federal government had been working in the Willamette Valley for many decades on matters that were not, at first, perceived to be relevant to water quality. By the early 1950s, however, it was increasingly apparent that the US Army Corps of Engineers' construction of dams on Willamette River tributaries would become an important component of abatement solutions. The corps of engineers had long been involved in managing the Willamette River. Civic leaders began dredging a channel through Portland Harbor and down the lower Columbia River to the Pacific Ocean in the 1860s to foster riverine transportation, but the corps of engineers soon took over this work.[52] Upriver of Portland, over the decades, engineers, farmers, business owners, and others removed obstacles from the river's main channel, closed-off alternate channels, built wing dams, reinforced embankments, and otherwise regularized the river.[53] These changes helped mold the river into a more useful transportation corridor, and aided in the reclamation of adjacent land, but it did not ameliorate flooding nor benefit waste disposal. To address the matter of flood control, the corps of engineers included a preliminary survey of the potential for developing reservoirs on Willamette River tributaries when they produced an engineering study of the Columbia River basin in 1932. Six years later, in response to a more thorough study (the corps's "308 Report"), the US Congress adopted a plan for developing the Willamette River water resources by building a number of dams (eventually thirteen) in tributary streams. The project's aim was to provide for flood control, navigation, irrigation, and hydropower, in addition to stream purification:

> Regulation of the main stream in the interest of navigation would
> tend to temporarily alleviate [the] present [pollution] situation
> by increasing present low-water flows. However, with continued
> development of the Willamette Basin, domestic and industrial wastes
> will eventually require partial or complete treatment before being
> discharged into the stream channels.[54]

In 1939, the Oregon legislature established the Willamette River Basin Committee to oversee project planning and allocate federal funds. The corps

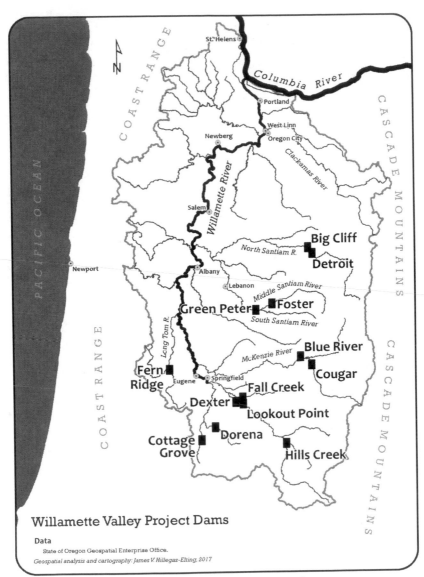

Fig. 5.1. Willamette Valley Project dams (James V. Hillegas-Elting).

of engineers built the first two flood control storage dams—Fern Ridge and Cottage Grove—at the south (far upstream) reaches of the watershed, and these were operational in 1941 and 1942, respectively. The remaining eleven dams were finished between 1949 and 1968.[55] In the aggregate, beginning in 1968, these corps of engineers–controlled dams regulated about 27 percent of the river system's flow.[56]

Seasonal water fluctuations became incrementally less extreme as the corps gradually completed these dams. River regulation led to some improvement in dissolved oxygen levels as early as 1953, with the completion of two high-head dams at Detroit (North Santiam River) and Lookout Point (Middle Fork Willamette River). Each of these represented a little more than 455,000 acre-feet of storage capacity, and together about 22.6 percent of the project's total storage capacity.[57] In practical terms, this meant that, beginning in 1953, flow in the main stem of the Willamette nearly doubled during the traditional seasonal low-water months of July through October.[58] By holding water in various tributaries, the dams modulated the annual flow variations in the main stem. This decreased the potential for winter and spring flooding, and also provided for higher average daily flows during the critical low-flow summer and fall months. Increased flow during these times meant that wastes put into the river were more likely to make their way more rapidly to the Columbia River and, thence, to the Pacific Ocean. This made the river better suited for year-round use as an integrated element of industrial and municipal waste management systems.[59] Sanitary engineering experts found that the gradual improvement of water quality in the mid-1950s *required* augmented flow from tributary reservoirs as much as it required Portland's sewage treatment system and modified pulp and paper industry waste disposal practices.[60]

Modifying river hydrology, whatever its benefits, also resulted in some unanticipated problems. Regularizing the river's flow by lowering winter/spring peaks and raising summer/fall troughs also meant that accumulated debris and sludge along the river bottom was less likely to be scoured and washed downstream during the winter months. An OSSA biologist recalled that "flood flows were especially important for flushing the Newberg Pool and the Portland Harbor zones" of the river.[61] These were precisely the zones that the NCSI's research had found to be most problematic in terms of maintaining adequate dissolved oxygen. In addition, managing river flows with Willamette Valley Project dams also raised the issue of water-rights allocations. The Oregon State Game Commission linked the September 1949 large-scale crawfish die-off near Scappoose, in part, to the over-allocation of water use rights that exacerbated seasonal low-flow conditions. Representatives of sportsmen's groups blamed the corps of engineers. These critics perceived that the corps did not release sufficient water because it was more concerned with maintaining water levels for recreationists at the Fern Ridge and Cottage Grove reservoirs.[62]

In early 1950, Governor Douglas McKay established an advisory committee on the state's natural resources to analyze the needs of parties with competing interests in Willamette Valley water allotments.[63] The same year, President Truman's National Water Pollution Control Advisory Board included flow regulation and rights allocation as important aspects of their water quality research.[64] In response to these state and federal initiatives, Oregon legislators created the State Water Resources Board in 1955 to develop and oversee a long-range, integrated management plan for "maximum beneficial use and control of the state's water resources," water quality being one element.[65] Because of the direct link between water allocations and Willamette River water quality, in 1960 the OSSA took a resolution to the water resources board to guarantee sufficient stream flow during the low-water months, thereby codifying such allocations as an integral pollution abatement measure.[66] Although Attorney General Neuner had warned pulp and paper industry executives during contentious hearings in 1950 not to rely upon augmented seasonal flows as part of their abatement plans, by 1960 state officials realized that reservoirs in the Willamette watershed were an essential part of multifaceted efforts to address water pollution.

Running as a thread through specific pollution abatement cases from the early 1950s into the early 1960s were four regional and national developments of critical importance. Willamette Valley population increases and further industrialization taxed sanitation infrastructure, causing undeniable environmental degradation. These effects were quantified in data collected by the City of Portland, OSSA, and National Council for Stream Improvement. Since these kinds of effects were becoming undeniable across the United States, federal legislators passed a series of water pollution control acts that increased funding for state-level pollution abatement and empowered the US Public Health Service to advance collaborative water quality research efforts among state agencies. Finally, as Willamette Valley Project dams came online, they significantly altered the river's character and positioned river regulation itself as an important pollution abatement tool. Each of these threads is evident to varying degrees in the cases of the "slime" scourge in the lower Columbia River, meat-processing wastes in the Columbia Slough, and the persistent shortcomings of Portland's sewage treatment infrastructure.

OTHER CONFOUNDING PROBLEMS: LOWER COLUMBIA RIVER "SLIME" AND COLUMBIA SLOUGH MEAT PROCESSING

Abel Wolman concluded in the 1930s that the lower Columbia River could assimilate 100 percent of Portland's untreated sewage without becoming degraded. He echoed expert consensus in determining that the very high river flows would be able to assimilate large amounts of municipal and industrial wastes while carrying these wastes to the vast Pacific Ocean, with no detrimental effects on fisheries, recreation, or public health. This consensus proved not to be reliable, however. Even by the late 1930s, a growing body of evidence indicated that the lower Columbia could not, in fact, assimilate all of the wastes that human activities were directing into it. The clearest example of this was the confounding "slime" problem.

Sport and commercial fishers noticed that lines and nets put into the lower Columbia were becoming fouled by long, stringy strands of green slime. The OSSA and its Washington State counterpart—the Washington Pollution Control Commission—collaborated on a joint study published in 1943. They found that the slime that coated fisherman's nets and suffocated juvenile fish was a strain of *Sphaerotilus*, "a filamentous, sheath-forming fungus." The fungus peaked in the warmer water temperatures of summer and early fall, and fed on phosphorous, nitrogen, and carbohydrates—carbohydrates of the kind found in lignins and other wood sugars from sulfite waste liquors. The fungus appeared to grow most rapidly when these wastes were concentrated at fifty parts per million (ppm) or greater.[67] Fishers and pulp and paper mill representatives held different opinions regarding the cause of the slime. The former asserted the problem was directly related to sulfite waste liquors, while the latter claimed that there was "little harm from the liquor for animal life." Industry representatives, echoing their counterparts in previous decades, opined that wastes from mills on the Willamette and lower Columbia rivers may even have "a cleansing effect upon the river"; there was, as yet, a lack of evidence either to prove or disprove industry assertions.[68]

Starting in 1948, the Oregon and Washington state water pollution authorities met every six months to address slime and other common water pollution issues. Based on their 1943 report and a US Fish and Wildlife Service survey in 1948, they developed a plan for mills to dilute sulfite wastes to concentrations of less than 50 ppm.[69] In 1950 the *Oregonian* observed this increased collaboration among state and federal officials and industry representatives and claimed that "a new era" in water quality was dawning.[70] The slime problem only intensified, however. Dam construction on the Columbia, more industrial development in the region, and increased water diverted for agricultural purposes had changed the river's character. By 1953, ten years after their initial report, Oregon and Washington water quality authorities found that diluting pulp and paper making wastes to fifty ppm did not alleviate slime growth, after all.[71]

On September 10 and 11, 1958, representatives from Oregon's sanitary authority, Washington's pollution control commission, industries, and other groups met in Portland at the third US Public Health Service–sponsored regional symposium to discuss the lower Columbia River. Conferees focused on two principal types of pollution: bacteria from insufficiently treated sewage and slime fed by pulp and paper mill wastes.[72] Research showed that the city of Portland contributed 89.6 percent of the bacterial pollution of the lower Columbia—and this in spite of the fact that since 1953 most of the city's population was being served by the primary sewage treatment plant. Another Willamette River pollution abatement practice that earlier seemed beneficial was also contributing significantly to pollution in the lower Columbia: Publishers' Pulp & Paper Company's practice of barging its Oregon City mill wastes to the Columbia contributed 12.5 percent of the nutrients necessary for slime growth.[73]

Following this, the Oregon and Washington sanitary agencies and the US Public Health Service announced a six-point compliance plan for the lower Columbia River that included a demand for cities and industries to cease discharging untreated wastes.[74] The *Oregonian* supported this plan and the collaborative, state-federal approach it represented, but not without qualification. Echoing observations made in the 1930s, editors claimed state laws were sufficiently strict but inadequately enforced.[75] The *Oregonian's* guarded optimism was justified, as it would take a combination of cooperation, increased enforcement, and scientific advances to address the *Sphaerotilus* problem. In the late 1960s, scientists discovered the fungus grew on wood fibers and created large, drifting flocs of slime that would get entangled in fishing nets.[76] With this evidence, the public health service supported the OSSA and Washington Pollution Control Commission's requirement that pulp and paper mills provide primary treatment of wood processing wastes by the end of 1967 to settle out these fibers and other solids.[77]

If expert opinion from Abel Wolman and other sanitary engineers in the 1930s was wrong about the Columbia River's ability to assimilate wastes, it proved correct about the Columbia Slough. These opinions influenced both the construction of the first phases of Portland's sewage treatment project in the late 1940s and the sanitary authority's approach to meat-processing facilities located along the slough in the 1940s and 1950s.

Part of the lower Columbia River and not the Willamette, the Columbia Slough differs drastically from both as it is much smaller and practically stagnant. The slough separates the North Portland peninsula from Hayden Island. Since the nineteenth century, European Americans had used the slough as a working stream, much as they used the Willamette and Columbia rivers. Barges transported log rafts by way of the slough, and shingle mills, meat-processing plants, animal feed lots, and other industries had long used the slough as a waste sink.[78] By the early 1940s, the effects of this practice were readily apparent in water that was unfit even to float log rafts.[79] During the previous decade, public health and sanitary engineering specialists realized that hydrological conditions made the slough an inadequate repository even for *treated* wastes. In 1939 Abel Wolman and his team expressed this conclusion clearly by contrasting the nonexistent assimilative capacity of the sluggish slough to the massive Columbia River. Whereas, they concluded, wastes should not be directed into the slough, from an engineering perspective the river could assimilate 100 percent of Portland's untreated municipal wastes—though they recommended against doing so to avoid "any conceivable evidence of sewage discharge."[80]

Degraded conditions and regular complaints had motivated the OSSA, from its first months in existence, to work with the sixteen meat industry companies located along the slough to find alternatives to dumping untreated wastes. Collectively, these processing plants, rendering facilities, and feedlots discharged a biochemical oxygen demand (BOD) population equivalent of more than 67,000 into the slough.[81] For years the sanitary authority had informed these firms of their obligation to abate pollution. On four occasions between December 1951 and October 1952, the authority granted abatement extensions to individual facilities because company representatives asserted that they were waiting on Portland officials to connect their facilities to the city's nearby Columbia Boulevard interceptor sewer.[82]

Connecting to the interceptor sewer might seem to have been the most simple and direct solution. Two issues complicated this option, however, one legal and the other technical. First, the meat industry facilities were located outside of the corporate limits of the City of Portland, therefore outside the

Fig. 5.2. Columbia Slough polluted with offal, blood, and other organic wastes flowing from meat-processing facilities, as well as trash and other solid wastes, Oct. 1970 (Oregon Historical Society Research Library, bb015682).

service area of the city's sewer system.[83] Second, the manager of Portland's Bureau of Sewage and Refuse Disposal testified that the city's sewage treatment works were not designed to deal with such wastes.[84] To these legal and technical complexities were added the sanitary authority's complicated enforcement process, its desire to work *with* companies to abate pollution, and its focus on more egregious polluters (i.e., the City of Portland and the pulp and paper industry). For these reasons, in October 1953 the sanitary authority gave these companies yet another sixty-day extension to develop abatement plans.

A few months later the OSSA's resolve strengthened and its patience waned. In response to measures that were still inadequate, in April 1954 Harold Wendel told industry representatives that he wanted action: "I, for one, am not going to quibble anymore."[85] By the end of the year, the authority considered progress satisfactory for fifteen of sixteen firms. Portland voters had, that November, also approved a charter amendment allowing the city to sign five-year, rather than just two-year, contracts. This change would enable the city to establish longer-term agreements with industries desiring to connect to the city's treatment system, thereby providing companies and the city with a longer-term planning horizon in which to operate.[86]

This apparent progress was to be short-lived, however, and Wendel's OSSA was forced to quibble yet again. In September 1955, the OSSA resolved to take legal action within sixty days against ten of sixteen Columbia Slough meat industry firms that had not yet connected to the Portland interceptor sewer or built their own wastewater treatment facilities. The authority initiated injunction proceedings against two companies in late 1955 for lack of adequate progress after years of extensions and warnings.[87] Threats of legal proceedings finally convinced even the slowest-moving firms. By 1958, the OSSA was able to get all but one facility to comply with Oregon law. This final firm—Pacific Meat Company—was not forced into compliance until the 1970s, even though in March 1961 the Oregon Supreme Court upheld a Multnomah County Circuit Court opinion supporting the OSSA's abatement order.[88]

PORTLAND'S INSUFFICIENT SEWER INFRASTRUCTURE (AGAIN)

"... aligned with the forces of evil"

Through the 1950s, editors at two of the state's prominent newspapers, the *Oregonian* and *Oregon Journal*, were variously praising and lamenting progress toward cleaning up the Willamette River. The *Oregonian*, for example, lamented the river was "clean, but not clean enough," putting pollution control "on [a] treadmill."[89] This stasis must not have been the worst of all possible worlds, however, because at about the same time, the *Oregon Journal* looked askance at water pollution abatement efforts in the Potomac River watershed as being appreciably behind efforts in Oregon.[90] By the late 1950s, however, with respect to Portland's still inadequate sewage treatment infrastructure, newspaper editors were in agreement on at least one thing: because the OSSA was advocating for health and welfare, it was "on the side of the angels," and Portland city leaders were "aligned with the forces of evil."[91]

More people began using the Willamette River for recreation in the mid-1950s as water quality gradually improved.[92] These improvements followed the completion of Portland's primary sewage treatment system, incremental abatement from the pulp and paper industry, and augmented flow from Willamette Valley Project reservoirs. David Charlton and others continued to press the sanitary authority to confront industrial polluters with more force, and OSSA members reacted to this pressure and evidence of environmental degradation while expanding their abatement efforts.

Portland's long-awaited primary sewage treatment plant in North Portland became operational in early October 1951 and was officially dedicated on October 27, 1952. The plant could process 60 million gallons of sewage per day, a capacity Portland City Commissioner William A. Bowes asserted would be sufficient until Portland's population numbered one million; the city's population in 1950 was only about one-third of this (373,000).[93] At the city's water quality testing laboratory, engineers measured a "record low" pollution level in late October 1953. This observation came after the last major

Fig. 5.3. Columbia Boulevard Wastewater Treatment Plant dedication, October 26, 1952. At microphone is Portland Mayor Dorothy McCullough Lee. Seated immediately behind is Commissioner Fred Peterson. Not visible behind Peterson is Commissioner William Bowes. Visible left to right are: past president of Associated General Contractors Ray N. Northcutt; OSSA Chair Harold F. Wendel; conservationist and former state game supervisor Edgar F. Averill; and chair of the board of project engineers Ben S. Morrow (City of Portland Archives, A2005-005.59.1283).

Fig. 5.4. Columbia Boulevard Wastewater Treatment Plant, July 1954, looking south with Columbia Boulevard in the background (City of Portland Archives, A2001-007.41).

eastside sewage lines serving the city's core were connected to the treatment plant and with the augmented flow provided by the Detroit and Lookout Point dams that had come online that year.[94] Portland's sewage treatment plant was, in the words of OSSA Sanitary Engineer Curtiss M. Everts, "the 'key' to the Willamette pollution puzzle."[95]

There were other important pieces of this puzzle as well. Until 1949, when primary sewage treatment plants in Junction City and Newberg came online, all cities along the Willamette River dumped untreated sewage into the river or its tributaries. Just eight years later, all cities on the river had operational primary treatment facilities after completion of Harrisburg's treatment plant.[96] Two Willamette Valley cities had secondary sewage treatment by summer 1960: Albany and Springfield.[97] Primary treatment neutralized bacteria and removed all visible solid matter and approximately 33 percent of the biochemical oxygen demand (BOD) of the sewage, while secondary treatment removed approximately 90 percent of the BOD.

As Portland's primary treatment plant came online and as other Willamette Valley communities built primary and secondary treatment facilities, attention turned to areas of population growth without sewer infrastructure in the suburbs and urban fringes. Local officials had begun to address the sewerage needs of the suburbs southwest of Portland in 1953 by creating the West Slope Sanitary District and contracting for engineering studies for the approximately six hundred square mile area. Although a bond issue to finance the district had failed in the November 1955 election, the Metropolitan Tri-Counties Health and Sanitation Committee continued "attacking the long-range problem."[98] In early February 1958, officials from Multnomah, Clackamas, and Washington counties agreed to a $27 million sewer plan for this suburban area.[99] Construction of the first phases of the Tri-County Sewerage System began by September 1959.[100] In the midst of this progress, however, Portland itself faced a "sewer emergency." City officials realized in late 1956 that extensive sewer line repairs and the need to expand infrastructure into neighborhoods at the city's fringes would take more than the $5 million bond issue city officials had proposed for voters that November. The *Oregonian* saw this lack of foresight and piecemeal approach as a prime example of the city council's weaknesses in formulating long-term infrastructure plans, a shortcoming that harkened back to city council failures in previous decades. [101]

In its comprehensive survey of river conditions in 1957, the sanitary authority found the river "still polluted to such an extent that it is not safe for certain recreational uses and occasionally in certain sections it is unsuitable for propagation or maintenance of fish life."[102] The authority's research led it to two conclusions. First, primary municipal waste treatment was no longer sufficient in the Willamette Valley. Second, the city of Portland still lagged behind other municipalities because it had yet to connect thirty-five raw

Oregon State Sanitary Authority Willamette Watershed Survey Data 1957

Sources	Treated Population Equivalent (1957)	Percent Total	Pre-treatment Population Equivalent	Percent Reduction (Since 1950)
Raw Sewage	42,700	1.8		
Treated Sewage	485,900	20.5	1,087,700	55.3
Sulfite Pulp Mills	1,518,000	64.2	3,850,000	60.6
Kraft Pulp Mills	158,000	6.7		
Other Industries	161,700	6.8	287,300	43.7
Totals	2,366,300	100		

Table 5.1. Oregon State Sanitary Authority data on Willamette River water quality, 1957 (Data: OSSA, "Interim Report on Status of Water Pollution Control in the Willamette River Basin," 1957).

sewage outfalls to the city's treatment plant, including those from Linnton and Guilds Lake (to the north), Tryon Creek (to the south), and other out-lying neighborhoods. With this evidence, at its January 1958 meeting the OSSA initiated another effort to compel compliance with Oregon's laws.[103] Supporting this cause, the *Oregon Journal* lamented that the goal of a clean river was still "far off," in spite of some recent achievements. The newspaper observed that the so-called "Willamette river cleanup" was a model for abate-ment programs in other areas, and State Sanitary Engineer Curtiss M. Everts had earned national recognition for his work. Population increases and indus-trial expansion in the valley, however, exceeded any previous gains from pri-mary sewage treatment and pulp and paper industry abatement practices.[104]

These conditions motivated civic organizations to re-engage in the clean stream fight. The Portland City Club turned its attention back to water pollu-tion for the first time since the mid-1920s. Its March 1958 municipal sewage report observed that untreated wastes from Portland neighborhoods con-tinued to be a health hazard for swimmers and drastically reduced dissolved oxygen levels during low-flow months. It endorsed the tri-county sewage plan and advocated for renovation of Portland's inadequate sewer infrastructure, while also observing the reluctance of citizens and elected officials to take action until faced with "extreme circumstances."[105] These conditions also spurred advocates to organize in Salem in July 1958 to begin lobbying all Willamette Valley cities to build secondary sewage treatment systems. The committee spearheading this campaign represented various interests and included David Charlton, Deputy State Sanitary Engineer Kenneth Spies, and Salem real estate developer Emmett Rogers.[106]

In the midst of this pressure, Portland's City Council prepared a measure for the November 1958 ballot. It bundled $5 million in capital improvements for Portland's sewer system with other projects into a $39,555,000 package. The OSSA opposed this strategy, predicting that voters would be loath to approve such a large amount. It reminded Portland Mayor Terry Schrunk that city officials were under legal obligation to keep sewage from the river; therefore, if voters did not approve the package, the sanitary authority would be authorized to take legal action.[107] As authority members feared, voters did not approve the measure.[108] In the aftermath of this failure and the author-ity's ultimatum to develop fiscal and construction proposals by January 15, 1959, the city council offered a seven-point plan.[109] Commissioner Bowes and other city officials asserted that they would do what they could to address their backlog of sewage system construction and upgrade projects, but only insofar as funds would allow. To correct these and other deficiencies, council members agreed to place yet another bond issue on the May 1960 ballot.[110] Notwithstanding city officials' claim of insufficient funds, in early March 1959 the sanitary authority rejected the city's proposal and filed suit in district court against the City of Portland for failure to comply in a timely manner.[111]

Newspapers sided with abatement advocates against Portland officials. In the *Oregonian*'s opinion, those not in agreement with the OSSA's efforts to clean up the river were on the side of evil. Editors urged city officials to comply with sanitary authority requests because it was "fighting for clean rivers, for motives of public health and welfare" and, therefore, was

automatically . . . on the side of the angels. One who opposes it plainly is aligned with the forces of evil, whether he likes it or not. The only smart way out of such a situation is to take the necessary steps to avoid getting into it.

To complement this advice, next to the editorial was a cartoon titled "Double Trouble" conveying the city council's problems in addressing water and air pollution issues.[112] There had long been elements of morality in the

Fig. 5.5. Editorial cartoon providing a comic twist on air and water pollution problems confronting Portland city officials in the late 1950s (*Oregonian*, Feb. 20, 1959, sec. 1, p. 30).

approach of pollution abatement advocates, particularly in some Izaak Walton League members and in *Oregon Journal* and *Oregonian* editorials. However, this particular editorial expressed a level of unambiguous moralizing that reflected a growing ethical argument against water pollution within and beyond Oregon.[113]

Responding to the suit in early March 1959, Mayor Schrunk accused the sanitary authority of deliberately impinging upon the policy of home rule. Calling upon the rhetoric of his predecessor George Baker from the 1920s, he also warned that the OSSA's goal of completely eliminating water pollution within Portland city limits would signal an end to all manufacturing concerns and a corresponding loss of jobs.[114] Portland's chief deputy city attorney said that city officials looked forward to the opportunity to challenge the state law empowering the OSSA—a threat to undermine the legal foundation of the OSSA itself.[115] On November 24, 1959, the authority refused Portland's request to dismiss its legal action. Authority members did, however, follow the recommendation of Circuit Court Judge Frank Lonergan to seek mediation and subsequently postponed the scheduled court hearing.[116]

Outside of administrative hearings, citizens and civic groups continued to pressure city leaders. The League of Women Voters at the national and state levels again became involved in the issue. In addition to conducting surveys and lobbying state legislators, the Portland League hosted Kenneth Spies at a luncheon program in late 1959. Spies's service with the sanitary authority since 1941 enabled him to provide both experiential and quantitative evidence of Willamette and Columbia river pollution. Harkening back to William Joyce Smith's 1940 film, luncheon guests also saw visual evidence to complement Spies's talk in the form of Sherman Washburn's color film *And On to the Sea*.[117] These efforts built upon the Portland City Club's report and lobbying the previous year as debate between the OSSA and Portland city officials continued. Authority members received a letter from Commissioner Bowes on December 14, 1960, outlining the steps that Portland was then taking to comply with the sanitary authority's order. These included treating wastes discharged into the Columbia River after primary treatment by installing a chlorine treatment facility no later than summer 1961; the city would also construct a secondary sewage system for the Tryon Creek area. On December 15, 1960, the OSSA found these and other actions satisfactory and dismissed its lawsuit.[118]

Forcing the City of Portland to move forward in constructing additional municipal waste infrastructure was not an insignificant achievement. For decades abatement advocates had repeatedly revisited the issue of Portland's discharge of inadequately treated wastes. City officials found that compliance with state water pollution regulations meant the investment of hundreds of millions of dollars in infrastructure. These investments would only be considered completely successful if the infrastructure functioned properly and became, thereby, an unseen part of the city, with pipes buried underground and treated effluent not polluting waterways. Even with funding available under the Federal Water Pollution Act, Portland city leaders had a hard time coming up with sufficient resources, not unlike the situation two decades previously. Real progress did not occur, however, until the OSSA, Izaak Walton League, and other groups confronted city leaders' intransigence.

POLLUTION IN PARADISE

As the sanitary authority was experiencing some successes in abating municipal and industrial water pollution, its purview was expanding. State legislators transferred the duties of the Oregon Air Pollution Authority to the OSSA in November 1959. For a few years after this merger sanitary authority members devoted less time during their monthly meetings to water pollution because they had to make room on their agenda for the air pollution topic.[119] Water pollution was no less pressing beyond the walls of the administrative hearing room, however: OSSA data from 1959 found that water pollution levels were only 16 percent lower than in 1939, in spite of progress abating both municipal and industrial discharges and the regulated seasonal water flow enabled by the reservoirs of the Willamette Valley Project.[120]

Motivated by abatement efforts that seemed to be on a treadmill, in the early 1960s local television correspondent Tom McCall became actively involved in water and air pollution abatement. With support from his boss at KGW-TV, he conceived, produced, and narrated a television documentary—*Pollution in Paradise*—that first aired on November 21, 1962. McCall had been in contact with David Charlton since at least 1954 and was familiar with the work of the IWLA and other local conservationists. In 1959, McCall received the Portland Chapter of the Izaak Walton League's Beaver Award recognizing achievements in natural resources conservation for his work producing television commentaries, particularly his 1958 documentary *Crisis in the Klamath Basin* about Indian rights and forestry practices in south-central Oregon.[121] *Pollution in Paradise* built upon his and Sherman Washburn's earlier works. Both of these films also marked an important shift in local grassroots sentiment that reflected changing perceptions of the environment throughout the nation.[122]

McCall characterized air and water pollution as complex—yet eminently solvable—issues that threatened economic vitality, public health, and natural resources. He also echoed the sentiments of David Charlton, the Izaak Walton League, and many other pollution abatement advocates in admitting that some level of air and water pollution was "inescapable in a society such as ours," but urging citizens to reconsider how much pollution was justifiable. Doing so was "central to a great debate," he continued—a debate "on which the shape of this region and this nation's future could well hang."[123]

McCall expressed faith in state authorities tasked with addressing the issue, such as the OSSA, state board of health, and attorney general. He also stressed the state's citizens played an important role.[124] His message resonated with many Oregonians. Sanitary authority members passed a resolution lauding McCall's documentary at their January 4, 1963, meeting, calling it "fair and unbiased" and congratulating KGW-TV for the film's role in public education. The citation read, in part, that "support for pollution control activities can only be gained when the people and the industrial leaders of the state have been accurately informed of what has been accomplished and what problems remain."[125] One Oregonian who lived near The Dalles in the Columbia River Gorge wrote McCall in early 1963:

> From what I read and hear on television I don't believe the pollution problem is over yet. When I look west from where we live a mess of smoke [from the Harvey Aluminum plant] often is coming out and no telling what goes into the river. . . . Now a paper mill is planning to move to town so there will be some more mess for the fish to take in.[126]

McCall's documentary and the sentiment expressed in this letter show that that even after nearly forty years of concerted efforts by pollution abatement advocates and recent OSSA achievements, water pollution remained unresolved. *Pollution in Paradise* heralded a new era for Tom McCall, Oregon politics, and the Willamette River.

Chapter 6
At Long Last?

In early August 1965, conditions below Willamette Falls threatened the fall Chinook salmon run. The year had been unusually dry, with water flow measured at the Salem gauge dropping as low as 5,000 cubic feet per second (cfs). The Oregon Water Resources Board calibrated their water quality management strategy to 6,000 cfs with regulated releases from the Willamette Valley Project dams, so a mere 5,000 cfs was near the absolute minimum flow to assimilate wastes and maintain fish life.[1] Less water without a corresponding decrease in pollutants meant higher relative amounts of wastes in the river, and as these wastes decomposed they consumed oxygen fish needed to survive. US Public Health Service officials measured a dissolved oxygen (DO) content of just 1.7 parts per million (ppm) at their Swan Island monitoring station in Portland Harbor. Well below the 5 ppm biologists targeted to sustain a healthy aquatic ecosystem, this was the absolute minimum DO salmon and trout could tolerate for brief periods before they had an "avoidance reaction." Much less than 1.7 ppm and not only would fish die but the river through Portland "would become septic, with resulting odor, appearance, and health problems."[2] Or, as Tom McCall narrated in *Pollution in Paradise* three years previously:

> Where these wastes are not treated in a safe manner, the effluent
> becomes an oxygen-gulping, slime-making scourge. It destroys fish
> life. It fouls fishing gear and fishing boats. Sometimes it churns at
> river's bottom, forming into rafts that rise to the surface as sluggish,
> foul-smelling masses of filth.[3]

Hot summer temperatures compounded these conditions. Warmer water retains less oxygen in solution than colder water; increased temperatures, therefore, drive DO levels down even further than decomposing mill wastes and festering, untreated human excrement by themselves. State Sanitary Engineer and sanitary authority member Kenneth Spies warned area residents "the river is not entirely safe for recreational purposes."[4] Echoing their refrain from the previous decade, *Oregonian* editors lamented the yearly pollution spikes during the low-water months that brought health warnings, fish kills, and repeated assurances from the sanitary authority, municipalities, and industries that the problem would soon be solved. These annual cycles— "Red Flag[s] on Pollution"—were "becoming monotonous."[5]

Three years before this environmental emergency, the monotony of the pollution treadmill had seemed to have been overcome. The broadcast of *Pollution in Paradise* in November 1962 had been an important catalyst in furthering the clean stream cause. In late March 1963, the added momentum from Tom McCall's documentary began to deliver tangible results. Democratic Portland senator Ted Hallock's Senate Bill (SB) 259—sponsored at the sanitary authority's request—passed by a wide margin. The bill significantly enhanced air and water quality enforcement mechanisms for the sanitary authority, city and county officials, and circuit courts. The Oregon State House approved the bill in mid-April, and shortly thereafter Governor Mark Hatfield signed it into law.[6] Advocates welcomed these legislative advances. However, as the alarming conditions during August 1965 indicated, they were not sufficient to forestall threatening water conditions. Complex natural systems were still beyond human capacity to predict and manage. Just as creation of the sanitary authority in 1938 did not, in and of itself, improve water quality, *Pollution in Paradise* did not, in and of itself, improve the river. Both of these events, however, introduced new eras in water pollution abatement. The new era Oregonians entered in early 1963 would be defined by the strengthening of both state and federal environmental laws, enforcement powers, and administrative reach. It would also be defined by the political career of the reporter turned secretary of state—and then governor—Tom McCall.

RECEPTION OF *POLLUTION IN PARADISE*

"... the shape of this region's and this nation's future"

Pollution in Paradise was groundbreaking as a public relations tool and an effective, concise statement of environmental degradation. After its first broadcast in November 1962, McCall's boss at KGW-TV, Tom Dargan, sent copies of the film to educators and state legislators. In January 1963, seeking to influence the legislative session, Dargan scheduled another television broadcast. The strategy worked, as Senator Hallock introduced the OSSA's legislative proposal and SB 259 became law a few months later.[7] *Pollution in Paradise* may not have done much to help the Chinook salmon in August 1965, but the positive response to McCall's documentary helped propel him, first, into the position of secretary of state in 1964, and then to the governorship two years later. The documentary contained the unmistakable flavor of Tom McCall's reportorial style and clearly conveyed his devotion to Oregon as a place as well as a state of mind. Though it bears these specific associations, it is only fully understood within a broader context. As Historian William G. Robbins aptly observes, the appearance of *Pollution in Paradise* was "serendipitous."[8]

Among the serendipities was that the early 1960s was a time in which the general public, news media, and politicians were becoming increasingly aware of, and concerned with, environmental topics. This stemmed from undeniable examples in Oregon and elsewhere that after decades of minimally regulated disposal practices, the environment had reached its capacity as a waste sink. By the late 1950s, environmental degradation was undeniably curtailing the accustomed activities and constraining future options of a growing number of people across race, class, gender, political, and geographic divisions.[9] More and more Americans were questioning destructive suburban development patterns, unregulated use of herbicides and pesticides, continued dam construction, and many other activities that before had largely gone unexamined.[10] Concurrently, advances in ecological science enabled humanity's effects on the environment to be understood in empirical and quantitative terms, and gradually this knowledge influenced federal environmental legislation.[11] Further, as society was becoming more affluent, Americans were increasingly coming to view nature as worthy of preservation for its own sake, not simply for conserving and then extracting resources.[12] Print and television media served as critical conduits for documenting and communicating environmental concerns to an increasingly receptive American audience. Historian Priscilla Coit Murphy characterizes this period as the "heyday of the television documentary," with a prime example of the positive power of this new medium being Edward R. Morrow's acclaimed 1960 CBS Reports exposé of the agricultural industry's mistreatment of migrant workers, *Harvest of Shame*.[13]

This era also saw environmental concerns widely and effectively communicated in print form. The most well-known example of this was Rachel Carson's groundbreaking work *Silent Spring*. It is another instance of serendipity that McCall's documentary appeared just a few weeks after publication of *Silent Spring* in book form. Carson compiled more than a decades' worth of scientific research on the devastating effects of pesticides on bird populations, and she succeeded in communicating this message well outside a small community of specialists in chemistry, ecology, entomology, and ornithology.[14] The medium of the printed word was critical to the influence and reach of Rachel Carson's research, just as the medium of the television documentary was critical in the reception of McCall's message. Carson used her book to cite scientific evidence exhaustively and make her case about the harm caused by DDT and other chemicals; these copious citations enabled her to counter charges of bias and defamation. Similarly, McCall built upon his years in radio and television to create a strong narrative backed by evidence comprised of graphic color images of polluted air and water. His background included award-winning work on the 1958 KGW-TV documentary *Crisis in the Klamath Basin* about looming environmental, social, and economic challenges that threatened this area of south-central Oregon after Congress voted to rescind federal recognition of ("terminate") the Klamath Tribe.[15] Additionally, Carson and McCall brought significant experience to their projects; both were recognized as skilled in their respective mediums; both were pursuing topics that were far from unknown but for which a critical mass of the general public was not yet sufficiently concerned; and both worked for people who gave them significant leeway and institutional insulation.[16]

In both broadcast and print form, stories drawing attention to environmental degradation increased throughout the 1950s and early 1960s, and the American public was growing more receptive to this message. A new way of understanding and interacting with the natural environment was gradually becoming discernable among some segments of American society. Historians and others call this new sensibility the modern environmentalist movement. It has deep roots, but can be differentiated from what came before by two key elements. The first is that it is based on an ecological approach to the study of nature, in which empirical, scientific methods are used to examine natural phenomenon as interconnected systems rather than discrete units. An example would be to study a wetland as an ecosystem in which flora and fauna exist within networks of interdependencies, rather than focusing solely on discrete plants or animals within the wetland. The second key element is an appreciation of nature for its amenity value as an intact system. This differs from the conservationist approach to natural resources, summed-up as the measured harvest of raw materials to ensure the greatest good to the greatest number for the longest time.[17] Organizations and writers who exemplified the environmentalist movement as it was coming into being in the early 1960s included the Sierra Club (under David Brower's executive directorship), Murray Bookchin, and Barry Commoner. With its application of ecological science and call to consider nature as a system, Carson's work shares much with this new sensibility. Both *Silent Spring* and *Pollution in Paradise* differed, however, from the products and activities of the emerging environmentalists. Carson and McCall framed complicated ecological issues in ways the general public could understand. They created their works for this broader audience, not necessarily those with sufficient time and money to travel to far-off wilderness areas or with the background and time to digest complex data and arguments. Carson and McCall fostered this understanding by drawing attention to specific episodes of undeniable environmental deterioration and placing these within a broader narrative: local examples illustrating a broader phenomenon.[18] *Pollution in Paradise* differed from *Silent Spring* in that McCall's perspective is best understood as somewhat of a bridge between the classic conservationist and emerging environmentalist views. McCall echoed Carson's findings in recognizing pesticides were "poisoning Man's environment" by "acting in much the same way as radiation lies in the soil and enters into living organisms" and observing insecticides "massacring birds, mammals, fishes and, indeed, every form of wildlife." He drew attention to the value of Oregon's natural areas as places of recreation and refuge while also stressing the need for balance between the inevitable consequences of modern industrial and urban society and the environment. As he narrates:

> Even before the coming of Man there was pollution of our air and
> water. Some pollution is inescapable, especially in a society such
> as ours whose commercial and domestic demands for water and
> air constantly accelerate. Thus some water, for example, must be
> employed for diluting pollution. But how much, and what priority
> should be given this use are central to a great debate on which the
> shape of this region's and this nation's future could well hang.[19]

While McCall was transitioning from KGW-TV into his first elected position as Oregon Secretary of State in 1964—and preparing to run for governor soon thereafter—the sanitary authority continued to do its work during the final years of the Mark Hatfield administration. The state legislature had strengthened the sanitary authority's enforcement powers in early 1963 in accordance with the authority's own recommendations. Not everyone was satisfied. David Charlton attended a hearing of the Legislative Interim Committee on Wildlife in May 1964 to speak on behalf of the clean water committee of the Oregon Division of the Izaak Walton League. As he had been doing since the 1940s, he asserted Oregon's water pollution laws were insufficient—the Willamette River overall had become cleaner than decades past, but industrial wastes continued to be a significant problem. Drawing from OSSA and US Public Health Service data as well as his own research, he found pulp and paper industry wastes within the Willamette watershed represented the biochemical oxygen demand equivalent of a city of 200,000. The sanitary authority itself still placed primary responsibility for pollution upon the valley's pulp and paper mills. According to its data, 58 percent of pollution in the Willamette's main stem was the result of wastes from five of the valley's seven pulp and paper mills: Western Kraft (Albany), Columbia River Paper Co. (Salem); Spaulding (Newberg); Crown Zellerbach (West Linn); and Publishers (Oregon City).

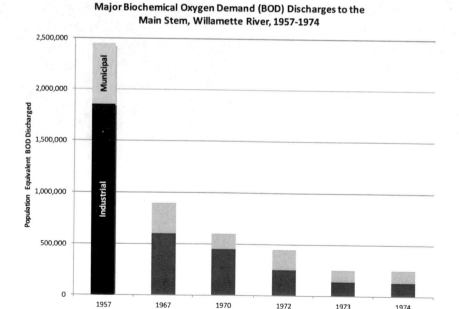

Fig. 6.1. Data collected 1957–1974 illustrates significant reductions in biochemical oxygen demand (BOD) discharges to the main stem Willamette River, with municipal wastes decreasing more rapidly than industrial wastes through the 1960s (Data: OSSA, *Ninth Biennial Report* (1950), 30; Oregon DEQ, *Proposed Water Quality Management Plan, Willamette River Basin: Appendices* (1976), C-43).

The Weyerhaeuser mill at Springfield and Crown Zellerbach's Lebanon mill also contributed pollutants, but the OSSA noted their waste treatment practices were better in comparison. Industry wastes were increasing in real and relative terms both because of expanded production and because cities in the watershed were building more efficient secondary and even tertiary treatment systems. The sanitary authority determined that upon completion of various municipal sewage plants over the next few years, mills would be spewing 82.6 percent of the pollution load within the watershed.[20] Even the City of Portland was making advances in their sewage treatment practices. The population equivalent pollution load from Portland city wastes during annual low-flow periods had decreased from 21.5 percent of the total in 1963 to 7 percent in 1965.[21] Figure 6.1 reflects these changes between 1957 and 1974. Backed by this evidence, Charlton and the Izaak Walton League called for legislators to increase the authority's staff and enforcement powers.[22]

As state legislators continued to receive complaints from David Charlton and others in 1964, Harold Wendel marked his twenty-five-year anniversary as OSSA chair. In an *Oregonian* interview, Wendel said that in spite of critiques he and the sanitary authority had received over the years, if he had to do it all over again he would make the same choice to head the OSSA. "A person can't just live in a city or state and 'take,'" he said, "and besides you develop a certain sense of pride and responsibility to help." Two of the frustrations Wendel noted were the "frequent refusal of voters to vote in sewer bonds," and the state court's lamentable record of enforcement. The article noted there was no record of the OSSA ever receiving fan mail, but that Oregon's waters and air quality had improved during Wendel's tenure nonetheless.[23]

Another occurrence in the years since Wendel started his tenure in 1939 was that water pollution had gotten much more complex, and more intertwined with other issues. A case in point was the threat to water quality in the Willamette and lower Columbia rivers by a proposal from officials in Southern California, Arizona, and Nevada to divert water from the Yukon, Fraser, and Columbia Rivers by way of a giant canal from Canada to Mexico. Proponents claimed the project would provide water to thirty-three US states, seven Canadian provinces, and three Mexican states. The Willamette River cleanup, as beholden as it was to regulated flow from Willamette Valley Project dams, would likely be jeopardized by this plan because water would be diverted out of the watershed. To what extent this grand scheme was likely to be implemented is not clear, but Governor Hatfield and Oregon legislators certainly thought enough of the threat that they agreed in March 1965 to appropriate the first installment of a four-year, $1 million project to survey the entire state's water resources and needs. Republican Senator Gordon McKay of Bend told his colleagues: "Let's not start filling swimming pools in California until we decide if our water is surplus."[24]

A few months later, during the dry, hot month of August 1965, it became apparent that Oregon did *not* have a surplus of water, at least in the Willamette Valley. W. W. Towne, director of the US Public Health Service's Columbia River Basin Comprehensive Project for Water Supply and Pollution Control, concluded in early August the Willamette River posed a threat both to public health and fish life. The 1965 fall Chinook run, which was just beginning, was particularly threatened. A low-water year resulted in a measurement of 5,000 cfs flow at the Salem gauge, causing the water to be abnormally warm. This compounded the effects of oxygen-depleting pollution in the river, with a DO of just 1.7 ppm measured in Portland Harbor. If DO levels fell further, Towne warned, fish would either turn away before they reached Willamette Falls, or large numbers of fish would die trying; with a DO of zero, fish and benthic life would die in the anaerobic conditions and the river would emit a putrid stench. Increasing the urgency of this situation was that the bare-minimum 5,000 cfs flow was being sustained only because the US Army Corps of Engineers had agreed to releases from Willamette Valley Project reservoirs (Figure 6.2 shows the lowest recorded river flows at the Salem Gage from 1927 to 1974).[25] In response, State Sanitary Engineer Ken Spies and his colleagues on the sanitary authority pressed the valley's pulp and paper mills to find ways to abate their effluents immediately. During this "present emergency," Spies had secured agreements from "each waste producer . . . to either improve housekeeping practices or reduce the flow of wastes until the critical period is passed."[26] At the OSSA's August 16 meeting, with little improvement in river conditions, authority member Barney A. McPhillips proposed to give the mills ten more days to "do some soul searching on this thing and come up with a practical, quick way to cut down the discharge of wastes into the river," or face closure orders. Representatives from the Northwest Pulp and Paper Manufacturers Association responded with pledges of full cooperation.[27]

Ten days later, in late August, the most critical period of the low-water emergency passed and salmon were making their way upriver—"but the danger is far from over," the *Oregonian* reported. The US Army Corps of Engineers had further cooperated with the electric utilities Portland General Electric and

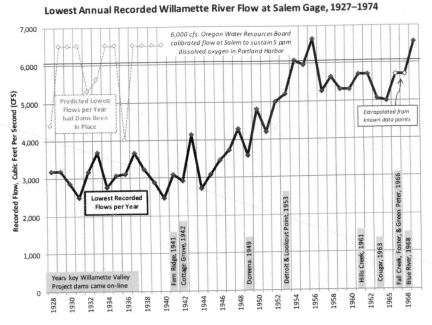

Lowest Annual Recorded Willamette River Flow at Salem Gage, 1927–1974

Fig. 6.2. Lowest annual recorded Willamette River flows measured at the Salem gage, 1927–1974. While clearly showing an upward trend as Willamette Valley Project dams came on-line, 1965 was the lowest-recorded flow in over a decade (Data: OSSA stream survey reports for 1951 and 1953; Clarence Velz's NCSI-funded river surveys of 1951 and 1961; and Oregon DEQ, *Proposed Water Quality Management Plan, Willamette River Basin: Appendices* (1976), C-43=.

the Bonneville Power Administration to lower Willamette Valley Project tributary dam levels by one-third to ensure survival of Chinook salmon. This higher flow on the river's main stem raised DO levels and allowed fish passage over Willamette Falls. Officials recognized another contributing factor improving water quality was "the pulp mills reducing their effluent output," but as Dr. L. Edward Perry of the US Bureau of Fisheries warned, "'There is no ground for relaxation.'" Because September typically witnessed the lowest monthly water flows, "dangers to fish runs have not been eliminated."[28]

With the catastrophe narrowly averted and the historically dry month of September yet ahead, OSSA members were not in a conciliatory mood when they met again with pulp and paper industry representatives. The sanitary authority unanimously approved McPhillips's motion that they take definitive steps to curtail pulp and paper industry pollution if they found the industry's actions during the crisis had caused "irreparable damage" to the spawning Chinook salmon.[29] This action highlights both the frustrations of McPhillips and his colleagues and the inadequacy of OSSA enforcement powers. The

sanitary authority had been successful by applying consistent pressure to cities, the pulp and paper industry, the meat-packing industry, and other polluters. Over time, this approach had incrementally resulted in positive results.
However, when faced with an immediate crisis such as existed in August
1965, the authority lacked both enforcement powers and support from the
state's executive branch to react promptly. It was not primarily the OSSA's
actions or industry measures that averted a large-scale die-off of the 1965 fall
Chinook salmon run—it was release of water impounded in Willamette River
tributary dams. These releases alleviated pollution sufficiently to save the fish,
but many observers were not pleased it had come to this. US Department of
the Interior Secretary Stuart Udall excoriated cities and industries along the
Willamette for their continued resistance to abatement. He said dams should
not be used every year for emergency abatement augmentation, and he
urged cities and industries to comply with OSSA mandates.[30] State Treasurer
Robert Straub, who would soon be the Democratic gubernatorial candidate
challenging Tom McCall, took advantage of these horrid conditions to blame
the incumbent, Governor Hatfield.[31] He did so at the time of the August crisis
and also at a hearing of the legislative interim committee on public health in
October. Straub critiqued Hatfield and called for the OSSA to be separated
from the board of health and made into a state agency. He held that if the
OSSA were an independent agency the likelihood of enforcement would
increase. He did not explicitly place blame on the OSSA or the Legislature for
shortcomings in Oregon's water pollution abatement, but on "budget-cutting
by the governor or his agents."[32]

SANITARY AUTHORITY FUNDING AS INDICATIVE OF STATE PRIORITIES

Conducting the day-to-day work of an organization charged with maintaining
water quality statewide was never going to come without appreciable financial costs. The sanitary authority came into being in late 1938 with a clear
public mandate but no funding. State officials subsequently allocated some
amount of funding, or else the authority could not have accomplished anything—and, as the preceding narrative has shown, the OSSA realized some
tangible achievements in the 1940s and 1950s. Robert Straub's critique that
insufficient funding had hindered the OSSA provides an opportunity to identify the financial constraints under which the authority operated. Analyzing
OSSA funding helps illuminate evolving priorities over the first twenty-five

years of its existence—during eight gubernatorial administrations and thirteen state legislative assemblies.

Though voted into existence by a significant majority in November 1938, and welcomed by many when it commenced work in February 1939, as reflected in its initial budget the sanitary authority did not get off to an auspicious start. Outgoing governor Charles Martin submitted his proposed 1939–1940 biennial budget to the Oregon legislature in October 1938, before voters created the OSSA.[33] Since the citizen's initiative did not include a funding mechanism, the sanitary authority was an unfunded mandate. Within its first days of operation, the sanitary authority requested $50,000 for the biennium. In the midst of the Great Depression, however, legislators were already facing a general fund deficit. When in early 1939 it became apparent that the legislature would not approve their desired amount, sanitary authority members and their allies lobbied the governor and legislators to allocate a minimum of $23,000. Incoming Republican Governor Charles Sprague requested $15,000. In the end, Oregon's Forty-first Legislative Assembly allocated just $10,000.[34] Oregon voters approved the OSSA by a two-to-one margin, indicating strong support for improved water quality. In deciding to fund it at only 20 percent of the amount authority members requested, the governor and legislators confirmed their priorities differed from voters. This raises the question: As the state's economy continued to struggle during the Depression, as state government faced a general fund deficit, and as the threat of war increased in Europe, what did state legislators consider worth funding?

Legislators allocated general fund appropriations to a variety of essential services, such as the state police, board of higher education, and department of forestry. The general fund also supported a recipient of much less importance than these other services, the Battleship *Oregon* Commission, an organization that maintained the decommissioned US Navy battleship USS *Oregon* (BB-3). The *Oregon*, once a world-class ship and a decorated combat veteran of the Spanish–American War of 1898–1899, was docked in Portland Harbor as a tourist attraction from 1925 to 1942.[35] During these years, legislators allocated $15,000 annually to the commission. During the same funding period in which the sanitary authority began operating—January 1, 1939, to June 30, 1941—legislators allocated $37,500 to the commission.[36] Thus, legislators decided that maintaining a stationary floating museum in Portland Harbor was more important than financing complex water pollution abatement work throughout the entire state when it authorized more than three times as much for the former as it did for the OSSA.

Another point of comparison with the sanitary authority's initial funding amount can be found outside Oregon. Washington's Governor Clarence D. Martin established the Washington State Pollution Control Commission (WPCC) in fall 1937.[37] This body, like the OSSA, was under the state department of health, but it was primarily a technical commission that conducted municipal and industrial pollution surveys and research. The WPCC was not, therefore, equivalent to the OSSA, because it lacked any kind of enforcement responsibilities or power.[38] Washington's Governor Martin allocated a $14,500 biennial budget to the WPCC in 1937.[39] Oregon's neighbor to the north allocated more funding for water pollution control—$4,500—but not drastically so; at the same time, however, the WPCC's purview was less expansive than the OSSA's.

Oregon voters overwhelmingly approved both the Water Purification and Prevention of Pollution initiative and the City of Portland sewer funding measure in November 1938. In the immediate aftermath, both state leaders and Portland city officials were unable to fund these pollution abatement approaches, in spite of voter support. In September 1939, a few months after the legislature set the OSSA's appropriation, Governor Sprague wrote to the sanitary authority recognizing it was "handicapped by a lack of sufficient funds to carry on an extensive program." One tangible example of the OSSA's handicap was in its inability to assist municipalities in quantifying water pollution and generating sewage treatment plans.[40] In mid-1940, the OSSA investigated ways to transfer money from the fish and/or game commissions, but reached no consensus with the respective commissioners about how this could be done legally.[41] Sprague's 1939 admission to authority members suggests he was not blind to the need for water pollution abatement; he did not, however, make the issue a priority in the midst of other budgetary demands (including upkeep of the floating museum USS *Oregon*).

If the OSSA suffered from inadequate funding during the first two years of its existence, did subsequent governors and legislators continue this trend? Previous chapters have identified some of the many ways in which water pollution was a complex and multivariate issue. It involved public sentiment and public policy as well as engineering approaches backed by empirical evidence, quantitative data, and comprehensible scientific models. Paying for pollution abatement offered yet another kind of complexity. Chapter 4 showed that

Oregon legislators authorized more funding for abatement in the 1940s and early 1950s than did the pulp and paper industry's National Council for Stream Improvement. This finding might not be surprising, since the former group fulfilled a public service mission and the latter operated only for financial profit. Since the early 1960s was an important transitional period in the Willamette River pollution control narrative, it is informative to analyze OSSA funding in depth to achieve some clarity in this area and at least provisionally answer an important set of questions: How did authority funding compare, year to year, to itself, to some other state agencies, and to the overall state general fund? In uncovering answers, some conclusions can be drawn about whether or not the OSSA was funded sufficiently to achieve its mandate.

In budgetary terms, the first ten years of the OSSA's existence can be characterized as a "coming to agreement" period. Figure 6.3 compares the state sanitary authority's biennial state appropriations between 1939 and 1949 with what the authority asked for and what governors requested in budgets submitted to the legislature. This figure (and the three to follow) shows how much the sanitary authority wanted, the extent to which the respective

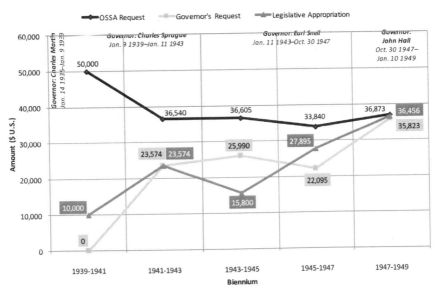

Fig. 6.3. Coming to agreement: OSSA biennial budget requests and legislative appropriations, 1939–1941 through 1947–1949 (Data: See biennial budget documents submitted by Oregon governors and published by the Budget Division of the Executive Department, Salem).

governor agreed, and how much, in the end, the legislature allocated.[42] In providing quantitative data about the extent to which the governor agreed with the OSSA, and the legislature agreed with the governor, this analysis provides direct evidence of the extent to which Oregon state leaders viewed the importance of the water pollution issue.

Figure 6.3 shows that the three respective parties—OSSA, governor, and legislature—began their relationship in 1939 with quite different expectations. Through the 1940s, these parties gradually came closer to agreement, until the 1947–1949 biennium, when they were more or less in accord. The OSSA had a budget of $36,456 to work with for that two-year period, a more than threefold increase over the $10,000 they started with at the beginning of the decade. The significant disparities between 1939 and 1945 is to be expected during the final years of the Great Depression and the emergency of World War II. Considering the momentum of the resounding 1938 vote, it is no surprise the OSSA reasoned it could request such a relatively high amount; given the constraints of the Great Depression, it is, likewise, no surprise that Sprague and state legislators provided only a small fraction of this request. The biennium in which the three figures begin to converge—1945–1947—was not long after the OSSA and other abatement advocates were finally successful in getting the City of Portland to fund its $12 million (eventually $15 million) primary sewage treatment system. Governor Earl Snell submitted his budget proposal to the Oregon legislature in late 1944, just a few months after Portland voters approved Robert Moses's *Portland Improvement* package. The city broke ground on this project in summer 1947 and the interceptor and sewage treatment plant came fully online in 1953. Governor Snell requested $36,823 for the OSSA for the 1947–1949 period.[43] The funding trend was, therefore, upward, and this is significant because state legislators approved the budget well before Congress passed the 1948 Federal Water Pollution Control Act providing financial support to states. A conclusion to be drawn from this trend during the 1940s is that increasingly more of Oregon's elected leaders were recognizing the need to enhance water pollution abatement resources—and this recognition in Oregon was occurring at the same time the issue was becoming a national concern.

The next ten-year period of the OSSA's work graphically illustrates the "treadmill" of water pollution control—in spite of significant increases in the authority's budget. Figure 6.4 extends the comparison of Figure 6.3 for the biennia 1949–1951 through 1957–1959. This period shows the OSSA, governors, and legislators more or less maintained the consensus established in the

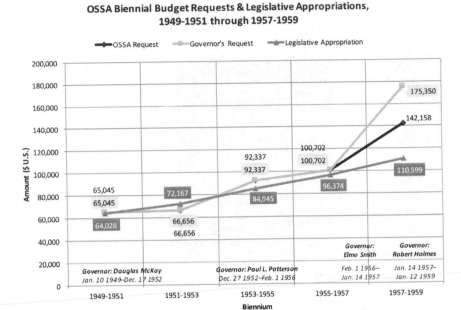

Fig. 6.4. Pollution control on a treadmill: OSSA biennial budget requests and legislative appropriations, 1949–1951 through 1957–1959 (Data: See biennial budget documents submitted by Oregon governors and published by the Budget Division of the Executive Department, Salem).

late 1940s about the merit of water pollution control. This period also shows how significantly OSSA funding grew during these years—from $65,045 for 1949–1951 to $110,599 for 1957–1959, an increase of 70 percent during this period and more than 1,000 percent since 1939–1941.[44]

There are two other important water quality-related funding sources not shown in Figure 6.4 that had begun to be a factor beginning in the late 1940s. In addition to state appropriations, the OSSA received federal funds for its own work from the 1948 Federal Water Pollution Control Act and 1956 amendments. For the period 1955–1957, these funds amounted to almost $22,000, or 16.9 percent of the authority's total available budget of $128,250. Federal funds were also increasingly becoming available to build sewage treatment infrastructure, which the OSSA passed through to municipalities and sewage districts throughout the state. For example, during the 1957 fiscal year, the OSSA disbursed slightly more than $700,000 for eleven sewage treatment projects, or more than twelve times the OSSA's funding for that year; in 1964 the authority disbursed nearly twice this amount ($1,387,983) for eighteen sewage treatment projects.[45] Figure 6.5 shows biennial state appropriations plus federal funds designated for the administrative, research, and enforcement work of the OSSA and its replacement agency, the Oregon Department of Environmental Quality (established in 1969); these amounts do not include federal funds the OSSA disbursed to build sewage treatment infrastructure.

From the 1949 through the 1959 general fund allocation periods, the sanitary authority consistently received the funding it requested, and it also received an increasing amount of federal dollars. These were the years in which the sanitary authority successfully compelled pulp and paper mills to initiate *some* kind of pollution abatement measures. However, this was also the period in which observers lamented water pollution control seemed to be on a treadmill. In spite of the steadily increasing amount of state and federal funds, the authority's work was merely keeping the Willamette River from getting worse.

The sanitary authority's budget began to get increasingly more complex after the 1957–1959 biennium, in two compounding ways. First, Governor Hatfield's budget documents do not track current and historical budget requests and appropriations in the same straightforward way as previous

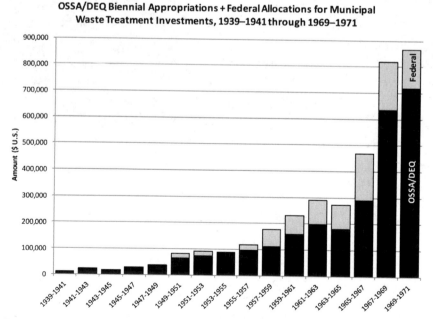

Fig. 6.5. OSSA/DEQ biennial appropriations and federal allocations for municipal waste treatment investments, 1939–1941 through 1969–1971 (Data: OSSA and Oregon DEQ biennial reports).

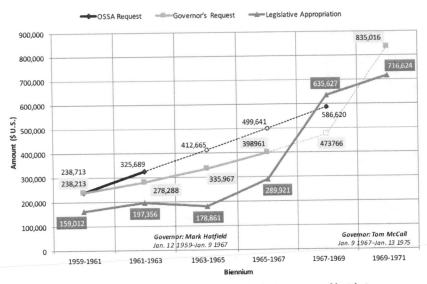

Fig. 6.6. Addressing pollution in paradise: OSSA biennial budget requests and legislative
appropriations, 1959–1961 through 1969–1971 (Data: See biennial budget documents submitted by
Oregon governors and published by the Budget Division of the Executive Department, Salem).

governors. Sanitary authority budgets are no longer differentiated within the
board of health figures, but subsumed under the broad category of "Sanitation
and Engineering." Within this category, OSSA programs received funds from
many sources, and the OSSA disbursed some funds to other bodies, such as
regional air pollution control authorities.[46] For these reasons, OSSA budget
numbers are much harder to track in a way that correlates with previous periods, making direct comparisons more difficult. In spite of these limitations,
available information allows for one final comparative graph that illustrates the
funding dynamic from the 1959–1961 biennium through the last of Governor
Hatfield's budgets (1965–1967) and into Tom McCall's tenure (1969–1971).

The period 1959–1971 was marked by significant increases in state legislative appropriations to the OSSA. Figure 6.6 shows appropriations growing
from $159,012 in 1957–1959 to $635,627 in 1967–1969, an increase of 299
percent during this period and about 6,356 percent since 1939.[47] McCall's
first budget request ($835,016 for the 1969–1971 biennium) is significantly
more than Hatfield's last request ($473,766) and the amount that Oregon
legislators approved in 1969 ($635,627). One trend from this figure is that
Governor Hatfield consistently requested less funding than the OSSA asked

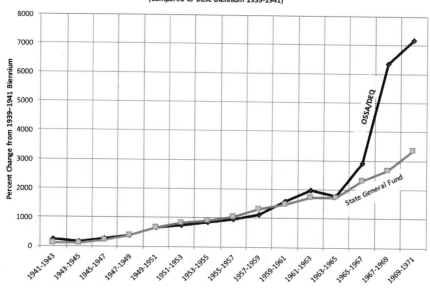

Fig. 6.7. Changes in OSSA/DEQ budgets and state general fund appropriations, 1941–1943 through 1969–1971 (Data: OSSA and Oregon DEQ biennial reports).

for, suggesting Straub's critique had some merit: as the data shows, unlike his immediate predecessors, Governor Hatfield did not support the authority's work by agreeing with its budget requests. Also supporting Straub's critique is clear evidence the legislature appropriated appreciably less than either the OSSA or the governor asked for throughout Hatfield's tenure. Finally, one can clearly see evidence both of the 1967 strengthening of the OSSA through additional staffing (afforded by more funds), and the significant increase in McCall's first budget request as governor, 1969–1971.

The data above shows OSSA funding in comparison to itself, but it is also instructive to compare this with the growth rate of the state general fund as a whole, of which the OSSA budget was a part. This enables conclusions to be drawn about whether the OSSA's budget grew along with increases to the general fund—or, if its growth rate differed appreciably, what this might say about state legislative priorities. Figure 6.7 compares the biennial OSSA budget growth with that of the state general fund as a whole, relative to where both budgets began in the 1939–1941 period. As this graph shows, OSSA funding grew at a larger rate compared to its baseline than did overall general fund appropriations between 1941 and 1947—during the same period when the legislature was appropriating significantly less than the OSSA had requested. From the 1951–1953 through the 1953–1955 biennia, when water pollution control was on a "treadmill," the state general fund grew at a larger percentage relative to its base than did OSSA funding. Beginning with the next biennium and continuing until 1961–1963, OSSA funding grew significantly faster than the overall state general fund. In spite of a significant drop in relative growth rates during Hatfield's tenure as governor, the authority's budget continued to grow faster than the general fund through the 1965–1967 biennium, before growing drastically, relative to the general fund, during Tom McCall's first four years in office.

Figure 6.8 shows the rate of growth of the OSSA/DEQ budget from the 1939–1941 through the 1969–1971 biennia compared to the rate of growth of the state general fund as a whole and to three other state programs, as measured against the 1939–1941 baseline. This graph illustrates clearly that OSSA/DEQ funding grew drastically during the 1960s, far outstripping the rate of change of the state general fund. During these years, OSSA/DEQ

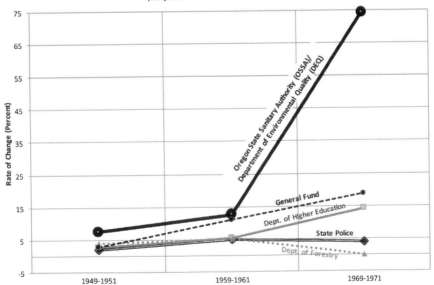

Fig. 6.8. Rate of change of general fund appropriations to selected state agencies relative to overall general fund, 1939–1940 through 1969–1971 (Data: See biennial budget documents submitted by Oregon governors and published by the Budget Division of the Executive Department, Salem).

funding increased at a higher rate than budgets for the department of higher education, state police, and department of forestry.

A final informative analysis comes from comparing the rate of change in OSSA/DEQ funding each biennium relative to inflation, because a rate of change less than the rate of inflation would amount to a decline in real terms. Biennial inflation averaged 3.36 percent between the 1939–1941 and 1969–1971 biennia, with a high of 9.9 percent during 1945–1947 and a low of 0 percent during 1953–1955.[48] As Figure 6.9 shows, in spite of two biennia during which legislators decreased OSSA funding, overall the growth rate far outpaced inflation. The long trough of the 1950s in Figure 6.9 reflects the period when water pollution control was on its treadmill.

The message to be drawn from the above data is that OSSA/DEQ funding grew significantly between 1939 and 1971. This growth was in real terms—as measured against inflation—and in relative terms compared to overall state general fund appropriations and funding of other state agencies. This leads to two important conclusions. First, these budgets illustrate Oregon governors and state legislators before Tom McCall were growing

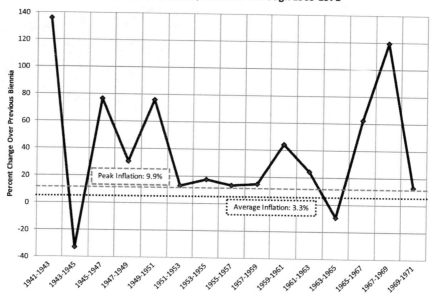

Fig. 6.9. Percent change in OSSA/DEQ general fund appropriations per biennia compared to inflation, 1939–1941 through 1969–1971 (Data: See biennial budget documents submitted by Oregon governors and published by the Budget Division of the Executive Department, Salem).

increasingly aware of the water pollution issue, and they were doing something about it. Measured against what occurred during the latter 1960s and beyond, the steps state leaders took were neither sufficiently stringent nor adequately funded, but Oregon was certainly not alone in this.[49] Second, even though OSSA/DEQ funding had grown in both real and relative terms, and even with funding through the Federal Water Pollution Control Act, the best the OSSA and pollution abatement advocates could do by the mid-1950s was put water pollution in stasis. The Willamette River was not becoming more degraded, but it was not getting much better, either. As the critical conditions of August 1965 had shown, this stasis *required* flow regulation from tributary dams. Environmental degradation from water pollution was a technical and scientific conundrum Oregon laws, administrative frameworks, and financial resources were struggling to address. It took a few years to be reflected in budget numbers and improved water quality, but McCall's *Pollution in Paradise* helped break through the gridlock of the early 1960s.

WILLAMETTE RIVER WATER QUALITY WITH TOM MCCALL AS GOVERNOR

Oregon voters elected Tom McCall secretary of state in November 1964 and governor two years later. During his eight years in office, Governor McCall took a prominent leadership position on a wide array of pressing environmental issues, including land-use planning, recycling, nuclear waste disposal, beachfront development, the Willamette River Greenway, and water pollution.[50] From his position of statewide leadership, his efforts to clean up the Willamette River would benefit from the ongoing work of many other Oregonians. Prominent among them were Oregon State Sanitary Engineer Kenneth Spies and his colleagues on the sanitary authority; chemist David Charlton; and the Oregon Division of the Izaak Walton League. Also contributing to abatement work was the City Club of Portland, League of Women Voters, and other groups. McCall's achievements would owe something as well to his allies in the state legislature—often Democrats—who helped him enact laws concerning land-use planning, beach protection, and water pollution control. He even benefited from support on many environmental issues from his opponent in the 1966 and 1970 elections, Democrat Robert Straub. In other words, McCall was not alone in his environmental advocacy generally, and in his push to clean the Willamette River in particular. David Charlton made this point when he lamented in a 1975 letter about "what might be called the McCall version [of the Willamette cleanup story] that appeared in the National Geographic" and how this version did not do justice to the many people long involved in the struggle.[51] Even with this admission, however, and in full recognition of the extensive work done by other Oregonians since the 1920s, Tom McCall's important role in cleanup efforts is a critical component of the story. In simplified renditions of the Willamette River pollution story, Tom McCall takes center stage. A more nuanced narrative helps contextualize McCall's role within the long history of abatement efforts. Beyond *Pollution in Paradise*, what makes McCall an important figure in the Willamette River cleanup narrative was that he was the first Oregon governor to care so deeply about a range of interrelated environmental issues, and to follow through with action on his sentiments.

McCall delivered his first speech as governor to the Oregon legislature the day he took office, January 9, 1967. In speaking about the OSSA and water quality, he observed the sanitary authority's budget was more than 60 percent below what the federal public administration service considered "desirable." While he did not commit to approving the authority's budget request in full, he did promise to strengthen its enforcement powers.[52] A few months later, in April, OSSA Chair Harold F. Wendel died of a heart attack at age seventy-four. The man who had been the sole chairperson of the sanitary authority for its entire existence had died after nearly thirty years of service.[53]

The *Oregonian* editorialized upon Wendel's death that he was "a gentle person for all his strong convictions, and in the many years under his guidance the Sanitary Authority followed a philosophy much like that of its chairman." This approach had been to apply "mild but steady pressure and persuasion in its efforts to win compliance" against industrial and municipal polluters, "rather than the more spectacular techniques of ultimatums and lawsuits." Wendel and his colleagues had "for many years fought the good fight before the fight became popular." During the 1940s and 1950s

> the Legislature was not at all keen about getting tough with industry and the cities, and the Sanitary Authority wisely did not try to get too far ahead of public sentiment in its activities. Perhaps it has not been as zealous in recent months as it should have been, but then the agency remembers how fickle the public can be . . .[54]

Many years later, biologist Glen D. Carter, who served with the OSSA under Wendel from 1956 to 1967, reflected on Wendel's approach by saying that he

> urged us to focus on the biggest water polluters first, because that would do the most good for society. He told us not to worry about the cost because that was the agency's concern. He also told us to seek voluntary correction of water pollution problems before resorting to legal enforcement. People outside the program may have perceived this as faint-heartedness, but it was really a matter of common sense and practicality; OSSA's budget allowed only $4,000 per biennium for legal services from the attorney general's office to take polluters to court.... In nearly all cases, however, OSSA relied on education and working cooperatively with polluters to solve problems in a fair and reasonable manner without causing economic hardship or ruin.[55]

Though the *Oregonian* and at least one of his former OSSA colleagues lauded his efforts, Wendel's approach of applying "mild but steady pressure and persuasion" was soon to be replaced by significantly more robust pressure from both the state's new governor and the federal government. Shortly after Wendel's death, McCall named himself interim OSSA chair. The *Oregonian* called McCall's decision a "surprise move." While McCall replied that a governor selecting himself as an agency chair was not unheard of in situations where the governor "wants to make things hum," such a move within the realm of water quality does appear to have been unprecedented. He said, "As governor of Oregon I know there is no more important task than to involve myself directly in the campaign against air and water pollution." The *Oregonian* reported McCall's step enabled the OSSA to begin "shoving aside the delaying alternatives which in the past have so appealed to the Authority," and OSSA members responded by "moving with briskness and new resolve through a long and vexsome agenda."[56] Long-time authority member Barney McPhillips recalled that "no governor before Tom ever cared much what we did. Tom was the only one to take an interest in what we were doing and push us. The only thing Tom didn't understand was that these things took time. He wanted action immediately."[57]

Considered in the context of McCall's 1962 documentary *Pollution in Paradise*, his decision to serve as OSSA chair was not entirely the "surprise move" the *Oregonian* made it out to be. In the documentary McCall articulated his position on water and air pollution. He firmly believed that Oregonians could have both economic prosperity and environmental health, and that state government should provide the leadership necessary to strike this balance. He was critical of industrial polluters who asserted they had done enough to abate pollution, and he trusted professional sanitary engineers, public health experts, and state administrators to enforce water quality standards.

With their governor as authority chair, in late May 1967 state legislators passed air and water quality bills for the OSSA that were to take effect January 1, 1968. These new laws included giving the OSSA permission to set more stringent standards and strengthening enforcement powers by enabling it to issue discharge permits. Legislative enhancements in 1963 allowed offenders to continue polluting while they challenged OSSA cease-and-desist orders in court; in contrast, the 1967 enforcement mechanisms required offenders to cease polluting while their case was being adjudicated. In other words, offenders could no longer continue to pollute for months or years while their case was in court.[58] While *Oregonian* editors largely lamented the results of the 1967 legislative session, the newspaper commended legislators for their environmental record:

> The most praiseworthy record of the 1967 session was written in the field of air and water pollution control. Taking recommendations of an interim committee as a starting point the Legislature adopted a comprehensive package of bills tightening up existing laws, strengthening the organization of the Sanitary Authority, and giving it a bigger enforcement budget. While the legislators disappointed many by their refusal to crack down on industrial polluters, they did provide incentives for river cleanup by offering tax credits to industry and state aid to cities for sewerage works.[59]

After only about three months as OSSA chair, in mid-July 1967, McCall stepped down. He originally stated he would remain chair until November 1, but justified this change for two reasons. The first was that he simply had many other responsibilities to consider, and so needed to delegate the role. A second reason was that the laws recently passed had both strengthened the authority's enforcement powers and added two new positions. With these improvements, McCall stated that "the next few months will mark the beginning of another critical phase of our war on pollution and only by starting as a unit can the restructured Sanitary Authority effectively implement this crucial legislation."[60] McCall's replacement was Washington County lawyer John Mosser. Mosser had served three terms as a state representative but did not run for reelection in 1966. For the first six months of McCall's governorship, Mosser served as his budget director. In naming Mosser to this position, McCall said he had "tried to get Mosser to take it" in April upon Wendel's death.[61] The *Oregonian* lauded McCall's choice, writing that Mosser was an expert in budget and finance. He was also said to have "an exceptional talent for quick and accurate analysis of complex situations, and has demonstrated many times in public office that he is not to be swayed by criticism or intimidation from what he regards as the proper course."[62]

Another of McCall's goals during his first term as governor was to reorganize and streamline governmental operations. The legislature approved these measures in early 1969, and implemented them on July 1 of that year. An *Oregonian* reporter characterized this as "the most extensive shake-up of state government in history" as they included creation of the Oregon Department of Transportation and other sweeping changes. Among these was McCall's creation of the Oregon Environmental Quality Commission (EQC) and Oregon Department of Environmental Quality (DEQ).[63] Republican Senator Victor Atiyeh and other legislators sponsored this proposal as Senate Bill (SB) 396.[64] It removed the sanitary and engineering divisions from the board of health, and placed these responsibilities—along with the solid waste program—under the EQC. The EQC was (and is) a five-member citizen panel appointed by the governor to provide policy and guidance for the DEQ.[65] McCall's decision to reorganize water pollution abatement efforts within the DEQ echoed recommendations that Robert Straub had made before the legislative interim committee on public health in 1965. Straub believed that taking the sanitary authority out of the board of health and creating an independent agency would be the best way to ensure enforcement.[66] Governor McCall signed SB 396 into law on June 16, 1969. Whereas the 1938 citizen's initiative creating the sanitary authority had been an unfunded mandate (which Governor Sprague and state legislators later funded at minimal levels), McCall and his legislative allies provided significant funding, increased staffing levels, and appreciably enhanced air and water quality statutes.[67] As Figure 6.7 shows, DEQ funding increased significantly in both real and relative terms during the initial years of McCall's leadership.

McCall's decision to take charge of the OSSA in 1967 and his collaboration with legislators to create a more effective environmental agency in 1969 illustrate a markedly different approach than that taken by his predecessor, Mark Hatfield. Both McCall and former OSSA member Barney A. McPhillips asserted that though Hatfield expressed a desire to clean up the Willamette, he did not follow through with substantive funding or leadership.[68] At a legislative committee hearing in October 1965, for instance, Robert Straub had blamed Hatfield's budget decisions for what he considered slow progress on abatement.[69] Analysis of OSSA funding data during the tenures of Hatfield and McCall support these claims (see Figure 6.6). One qualifying factor, however, resides in Wendel's testimony at the same October 1965 hearing. He said he was well aware that no agency ever gets all the funding it wanted, and "naturally we could use more money"—but he also admitted he had not approached the legislature directly for supplementary funds.[70] Though Wendel could have been more proactive in requesting funds, evidence still suggests that Governor Hatfield did not prioritize Willamette River pollution, and in this he shares more in common with his predecessors than with McCall, Straub, or future Oregon governors.

The first director of McCall's new DEQ was long-time state sanitary engineer Kenneth Spies. Spies took over for former OSSA chair John Mosser in July 1969.[71] In September 1971, the EQC replaced Spies with L. B. Day. Day had previously served as a business agent for a Salem cannery union, a Republican state representative from Salem in the 1960s, and then, from early 1970, as regional coordinator for the US Department of the Interior. One of his key tasks in this latter role was to oversee regional environmental protection efforts for estuaries.[72] With more funds, strengthened enforcement powers, and unprecedented support from the state's executive branch, Spies, Day, and the DEQ built upon the work of their predecessors. Biologist Glenn D. Carter was in a unique position to experience these changes directly. He began work with the OSSA in 1956, transitioned to the DEQ in 1969, and retired in 1988. In his experience, this administrative reorganization meant more than a change in name, it

> signified a much broader and more aggressive approach to solving
> environmental problems in Oregon. OSSA, functioning as an arm of
> the Oregon State Board of Health, had focused its attention largely
> on water-quality problems affecting public health. DEQ, on the other

hand, confronted a wide range of environmental issues, including air and water pollution, management of solid wastes, and land-use policies.[73]

One area of work the DEQ carried forward was municipal sewage treatment. In the early 1960s the sanitary authority threatened to bring suit against Portland city officials unless they expanded the municipal sewage system and provided more effective secondary waste treatment. This effort was part of a broader push to get all cities in the valley to provide at least secondary sewage treatment. The DEQ picked up where the OSSA left off, and by 1972 all cities in the Willamette Valley had secondary treatment facilities.[74] The DEQ also made progress addressing industrial point source pollution in the early 1970s. For decades, the sanitary authority faced budgetary and staffing limitations that forced it to prioritize actions against the most significant polluters before working down the voluminous list of culprits. The City of Portland's sewage and pulp and paper mill wastes topped this list, but other polluters included operations such as vegetable and meat processing, flax retting, tanning, mining, and logging. Two notable examples of the DEQ's work addressing industrial polluters include resolving a longstanding abatement struggle with a meat-processing facility along Columbia Slough in 1971 and, the following year, a dramatic showdown with the Boise Cascade pulp and paper mill in Salem.

The Pacific Meat Company case illustrates the way the OSSA worked down its list of polluters as well as the inadequacy of state water quality enforcement mechanisms prior to creation of the DEQ. The sanitary authority began pressing Pacific Meat and other firms along Columbia Slough to upgrade their waste management technologies in the 1940s. Solutions available included screens, retention ponds, or on-site treatment facilities—all of which were preferable to the traditional practice of routing wastes into Columbia Slough. By the mid-1950s, most firms availed themselves of the option of connecting to Portland's Columbia Boulevard sewer interceptor. Pacific Meat opted to purchase filtration screens for its effluents prior to dumping liquids into the slough; the company did not, however, install them. Instead, it continued to dump up to 150,000 gallons per day of untreated blood, feces, urine, and "animal particles" directly into the slough.[75] In October 1955 the OSSA asked the state attorney general to seek a court injunction against Pacific Meat. The court did not decide in the state's favor until April 1961. Following this, the company built lagoons in which their wastes settled somewhat before being dumped into the slough, an inadequate solution. As the authority admitted, "we won the case but did not stop the problem."[76]

The authority conducted periodic inspections of Pacific Meat's waste outfall but did not complete a comprehensive study of the wastes' effects on the slough until 1967. Even with this definitive information, the state attorney general still took the position it was better to work with the company than shut it down. Without the requisite enforcement authority to back up its actions, the OSSA and then the DEQ issued a series of discharge permits to the firm that allowed the status quo until June 1, 1970. The pollution did not stop, however, until Portland city officials agreed to annex the company's 8.2-acre tract early in 1971, which cleared the way for the firm to connect to the Columbia Boulevard sewer interceptor that August. By not implementing effective abatement measures in the 1950s, as the company fouled the slough it also externalized an estimated $150,000 in expenses it did not incur to deal with its own wastes, putting it at an economic advantage relative to the meat-processing firms that complied with state law.[77]

Abatement advocates had long been struggling against firms externalizing pollution's costs and effects. As governor, McCall was in an influential position to oppose pollution while advocating for economic growth, a perspective he articulated in *Pollution in Paradise* and during his gubernatorial campaign. He welcomed industrialization, so long as firms complied with environmental regulations. For this reason, he supported the siting of a new pulp mill in Halsey when the American Can Company of New York announced plans in 1967. The proposed mill would create pulp using the kraft (sulfate) process, which produced much less waste than the sulfite process; the firm's plans also included air and water pollution abatement technologies that would enable it to meet more stringent state standards taking effect in 1968. Considering the less-than-stellar record of the state's existing pulp and paper mills, significant public sentiment was against the mill, and the five-member OSSA was split two-to-two, until Chair John Mosser cast the deciding vote in favor. Significant factions in the legislature, press, and throughout Oregon disapproved of Mosser's vote, and McCall himself received criticism for this decision. However, the governor later reflected that American Can's Halsey mill ushered in a new era of environmentally compliant industrial development.[78]

If the mill at Halsey can be seen as an example of this new era, then the drama five years later surrounding Boise Cascade's pulp and paper mill in

Fig. 6.10. Aerial view (from the southwest) of the Oregon Pulp and Paper (later Boise Cascade) mill in Salem, 1949. Pringle Creek runs under the mill and railroad bridge at center, carrying untreated effluents into Willamette Slough, and then left (north) into the Willamette River. By the early 1970s, effluents were retained in ponds on Minto Island, west of the mill just beyond the frame at the bottom of the image (Willamette Heritage Center, P 2012.049.0115).

Salem shows the previous era had not been entirely superseded. As with the Pacific Meat Company, Boise Cascade's Salem plant had a long history of noncompliance with state water pollution law and unfulfilled promises to install abatement systems. Built as the Oregon Pulp and Paper Company in 1920, the mill produced pulp using the more-polluting sulfite process. Boise Cascade acquired the mill in 1962, and by the late 1960s the DEQ had evidence that the company's pulping liquors were seeping into the Willamette River from lagoons on Minto Island. Mill operations were also fouling Salem's air.[79] By summer 1972, state and company officials had failed to come to an agreement on the installation schedule for $6.5 million in abatement equipment. The company was supposed to have it installed by July 15. That the company was not making a good faith effort is supported by the fact that they did not apply to the DEQ for an extension; previous such instances involving the pulp and paper, meat-processing, and other industrial firms clearly illustrated the OSSA and DEQ were willing to extend deadlines if officials determined firms were making adequate progress.[80] In the face of Boise Cascade's intransigence, in late July 1972 the DEQ requested a Marion

County Circuit Court order that would enforce the stipulations of its waste discharge permits and shut down the mill. On July 24, the company, of its own volition, shut down the mill to install the equipment, and also agreed not to use its Minto Island lagoons until the seepage problems were fixed.[81] In spite of the company's decision to close temporarily, the DEQ proceeded with its suit. L. B. Day informed Oregonians that the state's objective was not to put people out of work, but to alleviate pollution for the benefit of all residents. Echoing his boss, Day said that "there is no reason why we can't have jobs without pollution."[82]

For about three weeks, this was not the case in Salem. The closure put hundreds of people out of work as the company installed the requisite abatement technologies. The mill reopened Monday, July 31, and was in full operation by the end of the week.[83] Abatement advocates were pleased the DEQ had succeeded in enforcing compliance from Boise Cascade. The brief closure did not please all Oregonians, however. On July 26 millworkers, union representatives, and others marched to the Capitol building to take their grievance directly to McCall and Day, whom they blamed for the shutdown. Some reportedly shouted, "Hitler, Hitler!" as McCall and Day spoke before the assembled crowd.[84] A letter writer to the *Oregonian* lamented in late July that "one man [McCall] can close down a manufacturing plant and send hundreds of people to the unemployment office." The writer claimed that "it took years and years for the rivers to get in their present shape, and yet the ecology zealots would have them cleaned up 'yesterday,' regardless of the hardship to the people."[85] This sentiment expressed a reactionary perspective that lacked both accuracy and historical context but likely reflected the fears of other Oregonians.

WILLAMETTE RIVER WATER QUALITY AT THE END OF MCCALL'S TENURE

Tom McCall was not an unreasonable "ecology zealot" and he did not single-handedly close down the recalcitrant Boise Cascade mill. He was not the sole person responsible for marked improvements in Willamette River water quality that were apparent by the early 1970s, but his contributions toward that end were significant. As his biographer Brent Walth writes, a critical part of McCall's personality contributed to his success as governor and helps contextualize his important contributions to cleaning up the Willamette: McCall "grabbed an idea, made it his own, and then fought for it with all his will."[86]

Tom McCall's importance to Oregon included his role as the catalyst for fundamental change in how the state government and—to a large degree—the state's population thought about and treated the environment. His importance in this regard can be observed in a wide array of issues that continue to resonate long after his tenure: land-use planning, recycling, hazardous waste disposal, and public access to ocean beaches, to name a few. In some of these he was at the vanguard: his leadership on land-use planning and his opposition to federal hazardous chemical weapons transport through the state in 1970 are examples. In some cases—such as bottle recycling—he played a pivotal role but did not originate the idea.[87] McCall also played an essential role in abating Willamette River pollution, though by no means did he originate the idea, nor did he have a decades-long record of involvement as had William Finley, David Charlton, Edgar Averill, Harold Wendel, and many others. McCall's contributions to the issue, however, are noteworthy. Foremost, he contributed passion, tenacity, and the power of the highest elected office in the state. He brought skills in investigative journalism and communicating with the public. One cannot overlook as well what Robbins calls the "serendipitous" time in which McCall reached the height of his career, just as Americans were becoming more aware of and anxious about environmental degradation and gaining new scientific methods and tools to detect, quantify, and address it. In summary, Governor McCall differs from previous abatement advocates in bringing to the issue a combination of passionate interest, political power, and public respect. McCall benefited from the successes and shortcomings of his predecessors and colleagues, and was able to catalyze these into something new for Oregon that served as a positive example for the entire nation.

Such achievements were what Ethel Starbird memorialized in her June 1972 *National Geographic* cover story, "A River Restored: Oregon's Willamette." Starbird wrote about the state's "amazing achievements in water pollution control" and characterized these as "a tribute to Oregonian tenacity in pushing through the most successful river rejuvenation program in the country." She wrote about the forty-plus-year pre-McCall era briefly before referring to the governor's documentary and the "protectionist sentiment [that had] snowballed" since the 1966 gubernatorial campaign between McCall and Straub. What emerged under McCall's leadership, she continued, "was a reasonable and economically feasible program to improve the Willamette."[88]

Oregon's progress in abating pollution within the Willamette watershed had already drawn national attention. The *New York Times* published an article in late 1969 providing an overview of a US Government Accounting Office (GAO) report lamenting national progress in water pollution abatement. Senator Mark Hatfield, in January 1970, wrote a letter to the editor of the *New York Times* critical of this report. Hatfield stated: "The GAO report as it relates to Oregon water quality control programs, and in particular the Willamette River, is grossly inadequate, and the conclusions drawn therefrom show a complete lack of understanding by the GAO accountants of water quality control problems and programs."[89] Not long afterward, both the US Attorney for Oregon Sidney Lezak and James Agee with the US Environmental Protection Agency agreed with Hatfield that "Oregon is doing one of the best jobs of any state in combating industrial water pollution."[90] A few months later, in September 1970, the *New York Times*'s E. W. Kenworthy wrote about the positive prospects for that years' fall Chinook salmon run in the river. "If the count is up," he wrote, "it will be an event of some national significance—visual proof that a Federal-state program to clean up the Willamette River, until a few years ago one of the nation's foulest, is continuing to make steady progress." He concluded that "at present they [Oregon officials] have 'a river on the way to restoration.'"[91] This phrase likely inspired Starbird's title two years later.

In May 1975, David Charlton wrote a letter to a Portland State University professor in which he bristled at "what might be called the McCall version" of the Willamette River cleanup narrative touted in the national press. Backed by his direct and longstanding involvement in the matter, he lamented that this simplified story neglected the contributions of many other people and organizations. Charlton continued by asking a question that was rhetorical to those who, like him, had long been active in the clean streams struggle: "Who were really concerned and took action leading to the first investigations of the problem and then to take corrective action?" He knew the answer. It was members of the state board of health, the game and fish commissions, and sporting and conservation groups—most notably Izaak Walton League members, beginning in earnest in the 1920s.[92]

The decades before "the McCall era" of water pollution abatement were more complex than Starbird's account and others that would follow. Water pollution challenges in subsequent decades would also prove to be complex, in ways both familiar and new. Tom McCall left office on January 14, 1975, having pushed the Willamette River pollution abatement effort further along than his predecessors—at least in terms of point source pollution. What he and his predecessors were not as prepared to address were nonpoint pollution sources and the hazards of persistent, bioaccumulative substances such as dioxins and heavy metals. By the mid-1970s it was becoming increasingly apparent that Oregonians would be facing these confounding issues next.

Chapter 7
Hydra-Headed Pollution

Oregonian reporter Don Holm provided one indication of improved Willamette River water quality in September 1976. The river "gets a setback every now and then through oil spills and other pollution incidents," he observed, "But if you want to know just how clean the river is, try crawdad [crawfish] fishing":

> Crawdads are among the first to go when a river begins to die of pollution. . . . Anglers who fish the Willamette above Portland know the crawdads have been coming back for several years . . . [and] We have caught them recently in Multnomah Channel. . . . The other day we decided to test the main river in the St. Johns area where there are sawmills, chemical plants, oil refineries and terminals and heavy ship concentrations.

Following the recommendations of a shipyard worker who "has been keeping himself supplied with fresh crawdads for the family for months, simply by hanging a trap under the oil dock" at the GATX rail terminal in the northwest Portland neighborhood of Linnton, he baited traps with salmon heads and waited. A few hours later the traps "were full of the biggest and healthiest crawdads" he had seen "for some years." He concluded: "Clearly, if the river is clean enough to support this kind of shellfish life, then progress has been good in restoring the river." Not only were they plentiful and easy to catch, "they turned out to be delicious" after being boiled, refrigerated for two days in a vinegar and salt brine, and dipped in hot sauce.[1]

Holm's experience harvesting crawfish in the industrialized lower Willamette provided experiential evidence that water quality had improved markedly as a result of fifty years of persistent efforts. By the mid-1970s dissolved oxygen levels had increased to support a broad range of aquatic and benthic life, even during the late summer and early fall low-flow period; the river had come a long way from the drastic crawfish die-off of September 1949. Improvement had come as a direct result of the work of the sanitary authority and department of environmental quality, the persistence of advocacy groups, and the beneficial effects of Willamette Valley Project dams. Portland's sewage, pulp and paper mill wastes, and most other significant point sources of pollution were largely being addressed, drastically reducing oxygen-depleting wastes. Dissolved oxygen (DO) levels were further improved after the final Willamette Valley Project dam was completed in 1968, and the river carried more water during its annual low-flow periods. Average flows measured at the Salem gauge during September 1972 were 12,000 cubic feet per second (cfs), having more than doubled from the 5,400 cfs recorded in 1966 and 5,000 cfs recorded during the water shortage crisis of August 1965.[2] DO improved along with decreased pollution and higher water levels: beginning in 1970, DO was consistently above minimum five parts per million, as indicated in Figure 7.1.

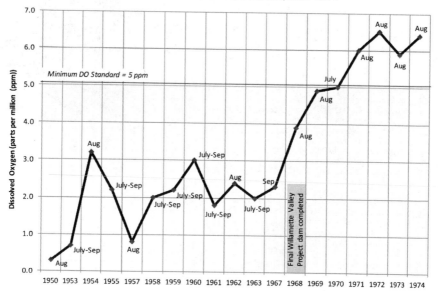

Fig. 7.1. Lowest dissolved oxygen (DO) levels recorded in Portland harbor during annual June–October low-flow periods, 1950–1974 (Data: OSSA stream survey report for 1953 and Oregon DEQ, *Proposed Water Quality Management Plan, Willamette River Basin: Appendices* (1976), C-43).

If DO was the only measurement of water quality, the Willamette River cleanup narrative would have ended with Holm's crawfish recipe. His non-rigorous statistical methods and unscientific sampling protocols could not, however, provide evidence of the stew of dioxins, polychlorinated biphenyls, pentachlorophenol, creosote, copper, chromium, oil, polycyclic aromatic hydrocarbons, arsenic, lead, and other chemicals and compounds permeating river sediments. After decades of minimally regulated industrial practices, the riverbed and banks were saturated with persistent toxic pollutants. David Charlton, Harold Wendel, Edgar Averill, Tom McCall, and the many others involved in pollution abatement in the 1950s and 1960s had been successful in getting the Willamette off the treadmill of point source pollution. In the 1970s it became apparent that previous successes brought into focus new pollutants and challenges that would require more time, money, and effort to confront. It would also require increased involvement from the federal government in the form of stronger environmental laws and administrative tools.

FEDERAL ENVIRONMENTAL LEGISLATION AND OREGON

"That's bad, make it good."

State officials and advocates in Oregon played the most pivotal roles in water pollution abatement into the 1970s. Two monumental developments in federal administration and law in the early 1970s fundamentally changed the state–federal dynamic, not only in Oregon but throughout the nation. The first was Republican President Richard Nixon's creation of the Environmental Protection Agency in 1970. The second was passage of the Clean Water Act two years later, over Nixon's veto. Nixon initiated the first of these sweeping changes in May 1969 when he established a citizens' advisory committee on environmental quality, in large part to counter political opposition. His likely Democratic opponent in the 1972 election, Senator Edmund Muskie of Maine, was gaining much popular support and political momentum for his work on water quality and other environmental issues.[3] To seize momentum, Nixon named three respected national figures to serve on his advisory committee: businessman and conservationist Laurance Rockefeller, famed aviator Charles Lindbergh, and Oregon Governor Tom McCall.[4] McCall's national reputation stemmed from a "refreshing candor that spoke of common sense, not obfuscation." This was apparent in his ability to identify a problem and implement solutions, a visible example of which was his leadership in prioritizing Willamette River pollution abatement and being a champion of a strengthened state role in ensuring water quality.[5] Input from this committee influenced what became the National Environmental Policy Act (NEPA) (Public Law 91-190), which Nixon signed into law on New Year's Day, 1970. It stipulated that federal agencies would endeavor to protect the environment, primarily through the medium of detailed, interdisciplinary analyses of the environmental repercussions of agency actions—reports known as Environmental Impact Statements (EISs). NEPA also established the President's Council on Environmental Quality as a permanent advisory committee on environmental matters of national import.[6] A few months later, in April 1970, the first Earth Day observation occurred. Shortly thereafter, Nixon asked Congress to create the Environmental Protection Agency (EPA) as part of his governmental reorganization plan. His proposal echoed the intentions of Governor McCall in 1969: Nixon wanted to centralize environmental oversight and enforcement within a new agency dedicated for this purpose. Congress concurred and the EPA opened for business in December 1970, for the first time bringing together air, water, and land pollution responsibilities formerly dispersed among an array of other agencies.[7] At the federal level, the EPA was pioneering and noteworthy for the same reasons the Department of Environmental Quality (DEQ) was for Oregon: it united under one body environmental oversight and enforcement. The EPA would come to play an important role in Oregon's water pollution abatement efforts, and the Clean Water Act of 1972 would come to serve as a primary mechanism through which it did so.

In the 1930s and 1940s, strong opposition thwarted the Izaak Walton League and other groups arguing to expand the federal governments' role in regulating water quality. Representatives from chemical manufacturers, pulp and paper companies, and other highly polluting industries did not want such laws because it was easier for them to influence disparate local and state governments than it was to influence the federal government. Political conservatives were also wary of any increase in federal power that, in their view, undermined states' rights.[8] Advocates' persistence combined with worsening pollution to shift debates in Congress. The 1948 Federal Water Pollution Control Act, though only a temporary measure lacking substantive enforcement mechanisms, provided the first step in a centralized national approach. Amendments in 1956 and 1961 established permanent federal research, outreach, and coordination functions. In 1965 Congress passed the

Water Quality Act (PL 89-234) that established the Federal Water Pollution Control Administration within the US Department of Health, Education, and Welfare.[9] By the late 1960s, state water quality programs ranged from relatively successful—as in Oregon—to functionally nonexistent, while water, air, and land pollution were becoming increasingly more widespread and confounding. When Congress passed another round of changes to water quality regulations, the law's title belied its significance: the Federal Water Pollution Control Act Amendments of 1972 (PL 92-500) was not another incremental change to previous laws but a drastic overhaul.

The Clean Water Act (CWA), as PL 92-500 is more commonly known, was 89 pages of fine print with an attached 424-page house committee explanatory report; an observer called it "one of the most complicated pieces of legislation ever to emerge from Congress."[10] It was complicated because it sought to address a slew of administrative, regulatory, and technological weaknesses apparent in previous laws. Many states had been slow to devise their own robust water quality programs as federal law evolved in the 1950s and 1960s—Oregon being one of the exceptions. The CWA increased federal oversight of regional and state water quality programs, mandated a detailed reporting process, and allowed for federal intervention into state actions and decisions. It established national goals and priorities, with adequate progress being defined as robust state pollution control programs that complied with EPA guidelines—the patchwork, state-by-state, implementation of varied water quality laws was no longer sufficient. If states were not making progress, the EPA could intervene and enforce CWA provisions.[11] Harkening to objections levied against proposed water quality legislation in the 1930s, critics charged the law trampled states' rights because it allowed the federal government to intervene on matters pertaining to intrastate surface waters, a jurisdictional boundary the federal government had not yet breached. While allowing for increased citizen participation and the ability to bring suits against polluters, the law also contained a "carrot" in the form of more than $20 billion in aid to build sewage treatment systems.[12] The CWA was more robust from a technical standpoint in that it moved away from ambient water quality standards and established effluent limitations: polluters would be required to install the best *practicable* technologies by 1977, the best *available* technologies by 1983, and achieve zero pollutant discharges by 1985. The CWA was also the first federal law to address nonpoint source water pollution, such as from agricultural runoff, construction projects, and parking lots.[13]

As with creation of the EPA, Congress's motivations in passing the CWA echoed what Oregonians had experienced. By empirical measurements such as DO and bacteria counts, water quality was improving, but these improvements brought attention to the complexities of pollutants from nonpoint sources and toxic substances. Municipal sewage treatment systems developed in the 1940s and 1950s were insufficient by the 1960s, and primary treatment had to be enhanced by costly secondary and tertiary systems. As McCall and Oregon legislators realized, addressing these shortcomings required a strengthened water quality agency, increased funding, and expanded infrastructure. State officials in Oregon made these changes in the late 1960s, and a few years later the federal government recognized the unavoidable need to address these issues at the national level. There were other important factors influencing Congress in 1972. Environmental advocates were becoming increasingly more creative and confrontational in their application of direct action, lobbying, lawsuits, and other tactics. Groups such as Environmental Action and Friends of the Earth, and individuals such as Ralph Nader and Barry Commoner, were increasingly successful in publicizing environmental and quality-of-life issues.[14] These efforts sometimes correlated with the work of the Izaak Walton League and other longstanding groups, but more often supplanted them. Highly publicized events such as the Santa Barbara oil spill of January–February 1969, the Cuyahoga River fire of June 1969, and the first Earth Day in April 1970 contributed momentum to arguments from this new generation of environmental advocates. In response, EPA lawyers and staff pledged to become more involved in water quality enforcement.[15] Oregon's DEQ came into existence in 1969, with funding and powers that eclipsed the sanitary authority and altered the dynamic of Willamette River pollution abatement. With creation of the EPA in 1970 and passage of the CWA in 1972, federal involvement in water quality within Oregon also changed in both quantitative and qualitative terms.

As comprehensive and stringent as the CWA was relative to regulations throughout the fifty states, when Congress passed the law in 1972, Oregon was already ahead of the curve. In October 1969 Federal Water Pollution Control Administration Commissioner David D. Dominick called Oregon's water pollution program among the most progressive and effective in the nation.[16] CWA provisions tasked the DEQ with a range of responsibilities. Some of these were direct carry-overs from previous federal or state legislation, such as distributing federal sewage infrastructure grants. CWA provisions also

required the DEQ to develop a permit program for point sources, but the DEQ had been doing this since 1967, in compliance with state law. Additionally, the DEQ established a water quality program with elements that predated—and in some cases inspired—the CWA, such as discharge permitting and setting a minimum requirement for secondary (rather than merely primary) municipal and industrial waste treatment.[17] A key change in Oregon under the Clean Water Act was that the DEQ had to answer to the federal EPA, rather than having full administrative oversight of the program.[18] The primary method for doing so was biennial reporting, mandated under Section 305(b) of the CWA. These "305(b) reports" summarized the status of water quality management throughout the state, provided updates on water quality programs, and communicated significant events that had transpired since the previous report.[19] The DEQ had been producing its equivalent to 305(b) reports for state officials since 1970. Its first such report did not address pollution from toxics, but as point source wastes from the Willamette Valley's largest polluters were gradually coming under control, the report identified three new and pressing water quality problems. The first was decreased flow caused by consumptive uses, such as irrigation, that contributed to increased water temperatures. The second was sediment caused by unregulated, nonpoint runoff from such sources as logging, road building, and mining. The third nonpoint sources of bacteria, such as from agricultural fields and urban sewer runoff (also known as combined sewer overflows, or CSOs).[20]

Nationally and in Oregon, the first decade of CWA implemcentered on devising and implementing a program to sustain and the integrity of surface waters. To do this, DEQ staff managed pr applied the best available technologies to regulate point sourc They were also primarily focused on the *chemical* aspects c ity integrity—water quality as measured by DO levels and t oxygen demand of effluents. Thus, their efforts through th expanded, were not drastically different from what they ha decades.[21] With this perspective, in 1976 the DEQ chara River water quality as "good overall," but still identifie the "bugaboo" of cleanup efforts. The assimilative cap its maximum level:

The main stem Willamette River water quality near minimum standard oxygen levels that a

waste treatment or control is apt to cause short term violations of the oxygen standards. The few DO violations witnessed since 1972 have been minor in duration and effect, but such violations stand as fair warning that there is no margin of river assimilation capacity with which to gamble.[22]

The combination of flow regulation from the Willamette Valley Project dams and point source abatement measures such as secondary treatment of both sewage and pulp and paper mill wastes was barely enough to support salmon and other aquatic life. This meant that "new or expanded BOD sources can be accommodated only by comparable reductions elsewhere in the system."[23] Since point source pollution was largely under control, the DEQ sought to address nonpoint source pollution. With federal financial assistance in the form of a $1.2 million grant, the DEQ conducted the state's first comprehensive inventory of nonpoint pollution sources between 1976 and 1978. Focused on pollution caused by water runoff from roads and ditches, poor forest management practices, overgrazing, agricultural chemicals, and other kinds of "human activity on large tracts of land," resultant data served as a baseline for future planning and abatement efforts.[24]

In continuing to gather data necessary to achieve state and federal mandates, DEQ Director Loren Kramer admitted abating nonpoint source pollution would be more difficult than meeting earlier water quality goals. "Our old point source pollution problems were very obvious," he told the *Oregonian*. "You just went to the end of some pipe, made tests, and told the owners, 'That's bad, make it good.'" However,

we're to the point where the additional improvements we can get by doing that may not be worth the additional cost. . . . We could spend millions trying to get the last ounce of pollution out of a pipeline, while just upstream we are getting millions of tons of silt and fertilizer washing into the river every time it rains.[25]

With its survey data as a foundation, in 1977 the DEQ adopted a set of for the state's twenty major river basins as part of the CWA compliance . These plans constituted the first step in setting state water quality ters under the CWA and set forth agency policies to establish water standards, protect unpolluted waters, and prepare for population

growth and economic expansion.[26] The plans identified areas where water quality suffered and proposed solutions to rectify these deficiencies. Back in 1973, DEQ had established its policy of mixing zones to recognize the fact that "individual point source waste discharges, even after receiving highest and best practical treatment, may cause a technical violation of water quality standards in the zone immediately adjacent to their outfalls."[27] Such zones were sections of receiving waters in which the DEQ would allow higher levels of pollutants to exist as long as these wastes were assimilated farther downstream. Four years later, DEQ found that even this policy of leniency was not always adequate. In accordance with the Clean Water Act, the agency began to designate certain sections of waterways as "water quality limiting" if the desired level of cleanliness could not be sustained even with the application of "best practicable treatment of point source wastes." The CWA required states establish Total Maximum Daily Load (TMDL) standards for each waterway both to quantify and limit the amount of pollutants put into the water.[28] TMDL is a measurement of the maximum amount of a given pollutant that a specific body of water can assimilate while still achieving overall water quality standards. TMDLs have been specified for a wide range of pollutants, including pathogens, nutrients, sediment, mercury, and pesticides.[29] In concept, once DEQ staff had set the aggregate TMDL for a given waste within a watershed, they then divided the load into natural source, nonpoint source, and point source categories. Allocated further to each polluter within a watershed, these limits then constituted the maximum daily amount a given company or municipality could contribute to the receiving waters.[30] The TMDL framework can be understood as the further refinement of watershed- and waste-specific assimilative capacity modeling that Clarence Velz of the National Council for Stream Improvement had done for the Willamette watershed in the 1950s. Velz's technique was, in turn, a more deliberate and quantitative refinement of the broad, taken-for-granted, assimilative capacity approach that predominated well into the twentieth century. This earlier approach was the model Abel Wolman used to provide suggestions for Portland's sewer system in 1939, that A. M. Rawn used in drafting engineering specifications for Portland's sewer system in the late 1940s, and that most other sanitary engineers applied throughout North America and Europe.

In spite of the large page count of the CWA and the lengthy explanatory report accompanying it, the law was not clear on how to establish TMDLs. Without this clarity, state agencies found it hard to establish limits, and without these there were no clear metrics to evaluate what constituted "pollution"; the OSSA faced the same fundamental issue in the early 1940s. In Oregon and elsewhere, this lack of clarity meant it often took a long time to establish TMDLs, and continuation of the status quo prevailed in the meanwhile. Advocates at the Northwest Environmental Defense Center (NEDC) at Portland's Lewis & Clark Law School sought to speed up the process when they brought suit against the EPA in December 1986 to compel the agency to develop TMDLs for the Tualatin River, one of the Willamette's tributaries.[31] The group petitioned the DEQ and EPA to establish TMDLs for eight years before bringing its suit. NEDC president J. Douglas Smith asserted, "The Tualatin River is one of the most egregious examples of DEQ's flawed statewide water quality management policies." He presented evidence collected since 1940 showing excessive algal growth caused by phosphorus and nitrogen from Washington County sewage treatment plants and raw sewage from combined sewer overflows. A co-plaintiff, Professor John Churchill of Portland State University, said that "time for additional study passed at least 10 years ago . . . what has been needed since, and what DEQ and EPA have refused to provide, is action."[32] In response, DEQ public affairs official Shirley Kengla commented that the NEDC, DEQ, and EPA all wanted the same outcome. "What we disagree about," she said, "is how quickly we go about it." She observed there were many factors contributing to pollution, and because cleanup costs would be significant, she warned, "If we rush this through, someone else is going to sue us because they're not going to believe what we tell them." Admitting that the DEQ had previously studied the Tualatin River, she said "This is something new in that we're looking at every single pollution source, including toxics, for which there has been no data." She was also concerned "with the introduction of all the new high-tech industry in Washington County," and noted the slow-moving Tualatin made the pollution problem worse.[33]

Negotiations continued through early 1987. On June 3, the parties signed a consent decree before the lawsuit went to trial. It established deadlines for setting TMDLs for eleven "water quality limited" watersheds. The process would begin with the Tualatin and end in June 1988 with the Grand Ronde River in eastern Oregon. The DEQ and EPA would establish point source TMDLs by looking at overall water quality in receiving waters, not merely the level of treatment of any given effluent. This revised approach would also be critical in incorporating nonpoint pollution into these metrics.[34] The decree also specified seventeen other "water quality limited" rivers, bays, sloughs, and aquifers for further study. Among these were the main stem of the Willamette River, the Coast Fork of the Willamette, and the lower Columbia.[35] J. Douglas Smith of the NEDC claimed the settlement would "revolutionize water management in Oregon" because the DEQ would have to administer its water quality management program "on the basis of water quality instead of sewage quality."[36] He was correct, as it shifted the focus of the DEQ's approach and helped clarify the EPA's interpretation of the Clean Water Act regarding TMDLs. With the Tualatin River case as precedent, other states initiated their own lawsuits against the EPA. The end result was that in July 1992 the EPA revised how it established TMDLs across the entire nation.[37]

The 1987 consent decree also helped propel efforts that had begun a few years earlier to expand the DEQ's scope of work from a focus on the *chemical* aspects of water quality to include *biological* and *physical* integrity, and also nonpoint pollution sources such as from streets, agricultural fields, and construction sites. Rather than pursuing *technology*-based approaches, the DEQ began to focus on *water quality*–based programs. In practical terms, this meant the DEQ would evaluate pollution sources based on their cumulative effects on the receiving waters, instead of on the level of treatment of each pollutant as it exited the pipe—or, in the words of the NEDC's J. Douglas Smith, evaluate "water quality instead of sewage quality."[38] With increased attention in the late 1980s and early 1990s, researchers in Oregon were finding that much nonpoint source pollution was generated during small-scale, everyday activities that were difficult, if not impossible, to regulate.[39] These conclusions complemented the EPA's findings after analyzing more than two thousand storm water runoff events in twenty-eight metropolitan areas over six years. This extensive project provided the first baseline data on pollutants washed into nearby streams, rivers, and lakes that included petroleum products, heavy metals, asbestos, herbicides, solid waste, and other materials.[40] Just as the Oregon State College Engineering Experiment Station's studies in the early 1930s gathered quantitative data necessary to enhance state water quality laws, Congress used the EPA's storm water runoff data to strengthen national water quality laws. Amendments made in 1987 to the CWA included a grant program to help address nonpoint source pollution.[41] In Oregon, Governor Barbara Roberts and legislators established the Watershed Health Initiative in 1993 to coordinate work on nonpoint source pollution through local watershed councils.[42] This initiative was one way that Oregon's Land Conservation and Development Commission, the DEQ, and other state agencies involved with environmental issues could spur voluntary water quality control efforts outside the regulatory framework. The DEQ also embarked upon a multiyear reorganization in 1994 to "put resources where they will be the most effective in protecting the environment and public health." This involved shifting staff from the Portland headquarters building to regional offices.[43]

In 1996, just a few years after the NEDC's successful suit and changes at the federal and state levels, NEDC members filed another suit in US District Court against the EPA, again for what it considered to be the state's still-inadequate TMDL program. Nina Bell, the NEDC's executive director, alleged that at current rates of progress, "cleanup could take until the year 3000 for the almost 870 stream segments now included on a state pollution list." Such a slow pace violated the Clean Water Act, the NEDC claimed.[44] The suit echoed other efforts seeking definitive deadlines for state agencies to determine stream pollution limits. A decision had given state officials in Georgia five years to finish the task; Bell said she hoped for a time frame of five to twenty-five years for Oregon. "Quickening the pace requires more money," Bell said, but she feared agricultural and other interest groups would lobby against allocating the necessary funding.[45]

State officials and agriculture industry representatives, in particular, were not pleased with the NEDC's stance. The lawsuit came in the midst of negotiations about the best ways to implement CWA-mandated cleanup processes, and Bell was among the negotiators. Abatement advocates felt compelled to take action because the state list of water quality–limiting stream segments had recently grown drastically, from 150 to 870. Most of this increase was because temperature levels exceeded state standards. Coming to agreement on TMDLs for these hundreds of segments was contentious because implementing methods to reduce stream temperatures was likely to increase restrictions of livestock grazing and other streamside operations.[46] Bell and the NEDC secured another consent decree in October 2000, replacing the one negotiated in 1987. It centered on a memorandum of agreement built around the Healthy Stream Partnership of 1996.[47] Governor John Kitzhaber, agriculture industry officials, and representatives of the environmental groups Oregon Trout and Water Watch forged this agreement to increase spending, foster cooperation, and facilitate water basin planning. Funding included a $25 million legislative appropriation every two years for water quality monitoring and generation of TMDLs.[48] The 2000 consent decree and the Healthy

Stream Partnership agreement evolved into the Oregon Plan for Salmon and Watersheds, a public-private effort to protect federally listed endangered species that soon evolved into a multifaceted statewide initiative to protect species, habitats, and "preserve and profit from Oregon's natural legacy."[49]

In response to these legal challenges and the increasing empowerment of local watershed-level planning and coordination, the DEQ's approach continued to evolve. The agency's efforts under the CWA had begun as an extension of its work in earlier years, focused on technologies and point source polluters. The NEDC's 1986 suit helped transition the DEQ from this more mechanistic approach to a focus on water quality integrity and non-point sources of pollution. Beginning about 2000, the DEQ's work began to emphasize a more inclusive approach. This involved research to understand watershed-level dynamics, sustain high water quality while also improving degraded water, and integrate requirements of the Endangered Species Act of 1973 (PL 93-205), Safe Drinking Water Act of 1974 (PL 93-523), and other federal and state environmental laws.[50] Just as pressure from the Izaak Walton League, League of Women Voters, and other groups in previous decades had compelled industries and state agencies to become more responsive and proactive, the NEDC's advocacy had been essential in bringing about critical changes to water pollution abatement policies and practices.

PORTLAND, CSOS, AND THE BIG PIPE

"We're only asking the city to do what is practical."

As the NEDC's campaigns illustrate, interpreting and applying complex state and federal laws to improve water quality continued to be a long and contentious process. Amidst the nuanced courtroom debates and technical administrative hearings, it is important to remember that the contention continued to center on a complex, dynamic natural system. Granted, Samuel Simpson's "Beautiful Willamette" of the mid-nineteenth century had become regulated, channelized, and largely "tamed" by the end of the twentieth century, but it was not completely under the control of human artifice. Willamette Valley Project dams were critical in raising water levels during the low-flow months and maintaining sufficiently high dissolved oxygen levels. Data from the mid-1990s provided further evidence of the effectiveness of these dams coupled with ongoing abatement work in that only 20 to 30 percent of Willamette River pollution came from point sources like municipal sewer outfalls and industrial effluents. This meant, however, that nonpoint sources caused up to 80 percent of the pollution, and these sources continued to be difficult to address.[51] Nonetheless, abatement advocates could take some solace in the knowledge that prior to the early 1970s nonpoint source pollution was not a primary concern and scientists had collected little or no data on it. Just as quantitative data collected in the 1930s was critical in defining the scope of the problem and identifying solutions, data collection on nonpoint source pollution was the requisite foundation for abatement. Notwithstanding such achievements, the Willamette River's ability to function as a natural system was still in question. Researchers in the mid-1990s characterized biological conditions in the upper reach as "good," the middle reach as "fair," and the lower reach—Wilsonville through Portland—as "poor."[52] A key reason water quality remained poor was that raw sewage from Portland continued to find its way into the water. In spite of millions of dollars the City of Portland had spent on its sewage infrastructure in the 1940s, 1950s, and 1960s, through the 1970s and 1980s raw sewage still spewed into the river from combined sewer overflows (CSOs).

CSOs result from release of excess liquid from sewer lines that carry a combination of sewage and storm water. Engineers designed these release outlets so sewer lines would not become overloaded and back up into basements and city streets during periods of excessive rainfall and other events. Into the 1960s, engineers built overflow sewer outlets into the Willamette River and Columbia Slough as integral parts of Portland's interceptor, pumping station, and treatment system.[53] This was standard engineering practice throughout North America and Europe.[54] Beginning in the early 1960s, the City of Portland began building separated sewage and storm water runoff systems for newly incorporated areas of the city, negating the need for the same kinds of direct-to-the-waterway release valves.[55] The city retained its combined system infrastructure in areas built before the early 1960s, however. In early 1990, city officials initiated efforts to reduce the incidence of CSOs with the launch of their Clean River Program. They were motivated, in part, by increasingly more stringent federal requirements and growing public demand, including the NEDC's successful suit to speed up TMDL allocations for Oregon's watersheds. Commissioner Earl Blumenauer, who oversaw the city's sewer program as head of the Bureau of Environmental Services, observed "We are in the midst of a revolution in environmental regulations" requiring the city to act. The solution promised to be expensive: Bureau of Environmental Services staff found that redesigning infrastructure to address CSOs could cost up to $1 billion over twenty years and double sewer charges for city residents.[56]

In April 1991 Portland officials faced additional public pressure in the form of a lawsuit from another in a long line of citizens' groups pushing to clean the Willamette, the Northwest Environmental Advocates (NEA). Citizens created the NEA in 1969 to oppose the Trojan Nuclear Power Plant in Rainier, Washington, and later expanded their advocacy to include water and air quality issues, wetlands protection, and wildlife habitat within the Columbia River basin.[57] NEA representatives observed that CSOs into the Willamette occurred an average of every four days during the rainiest months. CSOs also occurred periodically during the summer when many more people used the river for recreation. The group's position was that, in spite of their Clean River Program, city officials were not progressing quickly enough; as attorney Craig Johnston stated, "We're only asking the city to do what is practical. All we are saying is the approach they are taking right now is entirely too relaxed." Fifty-four outfalls from the city's 1,800-mile sewer system annually expelled 913 million gallons of wastes, which were a mixture of untreated sewage, storm water, industrial wastes, sludge, and garbage.[58] To convey the effect of these wastes on the river, advocates used tactics similar to those of previous generations. NEA members presented a "dirty message" to the city council in August 1991 composed of "a jar full of murky water" from Portland Harbor containing toilet paper and "'brown gunk' and substances too distasteful to describe."[59]

Such filth from 913 million gallons of wastes is not insignificant, and considered on its own seems to suggest a drastic failure of the city's sewer infrastructure. Compared to the amount of wastes expelled into the river and Columbia Slough before Portland's sewage treatment plant came online in 1952, however, a mere 913 million gallons actually shows how effective the infrastructure was. Prior to completion of the city's primary treatment plant and interceptor lines, the city had been expelling 50 million gallons of raw sewage *each day, every day,* heavy rains or not. This equates to about twenty times more—an incredible 18.2 *billion* gallons—than was released annually in 1991. While this significant decrease was occurring, the City of Portland's population increased 30.1 percent, from 373,628 in 1950 to 485,975 in 1990.[60] After nearly sixty years of abatement efforts, forty years of service from the treatment plant, and two decades under the Clean Water Act and other environmental laws, the Willamette suffered from significantly less untreated sewage than at any time before white settlers began arriving in large numbers in the mid-nineteenth century. As the twentieth century came to a close, however, Portland residents had grown accustomed to a river that was no longer the "open sewer" of years past. They demanded a much higher level of water quality than previous generations expected.

This demand is reflected in the DEQ's summer 1991 decision to approve the City of Portland's Clean River Program to rectify CSOs. The program was projected to cost $1 billion over its twenty-year construction schedule. Nicknamed the "Big Pipe" project, Commissioner Blumenauer characterized it as "the most ambitious project of its kind in the nation."[61] With the DEQ's approval, US District Judge Owen M. Panner dismissed NEA's lawsuit. Deferring to the expertise of state water quality officials, Judge Panner said, "The fact that it will take 20 years is regrettable, however, that is the course of action approved by DEQ."[62] NEA Executive Director Nina Bell said the group's failed lawsuit was not in vain, because it "lit a fire not only under the city but under DEQ."[63] The Big Pipe project was expensive, time-consuming, and complex. It included $146 million to separate sewer and storm water systems, install approximately three thousand sumps to collect storm runoff, disconnect building downspouts from sewer lines, and divert streams such as Tanner Creek that previous city engineers had directed into combined sewer systems.[64] The program also alleviated nonpoint source pollution through a combination of urban infrastructure and engineering design elements. These included installing bioswales to absorb storm water and replacing conventional building roof designs with "ecoroofs"—a layer of foliage and a growing medium—to decrease runoff.[65] To handle large flows experienced during heavy rains, the signature elements of the Big Pipe project were two very large pipes on the east and west side of the harbor, running generally parallel to the shoreline. The former was twenty-five feet in diameter, nearly six miles long, and averaged 150 feet deep; commenced in 2006 and completed five years later, it was the largest sewer construction project in the city's history. The latter was fourteen feet in diameter, 3.5 miles long, and 120 feet deep; engineers broke ground in November 2002 and completed it in September 2006. The pipes were built to serve as interceptor sewer lines as well as underground reservoirs to store excess liquids for incremental release to the Columbia Boulevard Waste Treatment Plant in North Portland.[66]

Over the twenty-year construction schedule, CSOs still plagued Portland Harbor. Well into the twenty-first century, heavy rains and operator accidents spurred health officials to warn against swimming in the Willamette. In August 2007 health risks from a CSO threatened the cancellation of Portland's first downtown triathlon, and after heavy rains in May 2008, officials from Portland's Bureau of Environmental Services inadvertently paraphrased a state board of health official in 1906 when the agency warned against engaging in any recreational activities in the Willamette "during which water could be swallowed."[67] The DEQ also fined city officials for unauthorized raw sewage discharges not related to the ongoing Big Pipe project. In November 2005, the DEQ levied a fine of about $450,000 for sixty-seven discharges, totaling about 1,875,000 gallons, into the Willamette, Columbia Slough, and Columbia River between March 2001 and November 2005. Channeling the same sentiments and almost the exact words of her predecessor Harold F. Wendel from the 1940s, 1950s, and 1960s, DEQ Director Stephanie Hallock said that "while we appreciate the progress made by the city in constructing a new system to control pollution, it is essential that the city make every effort to prevent raw sewage discharges to the Willamette and its tributaries." This was the DEQ's seventh fine against Portland since 1998 for these kinds of discharges.[68]

The city celebrated the conclusion of the Big Pipe project in late November 2011, on time and on budget, with a final price tag of $1.4 billion.[69] It resolved a shortcoming in the city's infrastructure that generations of Portland city officials had actively or passively resisted for nearly the entire twentieth century. The project also addressed the final and most significant cause of point source pollution in the Willamette watershed, nearly seventy years after Commissioner William Bowes, Sanitary Authority Chair Harold Wendel, and others turned the first shovels of dirt on Portland's sewage treatment system. Completion of the Big Pipe certainly was cause for celebration, but it did not do anything to address pollution from toxics. It would take another multimillion dollar effort and federal assistance to begin that part of the Willamette River cleanup.

The Superfund program began on December 11, 1980, when President Jimmy Carter signed into law the Comprehensive Environmental Response, Compensation and Liability Act (CERCLA). The EPA had existed for ten years before CERCLA's passage gave the agency authority to prioritize and clean up hazardous waste sites and spills. Legislators supported CERCLA because, as EPA Administrator Douglas M. Costle put it, "a rash of recent incidents resulting from improper disposal of hazardous wastes has made it tragically clear that faulty hazardous waste management practices, both past and current, present a grave threat to public health and to the environment." Most notable among the "rash of recent events" was the highly publicized

Love Canal incident of the late 1970s in Niagara, New York.[70] Superfund was an entirely different kind of law than the successive federal water pollution control laws and the National Environmental Policy Act, and was intended to address different kinds of pollution. The Superfund process is complex. EPA implements CERCLA by fostering involvement from state agencies and the affected communities, compelling responsible parties to contribute materially to the cleanup process, and overseeing remediation. CERCLA provides for a $1.4 billion trust fund to help finance the work.[71] Since 1980, Superfund projects have ranged from limited-duration events such as cleanup of the Space Shuttle Columbia disaster in 2003 to long-term identification and remediation of extensive hazardous waste contamination from accidental spills, illegal dumping, and unregulated industrial processes. The Portland Harbor Superfund Site is one of these latter projects.[72]

TOXICS: THE PORTLAND HARBOR SUPERFUND SITE

EPA designated the six-mile stretch of Portland Harbor between Swan Island and Sauvie Island a Superfund site on December 1, 2000. Observers had been expecting this announcement for months: Governor Kitzhaber had agreed to the listing in July, and three years previously EPA scientists had found high levels of tars, pesticides, polychlorinated biphenyls, and metals in river sediments. Further, since 1989 public and private parties had spent more than $56 million to clean up individual industrial sites, including two smaller, ongoing Superfund cleanups in the harbor at a creosote-treatment facility and a former battery recycling plant. Upon designation of the Portland Harbor Superfund Site, EPA set about notifying landowners along the river and commencing negotiations with these potentially responsible parties.[73] The agency also began community outreach. Beginning in the late 1930s, Oregon State Sanitary Authority meetings had been open to the public; CERCLA built upon this precedent by containing provisions to incorporate diverse public input into the cleanup process in a much more proactive way. An EPA directive from 1995 provided the framework for citizens to form community advisory groups. Impetus to form such groups was to come from the local communities—the EPA did not mandate their formation—but, when established, community advisory groups had a recognized seat at the table and channeled citizen input into all phases of the negotiation and planning processes.[74]

The EPA invited concerned Portland citizens to a meeting in March 2002, thereby initiating formation of the Portland Harbor Community

Advisory Group (PHCAG). The group's mission was to "ensure a Portland Harbor Cleanup that restores, enriches, and protects the environment for fish, wildlife, human health, and recreation, through community participation." The PHCAG includes representatives from neighborhood associations, environmental groups, businesses, and others. Represented at the March 2002 meeting were groups that had long been involved in Willamette River cleanup efforts, such as the League of Women Voters and the DEQ. It also included newer groups, such as Willamette Riverkeeper, founded in 1996 as "the only organization dedicated solely to the protection and restoration of the Willamette River" with advocacy efforts focused on improving wildlife habitat and water quality.[75] With the Superfund process still ongoing in early 2016, the PHCAG's work was ongoing as well, and included regular monthly meetings, special events (such as boat tours of Portland Harbor), cosponsored public education sessions, and a website providing historical and current information. The PHCAG engages with the EPA, DEQ, Port of Portland, City of Portland, the Grand Ronde tribe, and a consortium of fourteen "potentially responsible parties" known as the Lower Willamette Group.[76] "Potentially responsible party" is a legal definition under CERCLA referring to companies, municipalities, or other entities that may be held liable for cleanup costs based on the type and extent of contamination they contributed to a given site.[77] The Lower Willamette Group is composed of eighteen potentially responsible parties (out of more than eighty), such as the City of Portland, Port of Portland, Chevron USA, Bayer CropScience, Inc., Phillips 66 Company, and Union Pacific Railroad, among others.[78]

Among many pollutants the EPA, DEQ, and PHCAG are trying to address, the dioxin family of petroleum-based compounds are the most harmful to humans and wildlife. Groups of compounds within this family share a common chemical structure and toxic effects and include polychlorinated dibenzo dioxins (PCDDs), polychlorinated dibenzo furans (PCDFs), and polychlorinated biphenyls (PCBs). Manufacturers produced these chemicals in large volumes for a wide array of applications, such as herbicides, hydraulic fluids, and insulator fluids for heat exchangers and electric transformers. Dioxins were deposited in river sediment from decades of minimally regulated industrial processes within Portland Harbor, such as herbicides produced in the 1960s and 1970s for US military use during the Vietnam War.[79] Dioxins are present within and beyond the Superfund site from activities outside the harbor. The contaminant can be found in the particulates from incinerated solid waste and coal-fired power plants, pulp and paper mill effluents containing bleach, and herbicides applied in Oregon's managed forests.[80]

Prior to the 1970s, virtually no state or federal regulations governed dioxin use and disposal, with the result that by the second decade of the twenty-first century they are widely distributed throughout the world. Detectable levels of dioxins can be found in the tissues of most people on the planet, whether or not they live anywhere near an area where dioxins were manufactured, used, or deposited.[81] Dioxins bioaccumulate, which means that once they are in animal tissues they do not degrade or get removed, and any additional exposure adds to the total amount already present. The probability of cancer grows with increased dioxin accumulation in living tissue, and so does the chance for noncancer effects such as immune system suppression, hormone level changes, and fetal development alterations; in high doses, dioxins can also cause a skin disease known as chloracne, a broad term for a serious kind of dermatitis.[82] Specialists noticed the condition in some industrial workers in the late nineteenth century and linked it with the use of cutting oils and solvents as early as World War I. Similar to chloracne was "cable rash" or "Halowax dermatitis" found in World War II shipyard workers, a condition researchers in the mid-1940s also found among "those engaged in the manufacture of the chlorinated naphthalenes and diphenyls, as well as in men in other occupations in which these substances are employed."[83] Specialists learned of these effects gradually. Long before they knew about dioxins' persistence in the environment, they recognized evidence of harm in people who worked in shipbuilding and other industries that involved exposure to a new and growing array of petroleum-based solvents and other chemicals. It was not until the 1950s that experts directly linked such conditions to chlorinated hydrocarbons—dioxins—and it was not until the Vietnam War that people identified the propensity of these chemicals to accumulate and persist in the environment. In 1970, the US Department of Defense (DoD) stopped using the herbicide "Agent Orange" in Vietnam. Agent Orange was a mixture of the chemicals 2,4-D and 2,4,5-T, the latter of which contained the highly toxic dioxin 2,3,7,8-TCDD.[84] With this event, the word *dioxin* entered the broader public discourse for the first time in the form of news articles in the *New York Times, Oregonian*, and other newspapers.[85] In spite of the DoD's proscription for use in Vietnam, manufacturer Dow Chemical successfully lobbied against a federal ban of this herbicide in the US until the 1980s.[86]

The US Forest Service in Oregon had been using herbicides containing 2,4,5-T in the Siuslaw National Forest since the 1960s. The National Forest included a patchwork of lands that stretched along about 170 miles of Oregon's Coast Range from Tillamook in the north to Coos Bay in the south. Though most of these lands drained west directly into the Pacific Ocean, the National Forest boundaries included some streams that made their way east to the Willamette watershed. Dioxin-laden herbicide use in this area is tangentially linked to Willamette River pollution, but the larger narrative of the science and politics of dioxin detection and regulation relates directly both to the Portland Harbor Superfund Site and the valley's pulp and paper industry.

In 1977, Harvard biologists published a study indicating evidence of dioxin in the breast milk of a woman from the Five Rivers area near Waldport on the Oregon coast.[87] Scientists from universities and the chemical industry who sought to continue herbicide manufacturing and use questioned the study's accuracy. Taking the opposite view, area residents formed a citizen's group seeking to stop spraying in the nearby Siuslaw National Forest based on the Harvard study and previous concerns about dioxin-containing herbicide use. Later that year, the EPA's dioxin monitoring committee convened a conference to decide upon a plan for more environmental sampling and data gathering.[88] It would be ten years, however, before data about dioxin pollution in Oregon became public—and only in a roundabout way. In 1987, a federal magistrate in Eugene ordered the EPA to "release a document detailing levels of dioxin around the country—including those detected in Willamette River fish—or to state why it should legally be able to withhold release of the document." The judge was referring to an as-yet unreleased 1985 EPA study that had found "low levels" of dioxin at three Oregon locations, and he rendered his decision in response to a motion from writer and advocate Carol Van Strum. Van Strum claimed that information she received from an unnamed EPA scientist "reports that tissue samples taken from Willamette River fish show dioxin levels more than four times as high as levels detected in fish taken from the Great Lakes area."[89] The EPA countered that its data did "not constitute an environmental emergency." An EPA spokesperson stated the agency "had notified Oregon that dioxin had been found in 1985 in fish from the Willamette River in Portland, soil from rural Linn County, and creek sediments from a forest east of Salem." Dioxin found was the most toxic form: 2,3,7,8-TCDD. The highest concentration of this toxin from all three sites was 4.5 parts per trillion in two fish caught near the Southern Pacific Railroad

Bridge in Portland Harbor, upstream from St. Johns Bridge.[90] This stretch of the river is within what later became the Portland Harbor Superfund Site.

Oregonian correspondent Carmel Finley filed the newspaper's first-ever report that linked dioxins directly to the pulp and paper industry. In September 1987, Finley reported on EPA rule changes prompted by discovery of dioxin in fish downstream of the James River Paper Company mill at Wauna, Oregon, about seventy-five miles west of Portland on the Columbia River, and at four other US mills. Because of these findings, the EPA commenced a study of ninety mills nationwide, including three Oregon mills that used the kraft (sulfate) chemical bleaching process. Two of these three mills were within the Willamette and Lower Columbia river watersheds: the Pope & Talbot (formerly American Can Company) facility on the Willamette River near Halsey and the Boise Cascade operation in St. Helens at the confluence of the Columbia River and a tributary of the Willamette, Multnomah Channel.[91] The kraft process contributes less biochemical oxygen demand than the sulfite process because cooking liquors are more readily reused. However, bleached kraft pulp produces elevated amounts of a range of chlorinated organic compounds—including dioxins such as 2,3,7,8-TCDD—when molecular chlorine reacts with wood sugars (lignins). The discovery of dioxins in these wastes "was a new area of concern," as Finley quoted the chief of the EPA's Water Quality Analysis Branch.[92] In spite of the fact that TCDD specifically, and dioxins generally, are the most carcinogenic substances humans had yet created, Pope & Talbot Mill Manager Walt Sinclair downplayed the potential for contamination, stating that "the very minute amount they're finding almost defies science to detect it." At the same time, however, he admitted, "I don't think our society will stand to have any of this stuff in its water."[93] Carmel Finley's article also referred to a recently published report from Carol Van Strum and Paul Merrell, *No Margin of Safety*. It was based on records received from an anonymous EPA scientist suggesting a high-level agreement between EPA and pulp and paper industry executives to withhold information linking the industry with dioxin contamination.[94] *New York Times* correspondent Philip Shabecoff, reporting on the 1985 EPA study at the same time, wrote the EPA had found traces of dioxin in many paper products:

> The Environmental Protection Agency disclosed today that a Congressionally mandated study of dioxin contamination across the country had found less contamination than officials expected in land

and water. But data on water pollution from paper mills, gathered in the course of this broader study, led industry and Government researchers to focus on the possibility of paper contamination.[95]

Once this information was made public, EPA officials were put in a position of having to develop regulations regarding dioxin levels in pulp and paper industry wastes. Though these kinds of pollutants were not as visible as untreated sewage or industrial wastes, they were more harmful to humans, animals, and the environment. EPA's shift in focus from collecting data to being compelled to develop effective rules and standards influenced governments throughout North America and beyond to revise their pulp and paper industry regulations. It also spurred the industry, by the late 1990s, to reduce significantly their use of chlorine for bleaching pulp.[96]

The undaunted commitments of David Charlton, Edgar Averill, William Finley, Tom McCall, and many others enabled Don Holm to harvest crawdads in the mid-1970s in waters restored to a minimum level of dissolved oxygen and free from coliform bacteria. However, crawdads taken from the future Portland Harbor Superfund Site almost certainly contained dioxins and a long list of other toxic chemicals from bleached pulp processes, herbicide manufacture and application, shipbuilding, runoff from streets and fields, airborne particulates, and many other sources. Portland largely resolved its combined sewer overflow issue by completing its Big Pipe project in late 2011, but as of early 2018 remediating the Portland Harbor Superfund Site is still a work in progress and promises to be a long and complex challenge. Oregonians continue to face what *Oregonian* editors, in 1925, had fittingly called "Hydra-Headed Pollution."[97]

Conclusion
Speaking for Ourselves

The mirthful image of bass-drum-carrying Mayor Joseph Carson Jr., smiling with others at a Demand Clean Streams rally in Portland in late 1938 has become iconic in Oregon's environmental history. It speaks to the success of passing two significant water quality proposals after more than twelve years of grassroots efforts. This was no small feat. William Joyce Smith's color 1940 film, slime-coated nets in the Columbia River, and asphyxiated crawfish later in the 1940s illustrate how difficult it would be to translate good intentions into tangible, lasting results. Tom McCall's 1962 documentary *Pollution in Paradise* is another icon, helping refocus public attention on the state's degraded environment and putting pressure on state leaders to do more. It came after a decade in which the sanitary authority had made measurable progress against polluters but was still only able to put river pollution control on a treadmill. McCall, as governor, strengthened abatement efforts by taking an active public role and working closely with Republican and Democratic legislators to increase funding and reorganize the state's approach to environmental administration. When his second term ended in January 1975, Oregonians could reflect upon abatement work that alleviated the most egregious kinds of point source pollutants, such as untreated municipal sewage and pulp and paper mill wastes.

The achievements of McCall, his allies in government, and many other Oregonians in the 1960s and 1970s helped propel Oregon to the national forefront in such areas as land-use planning, recycling, public access to beaches, and environmental protection. Oregon's success in addressing water pollution has served as a national example, most notably encapsulated in Ethel Starbird's 1972 *National Geographic* article "A River Restored." The state generally, and Portland in particular, continues to have strong "green" credentials well into the twenty-first century. The twenty-year project to address Portland's combined sewer overflows and ongoing, expensive, and complex Portland Harbor Superfund cleanup program, however, illustrate that Willamette River water pollution has not gone away entirely. These examples suggest pollution of one form or another will remain a characteristic of the river for decades—if not centuries—to come. Neil Mullane's qualification of Starbird's article in the late 1990s is justified: "A River Restored?"

The vast majority of Oregon citizens involved in cleanup efforts into the 1970s wanted to create a healthier "working river," not a pristine, pre-contact watershed.[1] Even the most active members of the abatement community conceived of the river in terms of a collective resource to be used for a variety of purposes, and in this they shared a strong commitment to place. A broad spectrum of citizens supported these efforts to sustain human and piscatorial health and allow for future recreational and economic opportunities. David Charlton, William L. Finley, Edgar F. Averill, William Joyce Smith, Harold Wendel, John C. Veatch, and Kenneth Spies are a few among the many Oregonians long involved in the issue. They worked through the sanitary authority, board of health, and fish and game commissions, and also as members of groups such as the Izaak Walton League, Anti-Stream Pollution League, Advisory Committee on Stream Purification, and the Oregon Stream Purification League. Their motivations were diverse. Lifelong conservationists such as naturalist William L. Finley were as willing as chamber of commerce representatives to support pollution abatement using economic arguments. Editors at the *Oregon Journal* and *Oregonian* supported clean stream efforts with economic arguments but also, over the decades, used increasingly strong moral language to critique companies supplying the paper upon which their editorials were printed. Portland City Commissioner Ormond Bean spearheaded an initiative to fund city sewerage infrastructure in no small part to employ people during the Great Depression, yet Bean and other officials fought the sanitary authority in the early 1940s to avoid collecting the funds necessary to employ veterans after World War II. David Charlton provided leadership by forging cooperative relationships with a variety of professionals and state officials across the nation while concurrently lambasting his own state's sanitary authority for not *forcing* the pulp and paper industry to abate pollution immediately.

Charlton was foremost among abatement advocates criticizing the sanitary authority in the 1940s, 1950s, and 1960s. Decades later, some scholars still critique the authority's work. Brent Walth, for example, concludes that "overall the Sanitary Authority did a good job of convincing Oregonians that the Willamette was . . . growing cleaner all the time" when, in fact, it was not.[2] This interpretation implies authority members knew they were not making progress yet engaged in subterfuge by falsely claiming otherwise. As the preceding pages have shown, these implications are not valid, because the authority *was* making progress. It is easy, in hindsight, to deem as inadequate the pace and scale of this progress—and David Charlton and his Izaak Walton League colleagues certainly were not satisfied in the late 1940s—but considering the context within which the authority operated, it made identifiable progress. Historian William G. Robbins is more accurate when he argues that the sanitary authority did not achieve all it might have because of a lack of political will at the state level until Tom McCall became governor in the mid-1960s and increased funding significantly.[3] As earlier evidence indicated, however, Oregon was far from alone in lacking this political will. The authority's limited success into the 1960s was compounded by the constraints of legislation, funding allocations, and staffing levels. In such an environment it applied a cooperative, pragmatic approach to entreat and cajole polluters into following state law rather than coming down hard on them with threats of litigation. While not strong on enforcement, it was the approach of choice in Oregon and elsewhere because, in part, direct judicial oversight of water pollution had proven ineffective. Costly lawsuits redressed grievances retroactively but did not establish proactive water quality standards or administrative practices. Additionally, as was common among sanitary engineering professionals worldwide, the water pollution problem seemed solvable, given advances in scientific understanding and faith in expertise and technology.

Abatement methods and technologies evolved along with advances in scientific knowledge. Sanitary engineers through the 1960s—and to this day, in some respects—considered rivers, streams, and lakes as integral parts of urban sanitation infrastructure. Applying the "assimilative capacity" concept was the first attempt to provide a rational scientific framework to waste disposal and was a significant departure from the casual approaches that predominated through the nineteenth century. Making bodies of water serve multiple purposes, including diluting wastes, was a "reasonable use" of the resource. Advocates had to quantify the effects and sources of pollution before taking other steps. Complex technical and regulatory approaches required good supporting data: no steps could be taken until the issue, parameters, and possible solutions were put into quantitative, empirical terms. Such approaches were costly and so also required money and resources to implement. Research was needed to justify legislation and expenditures, but science alone was not enough to drive abatement forward. If the broader public did not perceive pollution as a nuisance until their accustomed economic and recreational activities were curtailed, then abatement advocates devised public education campaigns to convince them of the need to fund expensive measures. The many individuals and organizations involved in Willamette River water pollution abatement took advantage of the trials and successes in other locations. Events in Oregon proceeded with the benefit of—and contributed to—advances in the rest of the nation.

Funding sufficient to meet its many and ever-evolving challenges was a consistent issue for the sanitary authority. It is true that sufficient resources to move forward with large-scale infrastructure projects were not always easy to come by during the Great Depression and World War II, even considering New Deal funding opportunities. Lack of money may have stalled abatement work, particularly during the first decade of the authority's existence. However, a compelling question to ask is: What changed among a critical mass of people to get them to allocate the significant resources required to address pollution? In the case of water and other kinds of environmental pollution, degradation combined with ideology, morality, and economics to spur changes in individual and social behavior. The negative environmental effects constrained choices, which then caused discomfort, and the resulting discomfort led people to act for change.

With the knowledge and resources available to them, authority members found it easier to enact changes by getting cities to intercept and treat their sewage than it was to mandate that pulp and paper wastes be treated. Sewage represented more of a public health threat than pulp mill wastes, so it came under the purview of the earliest state water quality laws and was still a pressing concern when the authority began its work in early 1939; there were no such laws pertaining to mill wastes. Effective municipal sewage treatment systems were available by the 1910s, but technologies to treat or reuse sulfite wastes were still being developed into the mid-1950s. This meant the authority not only could call upon state law requiring sewage treatment, but its engineers could readily refer to system specifications and establish project budgets. Beginning in the late 1940s, the OSSA, through the federal government, was able to subsidize engineering, planning, and implementation of municipal sewage infrastructure, an active step that government was not empowered to do for private enterprises. Finally, these private enterprises sometimes used the threat of shutting down operations and moving elsewhere if they felt overly pressed by regulations and enforcement. Whether these threats were real or merely perceived, this kind of pressure was not something city officials were apt to ignore.

Many people suggest that modern civilization's propensity to reduce ecosystems into numbers and develop complex technologies to "control" the environment breaks the essential bonds humans have with nature. By using science, technology, and economic philosophies to transform nature into "organic machines," societies end up "seeing like a state" as they enforce digitization upon an inherently analog system. With such bonds compromised, humans then make decisions that undermine the long-term viability of the natural systems upon which they rely. By mechanizing, systematizing, and otherwise quantifying complex natural systems into simplified units, humans jeopardize the long-term viability of the species itself—or at least disadvantage those who cannot afford to avoid the negative consequences of pollution, public health threats, economic decline, and the like.[4] This dynamic has been a consequence of capitalist economic systems as much as it has been of command economies such as communism. While not without some merit, the Willamette River pollution abatement story calls into question the universality of such a simplistic view. Human actions led to the river's degradation, but human interventions were critical in cleaning it up. Advances in water quality science led to the quantification of oxygen-depleting wastes, which then served as the empirical data needed to establish laws to regulate waste disposal proactively; Willamette Valley Project tributary dams have contributed to declining Chinook salmon populations, but are essential in establishing minimum flow levels to maintain sufficient dissolved oxygen amounts, keeping water temperatures low, and carrying wastes downstream. The Willamette Valley is not the only location where such dynamics have played out, and water pollution is not the only environmental issue in which this kind of pattern can be observed, but these and other examples show how regulating and "taming" the Willamette has delivered benefits, such as helping clean up a seriously polluted river.

The Willamette River pollution abatement story is incomplete if it does not integrate the entwined scientific, technological, political, and economic factors that interacted over many decades within a dynamic, multivariate environmental system. It is also incomplete if it does not recognize explicitly the many dedicated people who worked through various official and unofficial organizations to force fundamental institutional changes in Portland city government and in pulp and paper industry waste disposal practices. Abatement advocates struggled to understand, identify, and rectify pollution sources and effects. They struggled to quantify water quality, they struggled to learn and apply abatement methods and technologies, and they struggled to understand the dynamic watershed. With a view of history that does not include such details, we are much less likely to perceive the vital importance of an engaged citizenry, the necessity of quantitative evidence to support environmental policies, and the requirement for complementary approaches at the local, state, national, and international levels. We also run the risk of wrongfully believing that fundamental economic, political, and environmental changes can be realized with minimal effort, or that a "Great Man" will rise from the mists of history on a white steed to rescue society from the brink of ruin.

What we learn from the Willamette River cleanup story is that both the environment and we actors upon its stage are much more complex than we generally suppose. Equipped with this knowledge, we can now stand at the top of Willamette Falls at Oregon City and perceive not just an industrialized landscape but the deep geologic events that laid the foundation for an environment that long supported Chinooks and Kalapuyans and called to European American immigrants. We can perceive as well the thriving eel and salmon fisheries at the base of the falls transformed into a death-dealing cesspool by the early twentieth century and then, gradually, transformed back into something approaching healthfulness by century's end. Trying to understand the complexity hidden in such a landscape is a requirement for informed decisions in the present that will, ideally, lead to future improvements. In claiming to speak for the river we are, in fact, speaking for ourselves. Welcome to this long and complex story about confronting pollution in Oregon's Willamette River watershed.

Appendix 1
Notable Willamette Valley and Lower Columbia River Pulp and Paper Mills in the Willamette River Pollution Story

This table summarizes data found in the sources referenced for each individual location, plus a number of general sources.[1]

Location Chemical Process	Established	Names & Owners
Oregon City, OR[2] sulfite	1866	Pioneer Paper Manufacturing Company (1866–1867)
	1908	Hawley Pulp and Paper (1908–1948)
		Publishers' Pulp and Paper Company (1948–1986)
		Jefferson-Smurfit Corporation (1986–2000)
		Blue Heron Paper Company (2000–2011)
Camas, WA[3] sulfite	1884	Columbia River Paper Company (1884–1905)
		Crown Columbia Paper Company (1905–1914)
		Crown Willamette Paper Company (1914–1928)
		Crown Zellerbach Corporation (1928–1986)
		James River Corporation (1986–1997)
		Fort James Corporation (1997–2000)
		Georgia-Pacific Corporation (2000–Present)

Location Chemical Process	Established	Names & Owners
West Linn, OR[4] sulfite	1889	Crown Paper Company (1889–1905)
		Crown Columbia Paper Company (1905–1914)
		Crown Willamette Paper Company (1914–1928)
	1889	Willamette Pulp and Paper Company (1889–1914) [founded as Willamette Falls Pulp and Paper Company]
		Crown Willamette Paper Company (1914–1928)
		Crown Zellerbach Paper Company (1928–1986)
		James River Corporation (1986–1990)
		Simpson Paper Company (1990–1996)
		West Linn Paper Company (1997–Present)
Lebanon, OR[5] sulfite	1889	Lebanon Paper Mill [founded as O'Neil's and Callaghan's Mill) (1889–1905]
		Lebanon Paper Company (1905–1906)
		Willamette Pulp and Paper Company (1906–1914)
		Crown Willamette Paper Company (1914–1928)
		Crown Zellerbach Paper Company (1928–1980)
Salem, OR[6] sulfite	1920	Oregon Pulp and Paper Company (1920–1962) [under holding company Columbia River Paper Company]
		Boise Cascade Corporation (1962–2007)
Vancouver, WA[7] sulfite	1921	Columbia River Paper Mills (1921–1962) [under holding company Columbia River Paper Company]
		Boise Cascade Corporation (1962–1996) [Pulp production ceased in 1969, paper production ceased in 1996]
St. Helens, OR[8] Sulfate (kraft)	1926	St. Helens Pulp and Paper Company (1926–1953)
		Crown Zellerbach Paper Company (1953–1964)
		Boise Cascade Corporation (1964–2008)
		Boise, Inc. (2008–2012)
		Cascades Tissue Group (2012–Present)

Location Chemical Process	Established	Names & Owners
Newberg, OR[9] sulfite	1927	Spaulding Pulp and Paper Company (1927–1965)
		Publishers' Paper Company (1965–1986)
		Smurfit Newsprint Corporation (1986–1999) [owned by Jefferson Smurfit Corporation]
		SP Newsprint Company (1999–2015) [owned by Southeast Paper Manufacturing Company (1999–2012); SP Fiber Technologies (2012–2015); WestRock Company (2015–Present)]
Springfield, OR[10] sulfate (kraft)	1949	Weyerhaeuser Timber Company (1949–2008)
		International Paper Company (2008–Present)
Albany, OR[11] sulfate (kraft)	1955	Western Kraft Corporation (1955–2002) [merged with Willamette Industries, Inc., 1973]
		Weyerhaeuser (2002–2008)
		International Paper Company (2008–2009)
Halsey, OR[12] sulfate (kraft)	1969	American Can Co. (1968–1981) [Pope & Talbot secured 50 percent share in 1978, full ownership in 1981]
		Pope & Talbot (1981–2008)
		Cascade Pacific Pulp, LLC (2008–2010) [owned by Wayzata Investment Partners (2008–2010), International Grand Investment Corporation, (2010–Present)]

1 W. Claude Adams, "History of Papermaking in the Pacific Northwest: I," *Oregon Historical Quarterly* 52:1 (March 1951), 21–37; W. Claude Adams, "History of Papermaking in the Pacific Northwest: II," *Oregon Historical Quarterly* 52:2 (June 1951), 83–100; W. Claude Adams, "History of Papermaking in the Pacific Northwest: III," *Oregon Historical Quarterly* 52:3 (Sept. 1951), 154–185; George W. Gleeson and Fred Merryfield, *Industrial and Domestic Wastes of the Willamette Valley, Bulletin Series, No. 7* (parts 1, 2, and 3) (Corvallis, Oregon State Agricultural College Engineering Experiment Station, [May] 1936); John H. Lincoln and Richard F. Foster, *Report on Investigation of Pollution in the Lower Columbia River*, Salem, Ore.: Interstate Technical Advisory Committee, 1943; and Raymond M. Miller, *The Pulp and Paper Industry of the Pacific Northwest, Part 1*. Portland: US Army Corps of Engineers, Office of the Division Engineer North Pacific Division, 1937.

2 "Blue Heron Paper Company: History," http://www.blueheronpaper.com/about_hist.html, accessed Nov. 13, 2012.

3 "Continuity and Change at the Camas Paper Mill," Center for Columbia River History, http://www.ccrh.org/comm/camas/millhist.php, accessed Nov. 13, 2011; Lori-Ann S. Young, "The Paper Mill and the City of Camas: Is Camas Still a Mill Town?" n.d. [ca. 2000], http://www.ccrh.org/comm/camas/student%20papers/camas%20mill.htm, accessed Nov. 13, 2012; "Camas Mill/Mill History." Georgia-Pacific, http://www.gp.com/camas/millHistory.html, accessed Nov. 13, 2012.

4 "West Linn Paper Company: History," http://www.wlinpco.com/company/history.aspx, accessed Nov. 13, 2012.

5 "Crown to Shut Down Old Lebanon Pulp Mill," *Oregonian*, July 11, 1980, A19.

6   "Columbia Paper Deal Completed," *Oregonian*, Apr. 16, 1962, sec. 1, 1; "Oregon Pulp Sells New Issue,"
    Morning Oregonian, July 27, 1927, sec. 1, 17; "Salem Investors Buy Key Waterfront Site in Oregon's
    Capital," *Oregonian*, Dec. 27, 2007.
7   Allan Brettman, "Boise Cascade's Columbia River Site Draws Looks," *Oregonian*, Sept. 17, 2002, B1;
    "Columbia Paper Deal Completed," *Oregonian*, Apr. 16, 1962, sec. 1, 1; "Oregon Pulp Sells New Issue,"
    Morning Oregonian, July 27, 1927, sec. 1, 17.
8   Kelly Moyer, "Will Boise close St. Helens mill?" *Portland Tribune*, Sept. 24, 2008 [updated Oct. 30, 2009];
    "St. Helens Deal Goes into Effect," *Oregonian*, June 2, 1953, sec. 2, 12; "St. Helens Mill Sold by Crown,"
    Oregonian, April 11, 1964, sec. 1, 1; "St. Helens Paper Plant Retools; Expands Adding 29 New Jobs,"
    Business Oregon, n.d., http://www.oregon4biz.com/story.php?storyID=212, accessed April 24, 2016; Shari
    Phiel, "Cascades Tissue Group Looks to Expand," *The St. Helens Chronicle*, Sept. 17, 2013; Amy Hsuan,
    "Halsey Pulp Mill Finds New Buyer in China," *Oregonian*, Feb. 4, 2010, Business Section; Amy Hsuan, "Hong
    Kong corporation buys Oregon pulp mill," *Oregonian*, Feb. 4, 2010; Amy Hsuan, "Millworkers Reapply for
    Their Jobs," *Oregonian*, June 19, 2008, C1; "Pope & Talbot Mulling Move." *Oregonian*, April 16, 1981, D11;
    "Pope & Talbot: Diversity Ends Adversity." *Oregonian*, Oct. 12, 1987, D10–D11.
9   "Firm Eyes Newberg Mill." *Oregonian*, March 16, 1965, sec. 1, 21; Jim Kadera, "Smurfit Millworkers Accept
    New Company Offer." *Oregonian*, Mar. 31, 2000, D2; "Smurfit Replacing Publishers in NW." *Oregonian*, June
    3, 1986, D13; David Bates, "New Owner Shutting Down Newberg Mill," *Yamhill Valley News Register*, Oct.
    19, 2015; and Allan Brettman, "Report: There May Be Hope Yet for Newberg Paper Mill," *Oregonian*, Oct.
    19, 2015.
10 "Weyerhaeuser Completes Sale of Assets to International Paper," Aug. 4, 2008, http://www.weyerhaeuser.
    com/Company/Media/NewsReleases/NewsRelease?dcrId=08-08-04_
    WeyerhaeuserCompletesSaleofAssetstoInternationalPaper, accessed Nov. 13, 2011.
11 Merlin Blais, "Business Briefs: Woods World," *Oregonian*, May 22, 1955, sec. 1, 42; Amy Hsuan, "Paper Mill
    to Close Near Albany, Putting 270 Out of Work," *Oregonian*, Oct. 23, 2009; "International Paper to Shut
    Down Three Mills: 270 Jobs from Albany, Oregon," Salem-News.com, Oct. 22, 2009; Steve Lathrop, "Final
    Shift Leaves Albany Paper Mill," *Albany Democrat-Herald*, Dec. 24, 2009; "Weyerhaeuser Completes Sale of
    Assets to International Paper," Aug. 4, 2008, http://www.weyerhaeuser.com/Company/Media/
    NewsReleases/NewsRelease?dcrId=08-08-04_WeyerhaeuserCompletesSaleofAssetstoInternationalPaper,
    accessed Nov. 13, 2011; "Weyerhaeuser to Thin Out Jobs," *Oregonian*, June 1, 2002, B1.
12 "Cascade Pacific Pulp, LLC: Mill History," www.cascadepulp.com/documents/history.html, accessed Nov. 13,
    2011.

Appendix 2
Willamette Valley Project Dams

This table centralizes information about each of the thirteen Willamette Valley Project dams, including their capacity, year they were first operational, and primary uses.[1] The "Storage Capacity: Volume" column provides figures in acre-feet; 3.07 acre-feet equals 1 million gallons. The "Storage Capacity: Percent" column provides figures as a percent of total volume of all thirteen Willamette Valley Project dams, 2,016,000 acre-feet.

Dam	Tributary	Storage Capacity: Volume	Storage Capacity: Percent	Year Operational	Primary Use(s)
Fern Ridge	Long Tom River and Coyote Creek	116,800	5.8	1941	Flood control storage
Cottage Grove	Coast Fork Willamette	32,900	1.6	1942	Flood control storage
Dorena	Coast Fork Willamette	77,600	3.8	1949	Flood control storage
Detroit	North Santiam	455,100	22.6	1953	High-head structure with considerable flood control and hydropower capacity
Lookout Point	Middle Fork Willamette	455,800	22.6	1953	High-head structure with considerable flood control and hydropower capacity
Big Cliff	North Santiam	n/a	0	1953	Smaller dam to help regulate downstream effects of large releases from upstream dam (Detroit)

Dam	Tributary	Storage Capacity: Volume	Storage Capacity: Percent	Year Operational	Primary Use(s)
Dexter	Middle Fork Willamette	n/a	0	1954	Smaller dam to help regulate downstream effects of large releases from upstream dam (Lookout Point)
Hills Creek	Middle Fork Willamette	355,500	17.6	1961	High-head structure with considerable flood control and hydropower capacity
Cougar	McKenzie	219,000	10.9	1963	High-head structure with considerable flood control and hydropower capacity
Fall Creek	Middle Fork Willamette	125,000	6.2	1966	Flood control storage
Foster	South Santiam	60,700	3.0	1966	Smaller dam to help regulate downstream effects of large releases from upstream dam (Green Peter)
Green Peter	South Santiam	28,100	1.4	1966	High-head structure with considerable flood control and hydropower capacity
Blue River	McKenzie	89,500	4.4	1968	High-head structure with considerable flood control and hydropower capacity

1 "Willamette Basin Reservoirs: An Overview of US Army Corps of Engineers Dams and Reservoirs on Willamette River Tributaries," n.d., www.oregon.gov/OWRD/docs/WillametteReservoirs.pdf, accessed Dec. 31, 2011; Douglas W. Larson, "Reservoir Limnology in the Pacific Northwest: Willamette River Basin, OR." *Lakeline* 21:4 (Winter 2001/2002), 11–16; US Army Corps of Engineers, "Willamette Valley Project," n.d., http://www.nwp.usace.army.mil/locations/willamettevalley.asp, accessed Dec. 31, 2011.

Appendix 3
North American Water Quality Entities as of 1949

The information in this table is taken largely from a 1948 article in the professional journal *Chemical Engineering Progress*.[1] The table shows that seventeen of fifty-seven possible jurisdictions—individual US states, the District of Columbia, and Canadian provinces—had enacted laws centralizing water quality oversight and enforcement powers within a single entity.

The two dates listed for Washington State reflect an evolution toward a centralized state-level entity. In fall 1937, Governor Clarence D. Martin established the Washington State Pollution Control Commission under the state department of health. It was primarily a technical commission that carried out surveys and research on both municipal and industrial pollution, but had no enforcement powers. A 1945 state law "invested in the Pollution Control Commission the power to adopt, prescribe, and promulgate the rules, regulations, and standards necessary to carry out the purposes of the Act."[2]

Jurisdiction	Year Enacted
Pennsylvania	1923
Connecticut	1925
Wisconsin	1925
Illinois	1929
South Dakota	1935
Manitoba	1935
New Jersey	1937
Oregon	1938
Arkansas	1941

Jurisdiction	Year Enacted
Indiana	1943
Minnesota	1945
Washington	1945 (1937)
Mississippi	1946
Virginia	1946
West Virginia	1946
Colorado	1947
California	1949

1 Anthony Abele and R. P. Kite, "Pollution Abatement: Appraisal of Current Regulations," *Chemical Engineering Progress* 44:1 (Jan. 1948), 3–16. Other important sources include: Benjamin Ross and Steven Amter, "Deregulation, Chemical Waste, and Ground Water: A 1949 Debate," *Ambix* 49:1 (March 2002), 51–66; Arn Keeling, "Sink or Swim: Water Pollution and Environmental Politics in Vancouver, 1889–1975," *BC Studies* 142/143 (Summer/Autumn 2004), 69–101; Washington State Pollution Control Commission, "Minimum Requirements for the Control of Industrial Wastes," *Sewage and Industrial Wastes* 22:4 (April 1950), 514–520; Washington State Pollution Control Commission, "The Seattle Sewage Treatment Problem" (Olympia: The Commission, 1948), http://www.washington.edu/uwired/outreach/cspn/Website/Classroom%20Materials/Curriculum%20Packets/Building%20Nature/Documents/27.html, accessed May 14, 2011.

2 Washington State Pollution Control.

Notes

INTRODUCTION: ONE RIVER, MANY VOICES

1 W. T. Burney, ed., *The Gold-Gated West, Songs and Poems by Samuel L. Simpson* (Philadelphia & London: J. B. Lippincott Company 1910), 19–20.

2 See, generally, Joel A. Tarr, *The Search for the Ultimate Sink: Urban Pollution in Historical Perspective* (Akron, Ohio: University of Akron Press, 1996).

3 Neil Mullane, "The Willamette River of Oregon: A River Restored?" in Antonius Laenen and David A. Dunnette, eds., *River Quality: Dynamics and Restoration* (Boca Raton, Fla.: Lewis Publishing, 1997), 65–76.

4 Coll Thrush, *Native Seattle: Histories from the Crossing-Over Place* (Seattle: University of Washington Press, 2007), 17–18.

5 Ethel Starbird, "A River Restored: Oregon's Willamette," *National Geographic* 141:6 (June 1972), 816–834; quote from 819. Note that Starbird incorrectly identifies the year of McCall's documentary as 1961. The essence of this popularized version of the narrative, with Tom McCall as the central figure, was repeated widely, including Lynton K. Caldwell, Lynton R. Hayes, and Isabel M. MacWhirter, *Citizens and the Environment: Case Studies in Popular Action* (Bloomington, Ind.: Indiana University Press, 1976), 90–93.

6 Thrush, *Native Seattle*, 19–20.

7 "Waltonians" is shorthand for members of the Izaak Walton League of America (IWLA).

8 David B. Charlton to Dr. Frank Terraglio, Portland State University, May 9, 1975, folder Water Pollution: Portland Area, box 18, David B. Charlton Papers (MSS 1900), Oregon Historical Society Research Library, Portland, Oreg. [hereafter Charlton Papers].

9 Richard White. *Organic Machine: The Remaking of the Columbia River* (New York: Hill and Wang, 1995), x–xi.

10 White. *Organic Machine*, x.

11 Matthew Klingle, *Emerald City: An Environmental History of Seattle* (New Haven, Conn.: Yale University Press, 2007), 9. For more on nature as a historical agent, see Ted Steinberg, "Down to Earth: Nature, Agency, and Power in History," *The American Historical Review* 107:3 (June 2002), 798–820.

12 Most treatments of this topic provide only the barest details—for example, Caldwell et al., *Citizens and the Environment*, 90–93. Elizabeth Orr and William Orr devote a chapter to pollution abatement in their book *Oregon Water: An*

Environmental History (Portland, Oreg.: Inkwater Press, 2005), but their treatment is cursory.

13 Iterations of this version of the narrative can be found in federal or state gray literature or in scientific reports; see, for example, William D. Honey Jr., *The Willamette River Greenway: Cultural and Environmental Interplay* (Corvallis, Oreg.: Oregon State University Water Research Institute, 1975).

14 George W. Gleeson, *The Return of a River: The Willamette River, Oregon* (Corvallis: Oregon State University Water Resources Research Institute, 1972), ii.

15 Gleeson, *The Return of a River*, 13.

CHAPTER 1: THE BEAUTIFUL WILLAMETTE

1 W. T. Burney, ed., *The Gold-Gated West: Songs and Poems by Samuel L. Simpson* (Philadelphia & London, J. B. Lippincott Company, 1910), 19–20.

2 W. W. Fidler, "Personal Reminiscences of Samuel L. Simpson," *Oregon Historical Society Quarterly* 15:4 (Dec. 1914), 264–276; Burney, ed., *The Gold-Gated West*.

3 Leslie M. Scott, ed., "John Work's Journey from Fort Vancouver to Umpqua River, and Return, in 1834," *Oregon Historical Quarterly* 24 (1923), 242, 249 [William G. Robbins, *Landscapes of Promise*, 31]; Robert T. Boyd, "Strategies of Indian Burning in the Willamette Valley," *Canadian Journal of Anthropology* 5 (1986), 66–67. William G. Robbins, "Narrative Form and Great River Myths: The Power of Columbia River Stories," *Environmental History Review* 17:2 (Summer 1993), 1–22.

4 Jim Mockford, "Before Lewis and Clark, Lt. Broughton's River of Names: The Columbia River Exploration of 1792," *Oregon Historical Quarterly* 106:4 (Winter 2005), 542–567.

5 See entries for April 2–6, 1806, in Gary E. Moulton, ed., *The Journals of the Lewis and Clark Expedition* (Lincoln: University of Nebraska Press, 1983–2001), http://lewisandclarkjournals.unl.edu, accessed May 3, 2010.

6 George W. Gleeson, *The Return of a River: The Willamette River, Oregon* (Corvallis: Oregon State University Water Resources Research Institute, 1972), 1–6.

7 David Alt and Donald W. Hyndman, *Northwest Exposures: A Geologic Story of the Northwest* (Missoula, Mont.: Mountain Press Publishing Company, 1995).

8 John Eliot Allen, Marjorie Burns, and Scott Burns, *Cataclysms on the Columbia: The Great Missoula Floods* (2nd ed.) (Portland, Oreg: Oooligan Press, 2009), 169–180.

9 Allen, Marjorie Burns, and Scott Burns, *Cataclysms on the Columbia*, 169–180.

10 Henry Zenk, "Notes on Native American Place-Names of the Willamette Valley Region," *Oregon Historical Quarterly* 109:1 (Spring 2008), 25–26. See entries for April 2–6, 1806, in Moulton, ed., *The Journals of the Lewis and Clark Expedition*.

11 Melinda Marie Jetté, "'At the hearth of the crossed races': Intercultural Relations and Social Change in French Prairie, Oregon, 1812–1843" (PhD diss., University of British Columbia, 2004); Melinda Marie Jetté, "'we have allmost Every Religion but our own': French–Indian Community Initiatives and Social Relations in French Prairie, Oregon, 1834–1837," *Oregon Historical Quarterly* 108:2 (Summer 2007), 222–245.

12 Henry B. Zenk, "Kalapuyans," in William C. Sturtevant, ed., *Handbook of North American Indians, vol. 7: Northwest Coast* (Washington, DC: Smithsonian Institution, 1990), 551.

13 Carl Abbott, *Boosters and Businessmen: Popular Economic Thought and Urban Growth in the Antebellum Middle West* (Westport, Conn.: Greenwood, 1981) and William G. Robbins, "Narrative Form and Great River Myths," 1–22. William G. Robbins, *Landscapes of Promise: The Oregon Story, 1800–1940* (Seattle: University of Washington Press, 1997), 179–187. On pastoralism and wilderness, see Richard W. Judd and Christopher S. Beach, *Natural States: The Environmental Imagination in Maine, Oregon, and the Nation* (Washington, DC: Resources for the Future, 2003), particularly x–xi, 4–7.

14 Frances F. Dunwell, *The Hudson: America's River* (New York: Columbia University Press, 2008), 54; 85–107.

15 Robert Bunting, *The Pacific Raincoast: Environment and Culture in an American Eden, 1778–1900* (Lawrence: University Press of Kansas, 1997), 40, 51–52; Robert T. Boyd, "Demographic History," in William C. Sturtevant, ed., *Handbook of North American Indians, vol. 7: Northwest Coast* (Washington, DC: Smithsonian Institution, 1990), 147.

16 Peter Boag, *Environment and Experience: Settlement Culture in Nineteenth-Century Oregon* (Berkeley: University of California Press, 1992), 4–9.

17 Michael Silverstein, "Chinooks of the Lower Columbia," in William C. Sturtevant, ed., *Handbook of North American Indians, vol. 7: Northwest Coast* (Washington, DC: Smithsonian Institution, 1990), 533–546.

18 Zenk, "Kalapuyans," 547–553.

19 Bunting, *The Pacific Raincoast*, 12–15; Boag, *Environment and Experience*, 10–17.

20 Robbins, *Landscapes of Promise*; Charles Wilkinson, *The People Are Dancing Again: The History of the Siletz Tribe of Western Oregon* (Seattle: University of Washington Press, 2010), particularly 57–130.

21 Boyd, "Demographic History," 135–148.

22 Stephen Dow Beckham, "History of Western Oregon Since 1846," in William C. Sturtevant, ed., *Handbook of North American Indians, vol. 7: Northwest Coast* (Washington, DC: Smithsonian Institution, 1990), 180–188.

23 Bunting, *The Pacific Raincoast*, 40, 51–52; Boyd, "Demographic History," 147.

24 Joseph E. Taylor III, *Making Salmon: An Environmental History of the Northwest Fisheries Crisis* (Seattle: University of Washington Press, 1999), 138–139; W. M. Hurst, et al., *The Fiber Flax Industry in Oregon: Its History, Present Status, and Future Possibilities* (Corvallis: Oregon State College Agricultural Experiment Station, 1953); William Lucas, *Canning in the Valley: Canneries of the Salem District* (Salem, Ore.: William Lucas, 1998); Bob Bourhill, *History of Oregon's Timber Harvests and/or Lumber Production: State Data, 1849 to 1992, County Data, 1925 to 1992* (Salem: Oregon Dept. of Forestry, 1994); Bunting, *The Pacific Raincoast*, 106–107; William G. Robbins, *Landscapes of Promise*, 72; Fred Merryfield, *A Preliminary Survey of Industrial Pollution of Oregon Streams* (Portland, Oreg., Oregon State Planning Board, [June] 1937).

25 Bunting, *The Pacific Raincoast*, 86.

26 US Department of the Interior, Census Office, Abstract of the Twelfth Census of the United States, 1900 (Washington, DC: Government Printing Office, 1904), 2, 470–471 [lxxxii–lxxxvi, 32, 40].

27 Quoted in Bunting, *The Pacific Raincoast*, 110.

28 Martin V. Melosi, *The Sanitary City: Urban Infrastructure in America from Colonial Times to the Present* (Baltimore, Md.: Johns Hopkins University Press, 2000), 90–

116; Jamie Benedickson, *The Culture of Flushing: A Social and Legal History of Sewage* (Vancouver: University of British Columbia Press, 2007), 83–85.

29 See, generally: Joel A. Tarr, *The Search for the Ultimate Sink: Urban Pollution in Historical Perspective* (Akron, Ohio: University of Akron Press, 1996), and also Melosi, *The Sanitary City*; Benedickson, *The Culture of Flushing*; Arn Keeling, "Urban Waste Sinks as a Natural Resource: The Case of the Fraser River," *Urban History Review* 34:1 (2005), 58–70.

30 Short, *Water*, 14–33. Bull Run water is still among the nation's cleanest water sources in the second decade of the twenty-first century. As was the case in Portland, efforts to establish municipal water supplies in other US cities were related to water quality issues as well as the Progressive Era push for administrative efficiency; see Melosi, *The Sanitary City*, 73–89, 117–147.

31 James V. Hillegas, "Working for the 'Working River': Willamette River Pollution, 1926–1962" (MA thesis, Portland State University: 2009), 20–21.

32 "An Act to Promote Drainage for Agricultural and Sanitary Purposes," *General Laws of Oregon, 1889*, 25–28; "An Act to Amend Section 1982 Title II, Chapter X, of the Laws of Oregon, as Compiled and Annotated by W. Lair Hill, relating to Crimes against the Public Health," *General Laws of Oregon, 1889*, 29.

33 "Oregon Department of Fish and Wildlife History, 1792–1990," http://www.dfw. state.or.us/agency/history.asp, accessed May 10, 2010.

34 See "Public Health Division: Agency History," *Oregon Blue Book*, http:// bluebook.state.or.us/ state/executive/health/healthhistory.htm, accessed Sept. 1, 2008; Duffy, *The Sanitarians*, 4.

35 "Oregon Department of Fish and Wildlife History, 1792–1990." Fifty dollars in 1903 was worth approximately $1,090 in 2017. This comparison is computed using the Gross Domestic Product (GDP) deflator formula; see the Bibliographic Essay and "Six Ways to Compute the Relative Value of a US Dollar Amount, 1774 to Present" for explanation of this method and a comparison calculator; http:// www.measuringworth.com/uscompare/.

36 "The Water Problem," *Morning Oregonian*, Feb. 28, 1906, 8.

37 "Germs in Rivers," *Morning Oregonian*, Aug. 17, 1906, 14.

38 Chapter 153, *General Laws of Oregon, 1921*, 267–303; Chapters 183 and 264, *General Laws of Oregon, 1919*, 264–266, 384–437. For agency reorganization, see "Oregon Department of Fish and Wildlife History, 1792–1990."

39 Neil S. Shifrin, "Pollution Management in the Twentieth Century," *Journal of Environmental Engineering* 131:5 (May 2005), 676–691; Jouni Paavola, "Interstate Water Pollution Problems and Elusive Federal Water Pollution Policy in the United States, 1900–1948," *Environment and History* 12:4 (2006), 435–465.

40 Albert E. Cowdrey, "Pioneering Environmental Law: The Army Corps of Engineers and the Refuse Act," *Pacific Historical Review* 44:3 (Aug. 1975), 331–349.

41 Paavola, "Interstate Water Pollution Problems," 436, 442, 447; Benjamin Ross and Steven Amter, *The Polluters: The Making of Our Chemically Altered Environment* (New York: Oxford University Press, 2010), 13–16.

42 Jennifer Read, "'A Sort of Destiny': The Multi-Jurisdictional Response to Sewage Pollution in the Great Lakes, 1900–1930," *Scientia Canadensis* 22–23:51 (1998–1999), 103–129; National Resources Committee, *Water Pollution in the United*

States: Third Report of the Special Advisory Committee on Water Pollution
(Washington, DC: US Government Printing Office, 1939), 17.

43 Notably, Oregon was excluded from fully participating in this recovery; see, generally, William G. Robbins, *Landscapes of Promise*.

44 Neil S. Shifrin, "Pollution Management in the Twentieth Century," *Journal of Environmental Engineering* 131:5 (May 2005), 676–691; Paavola, "Interstate Water Pollution Problems," 435–465.

45 Kendrick Clements, *Hoover, Conservation, and Consumerism: Engineering the Good Life* (Lawrence: University Press of Kansas, 2000).

46 Nicholas Casner, "Angler Activist," 536–537; Scarpino, *Great River*, 117–119; Thomas Howard Hayden (Paul E. Toren, ed.), *Citizen Leadership in Conservation: The Minnesota Izaak Walton League, 1922–1973* (St. Paul, Minn.: Izaak Walton League Minnesota Division, 2001). For more on the role of sportsmen's groups in nineteenth-century conservation efforts, see John F. Reiger, *American Sportsmen and the Origins of Conservation* (3rd ed.) (Corvallis: Oregon State University Press, 2001).

47 Dawn Merritt, "The Roaring 20s: A Call to Action," *Outdoor America* (Winter 2012), 25–23.

48 Clements, *Hoover, Conservation, and Consumerism*, 50–51, 71, 90–91.

49 Clements, *Hoover, Conservation, and Consumerism*, 50–51, 71, 90–91.

50 Clements, *Hoover, Conservation, and Consumerism*, 75–78.

51 Ross and Amter, *The Polluters*, 13–16.

52 Ross and Amter, *The Polluters*, 13–16.

53 Herbert Hoover, "A Remedy for Disappearing Game Fishes, Delivered before the Izaak Walton League of America, Chicago, Illinois, April 9, 1927" (Washington, DC: US Government Printing Office, 1927).

54 For Pennsylvania, see Nicholas Casner, "Polluter versus Polluter: The Pennsylvania Railroad and the Manufacturing of Pollution Policies in the 1920s," *Journal of Policy History* 11:2 (1999), 179–200; for Wisconsin, see Gregory Summers, *Consuming Nature: Environmentalism in the Fox River Valley, 1850–1950* (Lawrence: University Press of Kansas, 2006), particularly 19–30.

55 Lawrence M. Lipin, *Workers and the Wild: Conservation, Consumerism, and Labor in Oregon, 1910–30* (Urbana: University of Illinois Press, 2007), 51, 54–60.

56 Worth Mathewson, *William L. Finley: Pioneer Wildlife Photographer* (Corvallis: Oregon State University Press, 1986).

57 "Local Anglers Elect," *Morning Oregonian*, Dec. 16, 1922, 16; "Izaak Walton League Chapter Is Formed Here," *Oregon Sunday Journal*, Dec. 17, 1922, sec. 2, 2.

58 Ira Gabrielson served as the first head of the US Fish and Wildlife Service (1940–1946), and was one of the founders of the World Wildlife Fund. Jay Darling was a journalist and Pulitzer-prize-winning editorial cartoonist, and in the 1930s served as chief of the US Biological Survey, administered the Federal Duck Stamp Program, and helped establish the National Wildlife Federation. US Fish and Wildlife Service, "Ira Gabrielson," http://training.fws.gov/History/ConservationHeroes/Gabrielson.html; "The Papers of Jay Norwood 'Ding' Darling," http://www.lib.uiowa.edu/spec-coll/msc/tomsc200/msc170/MsC170_DarlingDing.html; both accessed April 30, 2010.

59 Worth Mathewson, *William L. Finley: Pioneer Wildlife Photographer* (Corvallis: Oregon State University Press, 1986), 6–17; Lipin, *Workers and the Wild*, particularly 51, 54–60. "Local Anglers Elect"; "Izaak Walton League Chapter is Formed Here."

60 Henry Ward would serve as Izaak Walton League national president, 1928–1930; see "Henry B. Ward Papers, 1885–1960" finding aid description, University of Illinois Archives, http://archives.library.illinois.edu/archon/?p=collections/controlcard&id=2392, accessed April 30, 2016; Jouni Paavola, "Interstate Water Pollution Problems and Elusive Federal Water Pollution Policy in the United States, 1900–1948," *Environment and History* 12:4 (2006), 435–465.

61 "Water Pollution Topic," *Morning Oregonian*, July 29, 1925, 11. The range "four to six millions of dollars a year" in 1925 was approximately $39 to $60 million in 2017.

62 "Sewage Survey Planned," *Morning Oregonian*, Aug. 19, 1925, 9.

63 "Willamette River Water To Be Given Test," *Oregon Daily Journal*, Feb. 2, 1926, 1, 15.

64 H. B. Hommon, Sanitary Engineer, US Public Health Service, "Report on the Pollution of the Willamette River," Feb. 26, 1926, file Stream Pollution, box 24, Master Fish Warden's Correspondence, Oregon State Archives, Salem, Oreg. Beyond this report, technical assistance from Hommon and his team also included designing and building the water sampling apparatus and helping city staff refine and interpret the data; see O. Laurgaard to A. L. Barbur, March 14, 1929, folder Sewage Disposal 1929 (26/5), accession A2001-008, City of Portland Archives, Oreg.

65 "Public Health Division: Agency History."

66 "Hydra-Headed Pollution," *Morning Oregonian*, July 30, 1925, 10. See also "The Cities Against Stream Pollution," *Morning Oregonian*, Feb. 1, 1926, 10. For specifics on the effect of Chicago's reversal of the flow of the Illinois river, see Craig E. Colten, "Illinois River Pollution Control, 1900–1970," in Lary M. Dilsaver and Craig E. Colten, eds., *The American Environment: Interpretations of Past Geographies* (Lanham, Md.: Rowman & Littlefield, 1992), 193–214.

67 City of Portland Dept. of Public Works [Olaf Laurgaard], *Report of City Engineer Pursuant to Resolution No. 11130 with Plans and Estimates for Water Front Project* (Portland, Oreg.: City of Portland, March 29, 1923).

68 William L. Lang, "One City, Two Rivers: Columbia and Willamette Rivers in the Environmental History of Twentieth-Century Portland, Oregon," in Char Miller, ed., *Cities and Nature in the American West* (Reno: University of Nevada Press, 2010), 96–111.

69 Olaf Laurgaard, *Treatise on the Design, Test, & Construction of the Front St. Intercepting Sewer and Drainage System in Portland, Oregon, Including Intercepting Sewer, Pumping Plant, & Concrete Bulkhead-Wall on Gravel Filled Timber Cribs* (Corvallis: Oregon State Agricultural College, 1933).

70 "The Beautiful Willamette," *Oregon Daily Journal*, Feb. 22, 1921, 6.

71 "127 Are Baptized," *Morning Oregonian*, July 31, 1914, 12; "Six Enter Trial Swim," *Morning Oregonian*, July 27, 1914, 8. See also "Some of Us Are Still Wearing Our Woolens but the Bathing Girls Have Opened Season," *Oregon Sunday Journal*, May 7, 1916, sec. 2, 3.

72 "Bathers Line River," *Sunday Oregonian*, Aug. 15, 1915, sec. 2, 14.

73 See "School Swimmers Carry Off Honors," *Sunday Oregonian*, July 31, 1921, sec. 2, 2.

74 See "City Quarantine on Dips in River May be Adopted," *Oregon Daily Journal*, July 24, 1924, 2, and "Windemuth Water Declared Impure," *Morning Oregonian*, July 16, 1924, 8.

75 US Bureau of the Census, *Census of Population: 1950 Oregon—Volume II, Part 37, Characteristics of the Population, Number of Inhabitants, Oregon*, 12, available at http://www.census.gov/prod/www/abs/decennial/1950.htm, accessed Oct. 6, 2008.

76 Melosi, *The Sanitary City*, 90–99.

77 Neil S. Shifrin, "Pollution Management in the Twentieth Century," *Journal of Environmental Engineering* 131:5 (May 2005), 676–691. See also Melosi, *The Sanitary City*, 161–165, and Keeling, "Urban Waste Sinks as a Natural Resource," 58–70.

78 "A River in a City," *Portland Telegram*, Nov. 8, 1922, 6. Geographer Arn Keeling has found a similar relationship of the residents of Vancouver, BC, to their urban river, the Fraser; see Keeling, "Urban Waste Sinks," 60.

79 For an analysis of the changing perceptions of another Pacific Northwest river, the Skagit of northwest Washington, see Linda Nash, "The Changing Experience of Nature: Historical Encounters with a Northwest River," *Journal of American History* 86:4 (March 2000), 1600–1629.

80 Daniel Schneider, *Hybrid Nature: Sewage Treatment and the Contradictions of the Industrial Ecosystem* (Cambridge, Mass.: The MIT Press, 2011), chapter 4. For more on the history of sewage as fertilizer, see the following: Sabine Barles and Laurence Lestel, "The Nitrogen Question: Urbanization, Industrialization, and Water Quality in Paris, 1830–1939," *Journal of Urban History* 33:5 (July 2007), 794–812; Nicholas Goddard, "'A mine of wealth'? The Victorians and the Agricultural Value of Sewage." *Journal of Historical Geography* 22:3 (1996), 274–290; J. A. Oleszkiewicz and D. S. Mavinic, "Wastewater Biosolids: An Overview of Processing, Treatment, and Management," *Journal of Environmental Engineering Science* 1 (2002), 75–88; John Sheail, "Town Wastes, Agricultural Sustainability, and Victorian Sewage," *Urban History* 23:2 (Aug. 1996), 189–210; and Joel A. Tarr, "From City to Farm: Urban Wastes and the American Farmer," *Agricultural History* 49:4 (Oct. 1975), 598–612.

81 Schneider, *Hybrid Nature*; Melosi, *The Sanitary City*, 172–173.

82 "Expert Advice Sought," *Morning Oregonian*, Sept. 17, 1925, 16, and "River Pollution Topic," *Morning Oregonian*, Sept. 18, 1925, 18. For support of this initiative by editors at the *Oregon Journal*, see "Dirty Rivers," *Oregon Daily Journal*, Aug. 27, 1926, 16.

83 "Willamette River Water To Be Given Test," *Oregon Daily Journal*, Feb. 2, 1926, 1, 15.

84 "Sewage Problem Hard," *Morning Oregonian*, Aug. 7, 1925, 17.

85 There were also three pulp and paper mills that came online in the following decades: Springfield (1949), Albany (1955), and Halsey (1969). See Appendix 1.

86 Burney, ed., *The Gold-Gated West*, 19–20.

CHAPTER 2: WHEN REASONABLE USE BECAME UNREASONABLE

1 "Fish Killed by Sewage," *Morning Oregonian*, May 5, 1926, sec. 2, 1.

2 "Stream Pollution Rouses Game Body," *Morning Oregonian*, May 11, 1926, sec. 2, 3; "Waste Disposal Argued," *Morning Oregonian*, May 21, 1926, sec. 1, 7.

3 "Fish Killed by Sewage," sec. 2, 1; Chapters 183 and 264, *General Laws of Oregon, 1919*, 264–266, 384–437.

4 "Fish Killed by Sewage."

5 "Edgar Averill Dies at Home," *Oregonian*, March 20, 1955, 15.

6 Conferees organized themselves under the name "Oregon Anti-Stream Pollution Committee." See "Minutes of the Meeting of the Oregon Anti-Stream Pollution Committee," May 20, 1926, file Stream Pollution, box 24, Master Fish Warden's Correspondence, Oregon State Archives, Salem, Oreg. [hereafter OR State Archives]. "River Pollution Ban Backed, But Remedy Lacking," *Oregon Journal*, May 20, 1926, sec. 1, 2; "Stream Pollution Rouses Game Body," *Morning Oregonian*, May 11, 1926, sec. 2, 3; Fred Merryfield, "History, Progress, and Problems of Stream Pollution in Oregon," *Commonwealth Review* 18:1 (March 1936), 80–86; "Sewer Fight Started," *Morning Oregonian*, May 14, 1926, 13.

7 "River Pollution Ban Backed, But Remedy Lacking," *Oregon Journal*, May 20, 1926, sec. 1, 2.

8 "Oil Waste Kills Fish," *Morning Oregonian*, July 20, 1911, 11; "Pure River Is Now Aim," *Morning Oregonian*, Dec. 8, 1911, 8.

9 "Foes of Pollution Organize," *Morning Oregonian*, Sept. 11, 1926, sec. 1, 6.

10 "Foes of Pollution Organize"; John C. Veatch, "Some Legal Aspects of Stream Pollution," *Commonwealth Review* 18:1 (March 1936), 91–93. Additionally, in the 1930s Veatch would serve as chair of the Oregon Fish Commission.

11 See, generally, Paavola, "Interstate Water Pollution Problems," 442–446, and Shifrin, "Pollution Management in the Twentieth Century," 680. For specific state examples, see: Nicholas Casner, "Polluter versus Polluter: The Pennsylvania Railroad and the Manufacturing of Pollution Policies in the 1920s," *Journal of Policy History* 11:2 (1999), 179–200, on Pennsylvania; Theodore Steinberg, *Nature Incorporated: Industrialization and the Waters of New England* (Amherst, Mass.: University of Massachusetts Press, 1994), 205–239, on Massachusetts; and Gregory Summers, *Consuming Nature: Environmentalism in the Fox River Valley, 1850–1950* (Lawrence: University Press of Kansas, 2006), particularly 19–30, on Wisconsin.

12 Executive Committee, Anti-Stream Pollution League, "A Bill for an Act to Provide for the Classification of the Streams in the State of Oregon Creating a Sanitary Water Board and Defining Its Duties," folder Sewage Disposal 1926 1 of 2, City Engineer's Historical Subject Records 8402-01, City of Portland Archives, Portland, Oregon.

13 "Drive to Purify Streams Starts," *Oregon Daily Journal*, Sept. 21, 1926, 1.

14 "Pollution League Drops Drastic Bill," *Morning Oregonian*, Jan. 15, 1927, sec. 1, 8. $35 million in 1927 was approximately $392 million in 2015.

15 C. C. Chapman, "Stream Pollution," *Oregon Voter*, Dec. 5, 1926, 507 (11).

16 J. C. Stevens and H. Loren Thompson, "The Portland Sewage Works Project," *Sewage Works Journal* 20:2 (Mar. 1948), 188.

17 For a brief statement of the continued lack of such information nearly thirty years later, see Richard Earle Noble, "The Willamette River Fishes as Biological Indicators of Pollution" (MS thesis, Oregon State College, 1952).

18 See, for example, Victor E. Shelford, "Fortunes in Wastes and Fortunes in Fish," *The Scientific Monthly* 9:2 (Aug. 1919), 97–124.

19 "River Pollution Ban Backed, But Remedy Lacking."

20 "Pollution Bill Defeated," *Sunday Oregonian*, Feb. 13, 1927, sec. 1, 10.

21 Since the Anti-Stream Pollution League (A-SPL) was an informal, nongovernmental group, quite often when the clean stream campaign made it into the *Oregonian* or *Oregon Journal* newspapers, these accounts specified individual members or their associated groups (such as Finley, Averill, or the IWLA), but not the league itself. Also common was for reporters to identify the league as the "Stream Pollution Committee" or the "anti-pollution league." To add further confusion for the researcher, in early 1933 a group of concerned Portland residents—many of whom had been associated with the A-SPL—formed the Portland Anti-Pollution Council to push city leaders to develop a sewage treatment proposal to submit for federal public works funding. Thus, while many individuals associated with the A-SPL continued working collaboratively on the topic, the A-SPL as a distinct group appears to have faded away in the late 1920s, to be "reconstituted" in activities if not in name in the work of the Portland City Club, Anti-Pollution Council, League of Oregon Cities, and other groups, until formation of the Oregon Stream Purification League in 1937 brought together many of the same advocates in support of a citizens' initiative to create the Oregon State Sanitary Authority. For the Portland Anti-Pollution Council, see "Council to Decide on Sewage Measure," *Morning Oregonian* June 8, 1933, sec. 1, 18.

22 Ellis Lucia, *The Conscience of a City: Fifty Years of City Club Service in Portland* (Portland, Oreg.: The City Club of Portland, 1966).

23 Public Health Section of the City Club of Portland, "Stream Pollution in Oregon," *The Pacific Engineer* 6:5 (May 1927), as quoted in George W. Gleeson, *The Return of a River: The Willamette River, Oregon* (Corvallis: Oregon State University Water Resources Research Institute, 1972), 13.

24 "Second Annual Commonwealth Conference," *University of Oregon Bulletin* 8:4 (Dec. 1910) [new series], 4.

25 O. Laurgaard to A. L. Barbur, March 14, 1929, folder Sewage Disposal 1929 (26/5), accession A2001-008, City of Portland Archives, Portland, Oreg.

26 Laurgaard to Barbur.

27 "River Pollution Deemed Curb to State Progress," *Oregon Daily Journal* March 29, 1929, 22; "Pollution Curb Begun," *Morning Oregonian*, March 23, 1929, sec. 1, 10.

28 "Committee on Resolutions, "A Brief Resume of the Conference on 'Stream Pollution' held at the University of Oregon, March 21–22, 1929," folder Sewage Disposal 1929 (26/5), accession A2001-008, City of Portland Archives, Oreg.; "Pollution Curb Begun"; "River Pollution Deemed Curb to State Progress," 22.

29 C. V. Langton and H. S. Rodgers, *Preliminary Report on the Control of Stream Pollution in Oregon, Bulletin Series, No. 1* (Corvallis: Oregon State Agricultural College Engineering Experiment Station, 1929).

30 Langton and Rodgers, *Preliminary Report on the Control of Stream Pollution in Oregon.*

31 H. S. Rodgers, C. A. Mockmore, and C. D. Adams, *A Sanitary Survey of the Willamette Valley, Bulletin Series, No. 2* (Corvallis: Oregon State Agricultural College Engineering Experiment Station, [June] 1930).

32 George W. Gleeson, *A Sanitary Survey of the Willamette River from Sellwood Bridge to the Columbia River, Bulletin Series, No. 6* (Corvallis: Oregon State Agricultural College Engineering Experiment Station, [April] 1936); Gleeson and Merryfield, *Industrial and Domestic Wastes of the Willamette Valley, Bulletin Series, No. 7* (parts 1, 2, and 3) (Corvallis: Oregon State Agricultural College Engineering Experiment Station, [May] 1936).

33 "City Quarantine on Dips in River May be Adopted," *Oregon Daily Journal*, July 24, 1924, 2, and "Windemuth Water Declared Impure," *Morning Oregonian*, July 16, 1924, 8.

34 William F. Willingham, *Army Engineers and the Development of Oregon: A History of the Portland District US Army Corps of Engineers* (Washington, DC: US Government Printing Office, 1983).

35 Olaf Laurgaard, *Treatise on the Design, Test, & Construction of the Front St. Intercepting Sewer and Drainage System in Portland, Oregon, Including Intercepting Sewer, Pumping Plant, & Concrete Bulkhead-Wall on Gravel Filled Timber Cribs* (Corvallis: Oregon State Agricultural College, 1933).

36 Willingham, *Army Engineers and the Development of Oregon*, 17–18.

37 Gleeson, *The Return of a River*, 18.

38 "A Debate on Sewage Disposal," *Sunday Oregonian*, Oct. 11, 1936, magazine section, 4–5.

39 Gleeson, *The Return of a River*, 18. One cubic foot per second equals about 7.5 gallons per second.

40 George Baker, untitled speech dated 1929, in Olaf Laurgaard Papers (AX 18), University of Oregon Archives, Eugene, quoted in E. Kimbark MacColl, *The Growth of a City: Power and Politics in Portland, Oregon, 1915–1950* (Portland, Oreg.: The Georgian Press, 1979), 286.

41 Laurgaard, *Treatise on the Design, Test, & Construction of the Front St. Intercepting Sewer*, 1–12.

42 Clements, *Hoover, Conservation, and Consumerism*, 39–45.

43 See, for example, Kendrick Clements, *Hoover, Conservation, and Consumerism*, and Jason Scott Smith, *Building New Deal Liberalism: The Political Economy of Public Works, 1933–1956* (Cambridge: Cambridge University Press, 2006).

44 Jason Scott Smith, *Building New Deal Liberalism: The Political Economy of Public Works, 1933–1956* (Cambridge: Cambridge University Press, 2006), 29–30; Philip W. Warken, *A History of the National Resources Planning Board, 1933–1943* (New York: Garland Publishing, Inc., 1979), 77–78. $3.3 billion in 1934 was approximately $496 billion in 2015.

45 Smith, *Building New Deal Liberalism*, 22–25.

46 James Butkiewicz, "Reconstruction Finance Corporation," EH.net, n.d., available at https://eh.net/encyclopedia/reconstruction-finance-corporation/, accessed Feb. 5, 2016.

47 Abbott, *Portland Planning*, 109–110.

48 See, for example, "Fund for Technical Work to be Urged," *Morning Oregonian*, March 23, 1933, sec. 1, 12, and "The Proposed City and County Bond Issues for

Unemployment and Material Relief: A Report by the Taxation Section of the City Club," *Portland City Club Bulletin* 13:1 (May 6, 1932), 1, 3–9.

49 The Portland Anti-Pollution Council organized in May 1933 to push for city officials to find a solution for municipal wastes. It was not affiliated with the similarly named Anti-Stream Pollution League of the late 1920s but contained many of the same members, including William L. Finley, Edgar F. Averill, and David Robinson. For more information, see, in particular, "Sewage Session Called," *Morning Oregonian* May 30, 1933, sec. 1, 2, and "Council to Decide on Sewage Measure," *Morning Oregonian* June 8, 1933, sec. 1, 18. The Civic Emergency Federation was a Portland-area group of unemployed persons organized in mid-1932 to advocate for measures to relieve unemployment. The group was highly active through 1934. For some background, see "Jobless Plan Picnic," *Sunday Oregonian*, Aug. 7, 1932, sec. 1, 11; "Unemployed to Unite," *Morning Oregonian*, Sept. 28, 1932, sec. 1, 12; and "Federation Status Up," *Morning Oregonian*, Jan. 27, 1933, sec. 1, 10.

50 "Projects Are Endorsed," *Morning Oregonian*, March 17, 1933, sec. 1, 10; "Fund for Technical Work to Be Urged," *Morning Oregonian*, March 23, 1933, sec. 1,12; "$11,700,000 Loan Sought by City for Five Self-Liquidating Projects," *Morning Oregonian*, June 16, 1933, sec. 1, 1, 4.

51 City Engineer to Ernest Cole, April 19, 1933, in folder Sewage Disposal 1933 (26/11), accession A2001-008, City of Portland Archives, Oreg.

52 City Engineer to L. G. Apperson, June 10, 1933, and O. Laurgaard to L. G. Apperson, June 10, 1933, both in folder Sewage Disposal 1933 (26/11), accession A2001-008, City of Portland Archives, Oreg.; City Engineer to R. E. Riley, July 10, 1933, folder Sewage Disposal 1933 (26/12), accession A2001-008, City of Portland Archives. One of the foremen the city hired for this work likely was Walter Baer; see O. Laurgaard to L. G. Apperson, June 10, 1933, folder Sewage Disposal 1933 (26/11), accession A2001-008, City of Portland Archives, Oreg.

53 "Engineer Board Named," *Sunday Oregonian*, June 11, 1933, sec. 1, 16; Roy E. Koon to Honorable Mayor and Council, July 19, 1933, folder Sewage Disposal 1933 (26/12), accession A2001-008, City of Portland Archives, Portland, Oreg.

54 Walter Baer and his allies also proposed their activated sludge plant would pay for itself through the sale of fertilizer from the sewage sludge. Though technically feasible, international experience had consistently shown the vast majority of sludge-to-fertilizer projects not to be economically feasible—fertilizer sales would not cover the costs of production. Thus, this aspect of Baer's proposal was not viable. For context, see Daniel Schneider, *Hybrid Nature*, chapter 4. $6 million in 1933 was approximately $90 million in 2015.

55 Ormond R. Bean, "Notes on the Portland Sewage Disposal Project, May 1933–June 26, 1936," undated typewritten manuscript [ca. June 1936], folder Willamette and Columbia River Pollution 1933–1934, box 2 Early Sewage Disposal System Data Sewage Systems (1922–1961), group Department of Public Works Bureau of Refuse Disposal (8890-02), City of Portland Archives, Portland, Oreg.

56 Board of Review (R. H. Corey, Wellington Donaldson, Carl E. Green, and Abel Wolman), "Report on the Collection and Disposal of Sewage, August 19, 1939," folder 8402-01 Sewage Disposal Project-Sewage Collection and Disposal Report

1939, box 1, Public Works Administration—City Engineer's Historical/Subject Records, City of Portland Archives, Portland, Oreg.

57 Smith, *Building New Deal Liberalism*, 35, 86.

58 "Harrison P. Eddy, Engineer, Is Dead," *New York Times,* June 16, 1937, 24.

59 Ormond R. Bean, "Notes on the Portland Sewage Disposal Project, May 1933– June 26, 1936," undated typewritten manuscript [ca. June 1936], 8–10, folder Willamette and Columbia River Pollution 1933–1934, box 2 Early Sewage Disposal System Data Sewage Systems (1922–1961), group Department of Public Works Bureau of Refuse Disposal (8890-02), SPARC.

60 Board of Review (R. H. Corey, Wellington Donaldson, Carl E. Green, and Abel Wolman), "Report on the Collection and Disposal of Sewage," Portland, Oreg., Aug. 19, 1939 [hereafter Wolman Report], 6, in folder 8402-01 Sewage Disposal Project-Sewage Collection and Disposal Report 1939, box 1, Public Works Administration—City Engineer's Historical/Subject Records, City of Portland Archives, Portland, Oreg. $461,000 in 1933 was approximately $6,890,000 in 2015.

61 Bean, "Notes on the Portland Sewage Disposal Project."

62 Bean, "Notes on the Portland Sewage Disposal Project."

63 Bean, "Notes on the Portland Sewage Disposal Project."

64 Bean, "Notes on the Portland Sewage Disposal Project."

65 "Wolman Report."

66 "Minutes of Meeting Sewage Disposal Committee August 3, 1936," folder 6 "Sewage Disposal," box 1, Edgar F. Averill Papers (AX 188), University of Oregon Archives, Eugene.

67 "River Looks Bad to Two Parties," *Morning Oregonian,* Aug. 5, 1936, 3.

68 "Minutes of Meeting Sewage Disposal Committee August 5, 1936." folder 6 "Sewage Disposal," box 1, Edgar F. Averill Papers (AX 188), University of Oregon Archives, Eugene.

69 "Minutes of Meeting on Sewage Disposal August 10, 1936," folder 6 "Sewage Disposal," box 1, Edgar F. Averill Papers (AX 188), University of Oregon Archives, Eugene.

70 "Pestilence from the Sludge Banks," *Oregon Daily Journal,* Aug. 8, 1936, 4.

71 Lawrence M. Lipin, *Workers and the Wild: Conservation, Consumerism, and Labor in Oregon, 1910–30* (Urbana, Ill.: University of Illinois Press, 2007), 71–3.

72 Chapter 350, *General Laws of Oregon, 1927,* 462–263; See "Attack Launched on River Cities," *Morning Oregonian,* Dec. 4, 1935, 1, and "Pollution Fight Reaches Court," *Morning Oregonian,* Dec. 6, 1935, 1. For details of the fish kill to which these fishermen were reacting, see "Pulp Mill Acid Slaughters Fish," *Morning Oregonian,* Oct. 19, 1935, 1, 4; "Pulp Mill Denies Polluting Water," *Sunday Oregonian,* Oct. 20, 1935, 6; and "St. Helens Supervisor Bans River Pollution," *Morning Oregonian,* Oct. 21, 1935, 2.

73 Smith, *Building New Deal Liberalism.*

74 State of Oregon, "Depression Era Public Works Web Exhibit," *Oregon Blue Book,* n.d., http://www.bluebook.state.or.us/facts/scenic/dep/depintro.htm, accessed Feb. 20, 2016.

75 The Works Progress Administration funded attorney John C. Ronchetto's work for the ACSP beginning in March 1936; see V. B. Stanbery, Consultant, OSPB, to

R. E. Koon, Chairman, ACSP, Feb. 25, 1936, folder Oregon State Planning Board 1936, box 20, Charlton Papers, OHSRL.

76 Merryfield, "History, Progress, and Problems," 85.

77 "Oregon State Planning Boards Records Guide, Agency History," http://arcweb. sos. state.or.us/state/planbd/hist/planhist.htm, accessed June 10, 2008.

78 Eve Vogel, "The Columbia River's Region: Politics, Place, and Environment in the Pacific Northwest, 1933–Present" (PhD dissertation, University of Oregon, 2007), 50–51. Also see Charles McKinley, *Uncle Sam in the Pacific Northwest: Federal Management of Natural Resources in the Columbia River Valley* (Berkeley: University of California Press, 1952), particularly 459–467 for the Pacific Northwest Regional Planning Council.

79 See Philip W. Warken, *A History of the National Resources Planning Board, 1933–1943* (New York: Garland Publishing, Inc., 1979), 1–4.

80 Merryfield, "History, Progress, and Problems," 85.

81 National Resources Committee Special Advisory Committee on Water Pollution, *Report on Water Pollution* (Washington, DC: US Government Printing Office, July 1935; reprint Feb. 1936), 6–8; 60.

82 ACSP members specified above include: Pacific Coast Association of Pulp and Paper Manufacturers representative George P. Berkey; Inman-Poulsen Lumber Company Vice President Henry Brooks Van Duzer; Crossett Western Company/Crossett-Watzek-Gates Industries Executive Aubrey Watzek; Oregon State College Civil Engineering Professor Fred Merryfield; and Portland Mayor Joseph Carson Jr. For full member list, see Merryfield, "History, Progress, and Problems," 85.

83 Other IWLA members on the ACSP included State Fish Commission Chair John C. Veatch and State Game Supervisor Frank B. Wire; see "1959 Roster Portland Chapter Izaak Walton League of America," folder Izaak Walton League, 1922–1958, series I, Personal Records, subgroup 1, Personal/Family Records, E. E. Wilson Papers, Oregon State University Archives, Corvallis, Oreg.

84 These were David Charlton, Ray E. Koon, Aubrey Watzek, and William Levin; see "Roster of Membership," *Portland City Club Bulletin* 17:9 (Sept. 4, 1936), 45–50.

85 John Ronchetto, *An Analytical Digest of Existing Legislation of Oregon and Other States Relating to Stream Pollution* (Salem: Oregon State Planning Board Advisory Committee on Stream Purification, Aug. 1936); Ronchetto, *Recommended Principles for Stream Purification Legislation in Oregon* (Salem: Oregon State Planning Board Advisory Committee on Stream Purification, June 1937); R. E. Koon, *Efforts to Reduce Stream Pollution in Oregon* (Salem: Oregon State Planning Board Advisory Committee on Stream Purification, Feb. 1937); and Fred Merryfield, *A Preliminary Survey of Industrial Pollution of Oregon Streams* (Portland: Oregon State Planning Board, June 1937).

86 R. E. Koon, *Efforts to Reduce Stream Pollution in Oregon*, 6.

87 "A Bill for an Act to Create a State Sanitary Commission of Oregon, for the Control of Water Pollution in Oregon, and for Other Purposes Thereto, and Making an Appropriation," folder Senate Bills #390-399, box 28, Secretary of State 1937 Legislature, Oregon State Archives, Salem.

88 Ronchetto, *An Analytical Digest of Existing Legislation*, v.

89 "A Bill for an Act to Create a State Sanitary Commission of Oregon."

90 C. C. Chapman, "Who's Who in the 1937 Legislature: Byron G. Carney," *Oregon Voter* 88:1 (Jan. 2, 1937), 22.

91 Chapman, "Who's Who in the 1937 Legislature." For Townsendism, see Rick Mayes, *Universal Coverage: The Elusive Quest for National Health Insurance* (Ann Arbor: University of Michigan Press, 2004), 24–25.

92 State of Oregon, *Official Voter's Pamphlet for the Regular General Election*, November 2, 1948, 72.

93 State of Oregon, *Official Voter's Pamphlet*, 72; A. E. Harding to Charles H. Leavy, May 17, 1939, box 7, Oregon Commonwealth Federation Records, University of Oregon Archives, Eugene; Monroe Sweetland to Harry Hopkins, May 3, 1939, box 7, Oregon Commonwealth Federation Records, University of Oregon Archives, Eugene.

94 "Senate Amendments to Senate Bill No. 392," folder Senate Bills #390-399, box 28, Secretary of State 1937 Legislature, Oregon State Archives, Salem.

95 "An Act for the Control of Water Pollution in Oregon," folder Senate Bills #410-419, box 29, Secretary of State 1937 Legislature, Oregon State Archives, Salem.

96 Floyd J. McKay. *An Editor for Oregon: Charles A. Sprague and the Politics of Change* (Corvallis: Oregon State University Press, 1998), 110.

97 Conrad, Bruce, & Co. advertisement, *Sunday Oregonian*, Nov. 14, 1934, 14; "Tax Discussion Slated," *Morning Oregonian*, Sept. 24, 1936, 9; "Sales Tax Issue Argued," *Morning Oregonian*, April 23, 1934, 7; "Two Bills Opposed by Security Owners," *Morning Oregonian*, March 9, 1935, 6; F. H. Young, "Danger Scented in Tax Measure," *Sunday Oregonian*, July 26, 1936, 17.

98 Conrad, Bruce, & Co. advertisement, 14; "Tax Discussion Slated,". 9; Laurel Gilbertson, "News of the Classes," *Old Oregon* 25:6 (Feb. 1944); "F. H. Young Dies at Game," *Oregonian*, July 3, 1954, 7. F. H. Young was the son of F. G. (Frederic George) Young. The elder Young was dean of the University of Oregon's School of Sociology, and was a founding member of the Oregon Historical Society where he served as secretary and editor of the *Oregon Historical Quarterly*; see "Grim Reaper Takes Dean of University," *Morning Oregonian*, Jan. 5., 1929, 14.

99 "Water Bill Held Pulp Plant Peril," *Oregonian*, March 2, 1937, 8.

100 *Oregon Business & Investors, Inc., Bulletin*, 64 (July 28, 1937), 2, in folder 12/4 Legislation, box 12 Stream Purification Records (4/10/4/5), Oregon State Planning Board Records, Oregon State Archives, Salem. Neither the Vinson-Barkley bill or similar proposals passed, but the substance of the bill became part of the 1948 Federal Water Pollution Control Act. For more, see Paavola, "Interstate Water Pollution Problems," 450, and Casner, "Angler Activist," 541, 545–546.

101 "Water Bill Held Pulp Plant Peril," *Oregonian*, March 2, 1937, 8.

102 Census numbers at US Bureau of the Census, *Census of Population: 1950 Oregon—Volume II, Part 37, Characteristics of the Population, Number of Inhabitants, Oregon*, 12–13, available at http://www.census.gov/prod/www/abs/decennial/1950.htm, accessed Oct. 6, 2008.

103 "An Act for the Control of Water Pollution in Oregon."

104 Journals of the Oregon Senate and House, 1937, pp. 245, 31, 292, 300.

105 "It Lost Its Teeth," *Oregon Daily Journal*, March 13, 1937, 4.

106 Gary Murrell, *Ironpants: Oregon's Anti-New Deal Governor, Charles Henry Martin* (Pullman: Washington State University Press, 2000).

107 Murrell, *Ironpants,* 112–163.

108 Murrell, *Ironpants,* 117.

109 Murrell, *Ironpants,* 158, 172, 193, 167–168.

110 Murrell. *Ironpants,* 129–130.

111 Governor Charles H. Martin to Honorable Earl Snell, Oregon Secretary of State, March 12, 1937, folder Senate Bills #410-419, box 29, Secretary of State 1937 Legislature 39th Session Records, OR State Archives. See also "State's Executive Denies Signature to Legislative Acts," *Oregonian,* March 13, 1937, sec. 1, 3. Tom McCall's biographer Brent Walth states simply that Governor Charles Martin vetoed the bill because it "went too far for [his] tastes"; see Brent Walth, *Fire at Eden's Gate: Tom McCall and the Oregon Story* (Portland, Oreg.: Oregon Historical Society Press, 1994), 138. $15 to 18 million in 1937 was approximately $197–237 million in 2015.

112 *Oregon Business & Investors, Inc., Bulletin* 64, 4.

113 "Attack Launched on River Cities"; "Pollution Fight Reaches Court"; "Pulp Mill Acid Slaughters Fish"; "Pulp Mill Denies Polluting Water"; and "St. Helens Supervisor Bans River Pollution."

114 "Officials Probe Death of Fish," *Morning Oregonian,* Oct. 20, 1936, 7.

115 "Lower Harbor River Water Devoid of All Oxygen," *Oregon Journal,* Aug. 6, 1936, 1, 14.

116 *Oregon Business & Investors, Inc., Bulletin* 60 (May 24, 1937), 1–2, in folder Oregon State Planning Board-Stream Purification Committee 1936, box 20, Charlton Papers.

117 "Papers Received by Organization," *Morning Oregonian,* July 22, 1935, 15. For context, see David T. Beito, *Taxpayers In Revolt: Tax Resistance During the Great Depression* (Chapel Hill, N.C.: The University of North Carolina Press, 1989).

118 "Cleaner Streams League Objective," *Oregonian,* May 9, 1937, 17.

119 "Cleaner Streams League Objective."

120 *Oregon Business & Investors, Inc., Bulletin* 60, 2.

121 "Initiative Objective of Stream Group," *Oregonian,* June 13, 1937, 4; see also *Oregon Business & Investors, Inc., Bulletin* 64, 2.

122 "Anti-Pollution Petition Work Completed," *Oregon Daily Journal,* Dec. 16, 1937, 3; "League Pushes Better-Stream Movement," *Oregon Daily Journal,* Dec. 15, 1937, 4.

123 Chapter 3, *Oregon Laws,* 1939, 9–14.

124 "Voters in Oregon Are Asked to Decide," *Oregon Daily Journal,* Dec. 22, 1937, 12; "United Against Stream Pollution," *Oregonian,* March 26, 1938, 10; C. C. Chapman, "Pure Streams: Constructive Policies," *Oregon Voter,* 90:9 (Aug. 28, 1937), 14–15, 230-231; "Stream Pollution Bill Approved," *Oregon Daily Journal,* Sept. 29, 1938, 6.

125 "League Writes Pollution Bill," *Oregonian,* Sept. 17, 1937, 9. See also "Anti-Pollution Bill Being Shaped by Business Firm," *Oregon Daily Journal,* Aug. 19, 1937, 5.

126 *Oregon Business & Investors, Inc., Bulletin* 64, 3; "'Clean the Rivers' Committee Formed," *Oregon Daily Journal,* Sept. 23, 1938, 13; Chapman, "Sewage Disposal:

Portland Again Asked to Authorize Program," *Oregon Voter* 95:3 (Oct. 15, 1938), 9–13; and Oregon State Sanitary Authority Meeting Minutes [hereafter OSSA Minutes], June 21, 1940, vol. 1, 51–84 , Oregon State Department of Environmental Quality, Portland; and "Complete Anti-Pollution Plan Favored," *Oregon Daily Journal*, Apr. 29, 1938, 28.

127 "River Looks Bad to Two Parties," *Morning Oregonian*, Aug. 5, 1936, 3.

128 "Dirty River Water Kills Fingerlings," *Oregon Daily Journal*, Oct. 12, 1938, 7; see also "Willamette Dip Fatal to Trout," *Oregonian*, Oct. 13, 1938, 1.

129 "Mayor Carson Dons Deep-Sea Diver's Suit for Personal Investigation of River Filth," *Oregonian*, Nov. 5, 1938, 1; quotes from a tour of the harbor a few weeks previously, see "River Pollution Seen First Hand," *Oregonian*, Oct. 11, 1938, 5.

130 "Dr. Dillehunt Heads Drive to Put Over Sewage Plan and Clean Polluted River," *Oregonian*, Aug. 11, 1938, 8.

131 "Dr. Dillehunt Heads Drive."

132 Quote from NRC report that Abel Wolman oversaw the research and writing of; see National Resources Committee Special Advisory Committee on Water Pollution, *Water Pollution in the United States* (Washington, DC: US Government Printing Office, 1939), 84. For more on Wolman's conservative approach, see Casner, "Angler Activist," 543–544.

133 "Finley Points Out Value of Clean River," *Oregon Daily Journal*, Oct. 19, 1938, 1. $12 million in 1938 was approximately $162 million in 2015, and $50 million was approximately $677 million. Newspaper editors often referred to other economic considerations—see, for example, Chapman, "Sewage Disposal," and Chapman, "Water Purification," *Oregon Voter* 95:6 (Nov. 5, 1938), 80.

134 "Argument Submitted by the Stream Purification League of Oregon, in Favor of the Water Purification and Prevention of Pollution Bill," *Proposed Constitutional Amendments and Measures Submitted to the Voters of Oregon, Regular General Election, November 8, 1938* (Salem, Oreg.) 38.

135 Gleeson, *The Return of a River*, 21–22; "Two Bills Strike Blows at Stream Pollution," *Oregon Daily Journal*, Nov. 10, 1938, 16. This account detailing Governor Martin's 1937 veto, organization of the Oregon Stream Purification League, and successful passage of the 1938 citizen's initiative revises the summary provided by historian William G. Robbins in his *Landscapes of Conflict: The Oregon Story, 1940–2000* (Seattle: University of Washington Press, 2004), 250.

136 Wolman Report, 6.

CHAPTER 3: SEWAGE AND THE CITY

1 William Joyce Smith, *Pollution in the Willamette*, film (Portland, Oreg., 1940) (viewable at http://oregonstate.edu/media/cbnnp, accessed Nov. 23, 2008). The Oregon State University Archives identifies this film as "Willamette River Pollution Film, circa 1940," but at an OSSA meeting Smith called it "Pollution in the Willamette"; see OSSA Minutes, Oct. 5, 1943, vol. 1, 267.

2 "Izaak Waltons Plan Confab," *Oregonian*, Sept 2, 1945, sec. 4, 2.

3 Harry C. Oberholser, "Organizations and Officials Concerned with Wildlife Protection: 1942," US Department of the Interior, US Fish and Wildlife Service, Circular No. 5, April 1943. http://www.archive.org/stream/fisherycircular05unit/fisherycircular05unit_djvu.txt, accessed Dec. 29, 2011.

4 OSSA Minutes, Dec. 13, 1940, vol. 1, 95–96. At its November 4, 1949, meeting, the OSSA allocated $900 to reprint the film and provide a soundtrack; see OSSA Minutes ,Nov. 4, 1949, vol. 2, 217, Oregon DEQ.

5 "Wildlife Meet Billed Friday," *Oregonian*, Oct. 11, 1940, sec. 3, 3.

6 See Appendix 3 for sources and further details.

7 Chapter 3, *General Laws of Oregon*, 1939, 9–14.

8 Gary Murrell, *Ironpants: Oregon's Anti-New Deal Governor, Charles Henry Martin* (Pullman: Washington State University Press, 2000), 179–186.

9 Jason Scott Smith, *Building New Deal Liberalism: The Political Economy of Public Works, 1933–1956* (Cambridge: Cambridge University Press, 2006), 155–156, 161–162.

10 Murrell, *Ironpants,* 179–186.

11 Charles A. Sprague, "Tenth Biennial Budget, State of Oregon, 1941–1943, Submitted to the Forty-first Legislative Assembly," Salem, Oreg.: Budget Division of the Executive Department, 1940, 146.

12 "Deficit Faced, Chapman Avers," *Oregonian*, Jan. 15, 1939, 12. $50,000 in 1938 was approximately $677,000 in 2015, and $63,116 was $854,000.

13 OSSA Minutes, Feb. 25, 1939, vol. 1, 1–2.

14 "Protests Voiced on Sanitary Fund," *Oregonian*, March 7, 1939, 10.

15 Final OSSA appropriation reflected in Charles A. Sprague, "Tenth Biennial Budget, State of Oregon, 1941–1943, Submitted to the Forty-first Legislative Assembly," 146. $10,000 and $21,290,778 in 1939 were, respectively, $137,000 and $291,000,000 in 2015. For additional comparison, the state's 1939 biennial budget was only about .5 percent of the state's 2009–2011 budget of $60.6 billion; see "Government Finance: State Government," http://bluebook.state.or.us/state/govtfinance/govtfinance01.htm, accessed Jan. 7, 2012.

16 Sprague, "Tenth Biennial Budget," 213.

17 OSSA Minutes, Sep. 12, 1941, vol. 1, 133.

18 OSSA Minutes, Dec. 12, 1941, vol. 1, 135–137.

19 OSSA Minutes, Feb. 25, 1939, vol. 1, 1–2.

20 "H. F. Wendel Dies at 74," *Oregonian*, April 18, 1967, sec. 1, 1, 16.

21 "H. F. Wendel Dies at 74."

22 OSSA Minutes Sept. 12, 1941, vol. 1, 129–130. It will be noted that Wendel expressed these sentiments nearly three years after the creation of the authority, but this succinct statement reflects his and the authority's basic approach throughout Wendel's tenure.

23 OSSA Minutes, July 7, 1939, vol. 1, 9.

24 OSSA Minutes, July 7, 1939, vol. 1, 10.

25 OSSA Minutes, Sep. 15, 1939, vol. 1, 13.

26 Adam Rome, "Coming to Terms with Pollution: The Language of Environmental Reform," *Environmental History* 1:3 (July 1996), 6–7.

27 Rome, "Coming to Terms with Pollution."

28 Martin V. Melosi, *The Sanitary City: Urban Infrastructure in America from Colonial Times to the Present* (Baltimore, Md.: Johns Hopkins University Press, 2000), 228–229.

29 George W. Gleeson, *The Return of a River: The Willamette River, Oregon* (Corvallis: Oregon State University Water Resources Research Institute, 1972), 11; C. V.

Langton and H. S. Rogers, *Preliminary Report on the Control of Stream Pollution in Oregon, Bulletin Series, No. 1* (Corvallis: Oregon State Agricultural College Engineering Experiment Station, [March] 1929), 15. To a somewhat lesser degree, water quality specialists were also concerned with turbidity, suspended solids, nutrients, major ions, color, and odor. See Neil S. Shifrin, "Pollution Management in the Twentieth Century," *Journal of Environmental Engineering* 131:5 (May 2005), 679.

30 Melosi, *The Sanitary City*, 228–229.

31 OSSA Minutes, March 1, 1940, vol. 1, 25–27. For the development of water quality standards generally, see Shifrin, "Pollution Management in the Twentieth Century," 682–683.

32 National Resources Committee Special Advisory Committee on Water Pollution, *Report on Water Pollution in the United States* (Washington, DC: US Government Printing Office, 1939), 39–41.

33 J. K. Hoskins, "A National Program for Water Pollution Control: Consideration of Some Elements in Its Formulation and Administration," *Sewage Works Journal* 7:3 (May 1935), 554.

34 OSSA Minutes, May 7, 1940, vol. 1, 32–33, 38–50.

35 John H. Lincoln and Richard F. Foster, *Report on Investigation of Pollution in the Lower Columbia River* (Olympia: Washington State Printing Department, 1943).

36 R. E. Koon, *Efforts to Reduce Stream Pollution in Oregon* (Salem: Oregon State Planning Board Advisory Committee on Stream Purification, Feb. 1937), 6–8, 11–12.

37 F. W. Mohlman, "Sewage Treatment in Large American Cities," *Sewage Works Journal* 12:1 (Jan. 1940), 168–171. Seattle provided 10 percent treatment for a population of four hundred thousand.

38 K. W. Brown, D. H. Caldwell, and H. E. Miller, "Metropolitan Seattle Sewerage and Drainage Survey: A Report for the City of Seattle, King County and the State of Washington on the Collection, Treatment and Disposal of Sewage and the Collection and Disposal of Storm Water in the Metropolitan Seattle Area," March 1958, Part 1, 9–11.

39 Ormond R. Bean, "Notes on the Portland Sewage Disposal Project, May 1933–June 26, 1936," undated typewritten manuscript [ca. June 1936], folder Willamette and Columbia River Pollution 1933–1934, box 2 Early Sewage Disposal System Data Sewage Systems (1922–1961), group Department of Public Works Bureau of Refuse Disposal (8890-02), City of Portland Archives, Portland, Oreg.

40 OSSA Minutes, Feb. 25, 1939, vol. 1, 1–2.

41 "Council Launches Rivers Clean-up," *Oregonian*, Nov. 17, 1938, 6; "New Sewage Charge Begins December 1," *Oregon Daily Journal*, Nov. 10, 1938, 34; "Sewage Project Gets Under Way in City," *Oregonian*, Nov. 11, 1938, 3. $5,000 in 1938 is approximately $64,000 in 2017.

42 Board of Review (R. H. Corey, Wellington Donaldson, Carl E. Green, and Abel Wolman), "Report on the Collection and Disposal of Sewage," Portland, Oreg., Aug. 19, 1939 [hereafter Wolman Report], 16–17, 21, in folder 8402-01 Sewage Disposal Project-Sewage Collection and Disposal Report 1939, box 1, Public Works Administration-City Engineer's Historical/Subject Records, City of Portland Archives, Portland, Oreg.

43 Wolman Report, 3–4. For Commissioner Beans's opinion of the Columbia Slough, see "Sewage Project Gets Under Way in City," *Oregonian*, Nov. 11, 1938, 3, and "Sewage Plan Ordinance is Prepared," *Oregon Daily Journal*, Nov. 11, 1938, 5.

44 Wolman Report, 4–7. $9 million in 1939 was approximately $123 million in 2017.

45 For an analysis of this practice, see geographer Arn Keeling's study of urban wastewater systems in Vancouver, British Columbia; Keeling, "Urban Waste Sinks as a Natural Resource: The Case of the Fraser River," *Urban History Review* 34:1 (2005), 58–70.

46 Wolman Report, 9–10, 14.

47 "Proceedings of Local Associations," *Sewage Works Journal* 10:4 (July 1938), 794–800.

48 Wolman Report, 16.

49 OSSA Minutes, June 21 1940, vol. 1, 61–62.

50 "Sewage Disposal Board Unchosen," *Oregonian* March 2, 1939, sec. 1, 9; "Council Rejects Enlarged Board," *Oregonian*, March 23, 1939, 4.

51 Jill Hopkins Herzig, "The Oregon Commonwealth Federation: The Rise and Decline of a Reform Organization" (MA thesis, University of Oregon, 1963). For more on Monroe Sweetland, see William G. Robbins, *A Man for All Seasons: Monroe Sweetland and the Liberal Paradox* (Corvallis: Oregon State University Press, 2015).

52 "Council Rejects Enlarged Board," *Oregonian*, March 23, 1939, 4.

53 OSSA Minutes, June 21, 1940, 61–62; OSSA Minutes, Oct. 5, 1943, vol. 1, 253. $275,000 in 1940 was approximately $3,760,000 in 2017, and $130,000 was approximately $1,780,000.

54 "River Cleanup Plan Stalled," *Oregonian*, Oct. 5, 1943, sec. 1, 10. $35,000 in 1943 was approximately $392,000 in 2017.

55 OSSA Minutes, June 21 1940, vol. 1, 52.

56 OSSA Minutes, Dec. 15, 1939, vol. 1, 17–18 ; "Council Launches Rivers Clean-up," *Oregonian*, Nov. 17, 1938, 6.

57 OSSA Minutes, May 7, 1940, vol. 1, 33–34, 37.

58 OSSA Minutes, June 21, 1940, vol. 1, 59–84. $7,000,000 to $10,000,000 in 1940 was approximately $94,600,000 to $135,000,000 in 2017.

59 OSSA Minutes, Dec. 12, 1941, vol. 1, 141–152; OSSA Minutes vol. 1, Sept. 17, 1943, 242–244.

60 OSSA Minutes, Nov. 13, 1942, vol. 1, 227–228. The OSSA was faced with a shortage of qualified sanitary engineers in the fall of 1941, before the United States' entry into World War II, because the US military had already begun enlisting men with this expertise; see OSSA Minutes Sept. 12, 1941, vol. 1, 133.

61 OSSA Minutes, March 12, 1943, vol. 1, 229–231.

62 "This Time, Let's Do It," *Oregonian*, July 14, 1943, 10.

63 OSSA Minutes, Oct. 5, 1943, vol. 1, 249–288.

64 OSSA Minutes, Oct. 5, 1943, vol. 1, 267.

65 "Raft Sitter Tells of Filth," and "River Cleanup Plan Stalled," *Oregonian*, Oct. 5, 1943, sec. 1, 10.

66 Quoted in "Advertising Pollution," *Oregonian*, Oct. 13, 1943, sec. 1, 8.

67 "Stream Purity Group at Work," *Oregon Journal*, Aug. 29, 1940, 20.

68 Ralph J. Staehli, "River Cesspool," *Oregonian*, Aug. 12, 1943, sec. 1, 10.

69 OSSA Minutes, Sept. 17, 1943, vol. 1, 242–245. Robert Moses, *Portland Improvement* (New York, William E. Rudge's Sons, Nov. 10, 1943), 13–15, 22, 43.

70 Carl Abbott, *Portland: Planning, Politics, and Growth in a Twentieth-Century City* (Lincoln: University of Nebraska Press, 1983), 137.

71 State of Oregon, *Progress Report of the Postwar Readjustment Commission*, May 15, 1943–Dec. 31, 1944. (Salem, Oreg., 1945), 34–36.

72 Abbott, *Portland*, 137–139.

73 Abbott, *Portland*, 137–139.

74 Val C. Ballestrem, "'In the Shadow of a Concrete Forest': Transportation Politics in Portland, Oregon, and the Revolt Against the Mount Hood Freeway, 1955–1976" (MA thesis, Portland State University, 2009); Moses, *Portland Improvement*, 13–15, 22, 43. $12 million in 1943 was approximately $135 million in 2017, and $24 million was approximately $269 million.

75 OSSA Minutes, Sep. 17, 1943, vol. 1, 247.

76 OSSA Minutes, Dec. 12, 1941, vol. 1, 137, 141–152.

77 Lew Wallace, "But We Still Have an 'Open Sewer,'" *Oregon Daily Journal*, Oct. 7, 1941, 12.

78 OSSA Minutes, Oct. 5, 1943, vol. 1, 253.

79 Harold Hughes, "Curmudgeon Bennett Fought Sacred Cows," *Oregonian*, April 4, 1970, 1, 15; "Jake Spoke for Many," *Oregonian* April 4, 1970, sec. 1, 14.

80 Harold Hughes, "Curmudgeon Bennett"; "Veteran State Legislator Dies After Heart Attack," *Oregonian*, April 4, 1970, 1. See also "Emergency Relief System Assailed," *Morning Oregonian*, April 11, 1933, sec. 1, 4; "Relief Board Ouster Up," *Sunday Oregonian*, April 23, 1933, sec. 1, 12.

81 OSSA Minutes, June 21, 1940, vol. 1, 67.

82 OSSA Minutes, Oct. 5, 1943, vol. 1, 266–267.

83 "Bowes Selected for Council Job," *Oregonian*, June 1, 1939, sec. 1, 1; "Bowes Declares His Hands Untied," *Oregonian*, June 1, 1939, sec. 1, 17.

84 "Seizure Takes Life of William Bowes," *Oregonian*, Oct. 19, 1969, sec. 1, 1, 31.

85 "Seizure Takes Life of William Bowes."

86 E. Kimbark MacColl, *The Growth of a City: Power and Politics in Portland, Oregon, 1915–1950* (Portland, Oreg.: The Georgian Press, 1979), 565, 584, 587, 591–592.

87 Kimbark MacColl, *The Growth of a City*, 565, 584, 587, 591–592.

88 OSSA Minutes, Oct. 5, 1943, vol. 1, 268.

89 OSSA Minutes Dec. 15, 1939, vol. 1, 17–18. The Portland city council authorized a $5,000 loan for plans and specifications on November 16, 1938; see "Council Launches Rivers Clean-up," *Oregonian*, Nov. 17, 1938, 6. See also "New Sewage Charge Begins December 1," *Oregon Daily Journal*, Nov. 10, 1938, 34; "Sewage Project Gets Under Way in City," *Oregonian*, Nov. 11, 1938, 3. $5,000 in 1938 was worth approximately $67,000 in 2017.

90 Abbott, *Portland*, 118–119, 141–142.

91 Harold Wendel to George Neuner, Dec. 17, 1943, and George Neuner to Harold Wendel, Dec. 31, 1943, both in folder S-9-1 Legal Opinions & Decisions 1943–

1955, box Health Board Sewage Systems Correspondence 1938–1961 (#471-499), Department of Human Resources, Oregon State Archives, Salem.

92 Wendel to Neuner, Dec. 17, 1943, and Neuner to Wendel, Dec. 31, 1943.

93 "Bond Issue for Sewage Plant Urged," *Oregonian*, Jan. 21, 1944, sec. 1, 1, 10. $12 million in 1943 was approximately $120 million in 2017.

94 "Sewage Group Elects Heads," *Oregonian*, Feb. 29, 1944, sec. 1, 11.

95 "City Council Defers Action on Engineer's Sewer Plans," *Oregonian*, Feb. 17, 1944, sec. 1, 11.

96 "Sewage Disposal Angle Explained," *Oregonian*, Feb. 11, 1944, sec. 1, 5.

97 OSSA Minutes, Sep. 17, 1943, vol. 1, 242–247.

98 "Citizens Plan for Economy," *Oregonian*, Feb. 14, 1942, 8.

99 "Incumbent Judges and County Office Holders Lead in Race for Re-Election," *Morning Oregonian*, Nov. 9, 1932, 1; "Pay Cuts at Top Backed," *Oregonian*, April 14, 1932, 3; "Former Legislator Frank Hilton Dies," *Oregonian*, March 22, 1973, sec. 3, 9; "G.O.P. Men Top Legislative Race," *Oregonian*, Nov. 9, 1938, 1.

100 "Petty theft Brings Terms in County Jail," *Oregonian*, June 15, 1937, 12.

101 Herbert Lundy, "Candidates In Annual Filing Rush," *Oregonian*, March 31, 1940, 1; "Political News In Summary," *Oregonian*, Oct. 15, 1940, 6; Paul Laartz, "Portland Apartment News Briefs," *Sunday Oregonian*, Nov. 24, 1940, sec., 2, 1; Dan E. Clark, II, "Survey Shows Riley Leads Hilton in Race for Mayor," *Oregonian*, Oct. 22, 1940, 1; Dan E. Clark, II, "Front Door Ballot Box Poll Indicate Mayoralty Race to Be Close, Show Roosevelt Ahead Here," *Oregonian*, Nov. 4, 1940, 10; "Attention, Voters," *Oregonian*, Nov. 4, 1940, 10; "Riley Victorious in Mayor Race," *Oregonian*, Nov. 8, 1940, 1. As of 1973, Hilton received the most votes of any unsuccessful Portland candidate for mayor; see "Former Legislator Frank Hilton Dies," *Oregonian*, March 22, 1973, sec. 3, 9.

102 "Meier Hits Lunch Clubs," *Morning Oregonian*, Sept. 30, 1930, 12.

103 "Sales Tax Debate Set," *Sunday Oregonian*, April 22, 1934, sec. 5, 5. For an analysis of mid-twentieth-century party politics in Oregon, see Robert E. Burton, *Democrats of Oregon: The Pattern of Minority Politics, 1900–1956* (Eugene: University of Oregon Books, 1970).

104 "City News In Brief: Townsendites to Meet," *Oregonian*, April 18, 1939, 7; "Measures Pour Into Law Mill," *Oregonian*, Jan. 31, 1939, 1.

105 "Death Removes Harry L. Gross," *Oregonian*, March 22, 1938, 9.

106 OSSA Minutes Oct. 5, 1943, vol. 1, 260–261.

107 "Representative Portlanders Form Anti-Sewage Group," *Oregonian*, Feb. 22, 1944, sec. 1, 1, 11.

108 "Sewage Group Elects Heads," *Oregonian*, Feb. 29, 1944, sec. 1, 11.

109 "Izaak Walton League Chapter Is Formed Here," *Oregon Sunday Journal*, Dec. 17, 1922, sec. 2, 2.

110 For his IWLA connection, see "Izaak Walton League Chapter Is Formed Here."

111 Eve Vogel, "The Columbia River's Region: Politics, Place, and Environment in the Pacific Northwest, 1933–Present" (PhD dissertation, University of Oregon, 2007), 49.

112 Vogel, "The Columbia River's Region," 37–40.

113 Vogel, "The Columbia River's Region," 49–50.

114 "City Moves to Further Sewer Plan," *Oregonian*, Feb. 25, 1944, sec. 1, 1, 11.

115 "Shingle Men Tell of Filth." See also Lawrence Barber, *Columbia Slough* (Portland, Oreg.: Columbia Slough Development Corporation, Oct. 1977).

116 "Bond Issue to be Put on Ballots," *Oregonian*, March 3, 1944, sec. 1, 1, 7.

117 Carl Abbott, *The Metropolitan Frontier: Cities in the Modern American West* (Tucson, Ariz.: University of Arizona Press, 1993), 37.

118 Wolman Report.

119 In 1943 the Oregon State Legislature authorized a biennial state budget of $24,164,706; see Earl Snell, "Twelfth Biennial Budget, State of Oregon, 1945–1947, Submitted to the Forty-third Legislative Assembly," 73.

120 John A. Guthrie, *The Economics of Pulp and Paper* (Pullman: State College of Washington Press, 1950), 9.

121 William G. Robbins, *Landscapes of Conflict: The Oregon Story, 1940–2000* (Seattle: University of Washington Press, 2004), 10–11.

122 "L.A. Engineer Here Advising," *Oregonian*, July 20, 1944, sec. 1, 9. A. M. Rawn was also involved in the redesign of Vancouver, BC's, sewer infrastructure; see Keeling, "Urban Waste Sinks," 62.

123 This is identified as "Project (2)" of the six proposed options; see Wolman Report, 16–17, 21.

124 "Bottleneck in Materials Delaying Progress on New City Sewer System," *Oregonian*, Sept. 19, 1948, sec. 1, 28; see also OSSA Minutes, Jan. 23, 1948, vol. 2, 82–86; "Sewer Plant Site Chosen," *Oregonian*, Sept. 7, 1944, sec. 1, 6.

125 These reports included Fred Merryfield, "Stream Pollution Studies–1944: Preliminary Report Submitted to Governor's Special Committee, Oregon State Sanitary Authority" (Corvallis: Oregon State Agricultural College Engineering Experiment Station, [Jan.] 1945), R. E. Dimick and Merryfield, *The Fishes of the Willamette River System in Relation to Pollution, Bulletin Series, No. 20* (Corvallis, Oreg.: Oregon State Agricultural College Engineering Experiment Station, [June] 1945), and Merryfield and W. G. Wilmot, *1945 Progress Report on Pollution of Oregon Streams, Bulletin Series, No. 19* (Corvallis: Oregon State Agricultural College Engineering Experiment Station, [June] 1945). For Governor Snell's approval of this funding, see Merryfield and Wilmot, *1945 Progress Report*, 7–8.

126 "Sewage Plans Revised to Fit," *Oregonian*, March 27, 1945, 5. For analyses of attempts to use sewage treatment byproducts in the United States, United Kingdom, and France in the nineteenth and twentieth centuries, see Joel A. Tarr, "From City to Farm: Urban Wastes and the American Farmer," *Agricultural History* 49:4 (Oct. 1975), 598–612; Nicholas Goddard, "'A Mine of Wealth'? The Victorians and the Agricultural Value of Sewage," *Journal of Historical Geography* 22:3 (1996), 274–290; John Sheail, "Town Wastes, Agricultural Sustainability, and Victorian Sewage," *Urban History* 23:2 (Aug. 1996), 189–210; and Sabine Barles and Laurence Lestel, "The Nitrogen Question: Urbanization, Industrialization, and Water Quality in Paris, 1830–1939," *Journal of Urban History* 33:5 (July 2007), 794–812.

127 Virgil Smith, "Oregon Waters: Crystal Ball Shows Them Clean Again," *Oregonian*, Jan. 21, 1945, magazine section, 5. For an expression of other questions about the project, see "Sewage Plant Opinion Asked," *Oregonian*, Oct. 26, 1944, sec. 2, 3.

128 Smith, "Oregon Waters."

129 Smith, "Oregon Waters." See also Robbins, *Landscapes of Conflict*, 250–251.

130 Ron Moxness, "Progress Report Made on $16,000,000 Project," *Sunday Oregonian*, July 31, 1949, sec. 2, 6; "Work Begins on Sewage Disposal System Authorized by Portland Voters in 1944," *Oregonian*, July 18, 1947, 14.

131 "Work Begins on Sewage Disposal System."

132 "Oregon Cities Push Work on Sewage, 15 Disposal Plants Under Construction," *Oregonian*, Oct. 26, 1947, 20; OSSA Minutes, July 11, 1947 vol. 2, 60–65.

133 In late October 1947, fifteen sewage treatment plants were under construction in communities along the Willamette River; see "Oregon Cities Push Work on Sewage."

CHAPTER 4: POSTWAR EFFLUENTS

1 Walter Mattila, "Millions of Crawfish Perishing; Pollution Hits Hardy Bullheads," *Oregon Journal*, Sept. 18, 1949, A9.

2 "Willamette River Pollution: A Report of the Stream Purification Committee—D. B. Charlton, Chairman," Oregon Division, Izaak Walton League of America, Dec. 4, 1948, Folder 4: Willamette River Pollution Dec. 4 1948, Box 23, David B. Charlton Papers (MSS 1900), Oregon Historical Society Research Library, Portland, Oreg. [hereafter Charlton Papers]. "Stream Pollution Program Draws Walton League Rap," *Oregonian*, Dec. 21, 1948, 12. The Oregon IWLA's 1948 report is also discussed in Robbins, *Landscapes of Conflict*, 254–255. For IWLA's opposition to dam building, see Robbins, *Landscapes of Conflict*, 240–247.

3 "Willamette River Pollution," Dec. 4, 1948, Charlton Papers.

4 For Charlton's materials to Sanitary Authority Chairman Harold Wendel, see Oregon State Sanitary Authority Meeting Minutes [hereafter OSSA Minutes], Aug. 20, 1948, vol. 2, 119 120, Oregon State Department of Environmental Quality, Portland, Oreg.

5 Mervin Shoemaker, "Test Pilots for Industry," *Oregonian*, June 13, 1954, Northwest Roto Magazine section, 7.

6 OHS finding aid to David B. Charlton papers, Mss 1900; Reuben O. Norman, ed., *Capitol's Who's Who for Oregon, 1948–1948* (Portland, Ore.: Capitol Publishing Company, 1948), 107.

7 Charlton to "Dear Al," April 9, 1975, folder Correspondence re: Water Pollution 1930–1950, box 18, Charlton Papers.

8 "Past Chiefs Gain Oscars," *Oregonian*, Nov. 17, 1960, 3L.

9 Bill Hulen, "Waltons Heat Talk by Holmes," *Oregonian*, Dec. 9, 1956, sports section, 1, 4.

10 David B. Charlton to Clarence W. Klassen, Sept. 1, 1948, folder Correspondence re: Water Pollution 1947–1948, box 18, Charlton Papers. For Klassen's contribution to national industrial pollution management practices between the 1930s and 1970, see Hugh S. Gorman, "Laying the Foundation for the Control of Industrial Pollution, 1930–1970: Two Canals, A Refinery, and Clarence W. Klassen," *Journal of Illinois History* 8:3 (2005), 182–208.

11 Charlton to Dr. W. H. Stark, Vickers-Vulcan Process Engineering Co. Ltd., Montreal, Quebec, n.d. [ca. 1949], folder Correspondence re: Pollution 1949, box 18, Charlton Papers. Readers will find that I use the spelling "sulfite" in my own narrative, but, when providing direct quotes, I use the period-appropriate spelling "sulphite" if this was the word that the speaker or writer used at the time.

12 See Gregory Summers, *Consuming Nature: Environmentalism in the Fox River Valley, 1850–1950* (Lawrence: University Press of Kansas, 2006), for an analysis of local Izaak Walton League of America (IWLA) involvement in water pollution abatement efforts along Wisconsin's Fox River.

13 V. J. Meunch, President, Wisconsin Division, IWLA, to Charlton, Dec. 1, 1949, folder Correspondence re: Pollution 1949, box 18, Charlton Papers.

14 William Voigt Jr., Executive Director, IWLA, to Charlton and Meunch, Dec. 21, 1949, folder Correspondence re: Pollution 1949, box 18, Charlton Papers.

15 Charlton to Klassen, Sept. 1, 1948, folder Correspondence re: Water Pollution 1947–1948, box 18, Charlton Papers.

16 Abatement advocates in other states were also not pushing for a "pristine" river. For this kind of advocacy in Pennsylvania, see Nicholas Casner, "Angler Activist: Kenneth Reid, the Izaak Walton League, and the Crusade for Federal Water Pollution Control," *Pennsylvania History* 66:4 (1999), 535–553; for Minnesota, see Thomas Howard Hayden (Paul E. Toren, ed.), *Citizen Leadership in Conservation: The Minnesota Izaak Walton League, 1922–1973* (St. Paul, Minn.: Izaak Walton League Minnesota Division, 2001) and Philip V. Scarpino, *Great River: An Environmental History of the Upper Mississippi, 1890–1950* (Columbia: University of Missouri Press, 1985).

17 See, for example, Kenneth A. Reid, "Editorial: The Water Policy of the League," *Outdoor America*, January 1947, 3, and Casner, "Angler Activist."

18 George H. Risley, Secretary, Portland Chapter, IWLA [membership solicitation form letter], n.d. [ca. 1943], folder Portland Chapter-Izaak Walton League 1943, box 8, Charlton Papers.

19 Charlton to "The Editor," *Springfield News*, Aug. 3, 1951, folder Correspondence re: Water Pollution 1951–1958, box 18, Charlton Papers. For this same sentiment from pollution abatement advocates along Maine's Androscoggin River, see Richard W. Judd, "The Coming of the Clean Waters Acts in Maine, 1941–1961," *Environmental History Review* 14:3 (Fall 1990), 55.

20 John A. Guthrie, *The Economics of Pulp and Paper* (Pullman: State College of Washington Press, 1950), 28–30.

21 Guthrie, *The Economics of Pulp and Paper*, 28–30.

22 Guthrie, *The Economics of Pulp and Paper*, 28–30.

23 Guthrie, *The Economics of Pulp and Paper*, 28–30. Two additional processes represented a small segment of total industry production as of 1950. The *semi-chemical* process produced a pulp often used to make board and coarse heavy papers, and involved partially digesting woods with solvents and then reducing this material to pulp with mechanical pulverizers. A few mills in the United States recycled paper and rags, and used wheat, oat, and rye straw to produce pulp; as of 1950, outside of California, the use of straw for pulp was rare on the Pacific Coast.

24 John A. Guthrie, *An Economic Analysis of the Pulp and Paper Industry* (Pullman: Washington State University Press, 1972), 101–103.

25 Guthrie, *An Economic Analysis of the Pulp and Paper Industry*, 101–103.

26 Guthrie, *An Economic Analysis of the Pulp and Paper Industry*, 101–103.

27 Examples of this consumption included evaporative losses from paper machines and chemical recovery processes. Guthrie, *An Economic Analysis of the Pulp and Paper Industry*, 101–103.

28 Guthrie, *The Economics of Pulp and Paper*, 41–45.

29 Guthrie, *The Economics of Pulp and Paper*, 9.

30 Guthrie, *The Economics of Pulp and Paper*, 41–45, 81.

31 Nancy Kane Ohanian, *The American Pulp and Paper Industry, 1900-1940: Mill Survival, Firm Structure, and Industry Relocation* (Westport, Conn.: Greenwood Press, 1993), 1–2.

32 Ohanian, *The American Pulp and Paper Industry*, 27.

33 Guthrie, *The Economics of Pulp and Paper*, 9. For timber industry waste utilization, see John L. Denny, "Oregon Lags in Use of Lumber Waste, According to Trade Magazine Report," *Sunday Oregonian*, Nov. 10. 1957, sec. 1, p 64.

34 National Resources Committee, Water Pollution in the United States: *Third Report of the Special Advisory Committee on Water Pollution* (Washington, DC: US Government Printing Office, 1939), 17. $18,240,000 in 1939 was approximately $249,000,000 in 2017.

35 "One Really Bright Spot," *Oregonian*, Sept. 19, 1957, 18. Individually, Washington State ranked first in the nation, and Oregon twelfth.

36 "Paper Statistics," *Sunday Oregonian*, Sept. 22, 1957, sec. 1, p 33. In comparison, Washington State had twenty mills employing 14,333, $75,000,000 annual payroll, and shipped more than $450,000,000 worth of products in 1956. $22 million in 1957 was approximately $145 million in 2017, and $126 million was approximately $832 million.

37 "One Really Bright Spot."

38 There are many sources on this in professional journals through the 1940s and 1950s, but see, for example, Harold R. Murdock, "Industrial Wastes," *Industrial and Engineering Chemistry* 42:6 (June 1950), 71A–72A.

39 Ohanian, *The American Pulp and Paper Industry*, 41–45.

41 World Bank Group, "Introduction to Wastewater Treatment Processes," http://water.worldbank.org/shw-resource-guide/infrastructure/menu-technical-options/wastewater-treatment, accessed Feb. 17, 2016.

42 Oregon State Sanitary Authority, "Fifth Biennial Report for the Period July 1, 1946 to June 30, 1948," 8; Oregon State Sanitary Authority, "Stream Survey Report: Willamette River, 1953," 3; US Bureau of the Census, *Census of Population: 1950 Oregon—Volume II, Part 37, Characteristics of the Population, Number of Inhabitants, Oregon*, 12–13, available at http://www.census.gov/prod/www/abs/decennial/1950.htm, accessed Oct. 6, 2008.

43 "River Pollution Ban Backed, But Remedy Lacking," *Oregon Journal*, May 20, 1926, sec. 1, p. 2.

44 George W. Gleeson and Fred Merryfield, *Industrial and Domestic Wastes of the Willamette Valley, Bulletin Series, No. 7* (Parts 1, 2, and 3) (Corvallis: Oregon State Agricultural College Engineering Experiment Station, [May] 1936), 11.

45 For an overview of the pulp and paper industry's lack of interest in pollution abatement prior to this time, see Richard A. Bartlett, *Troubled Waters: Champion International and the Pigeon River Controversy* (Knoxville: University of Tennessee Press, 1995).

46 David C. Knowlton, "Progress in Pollution Abatement and Water Conservation," *Tappi* 47:11 (Nov. 1964), 16A.

47 Wallace Scot McFarlane, "The Limits of Progress: Walter Lawrance and the Shifting Terrain of Science, Pollution, and Environmental Politics on Maine's Androscoggin River, 1941–1977" (Senior thesis, Bowdoin College, 2009), 27.

48 Hervey J. Skinner, "Waste Problems in the Pulp and Paper Industry," *Industrial and Engineering Chemistry* 31:11 (Nov. 1939), 1331.

49 Benjamin Ross and Steven Amter, *The Polluters: The Making of Our Chemically Altered Environment* (New York: Oxford University Press, 2010), 14–16, 142–143.

50 Knowlton, "Progress in Pollution Abatement and Water Conservation," 16A.

51 "The National Council for Stream Improvement," *Science* 99:2582 (June 23, 1944), 508; Russell L. Winget, "A Definitive History of the National Council for Stream Improvement (of the Pulp, Paper, and Paperboard Industries), Inc., 1943–1966," National Council for Stream Improvement, 1966.

52 "The National Council for Stream Improvement"; W. Rudolfs et al., "A Critical Review of the Literature of 1946 on Sewage and Waste Treatment and Stream Pollution," *Sewage Works Journal* 19:2 (Mar. 1947), 232; Winget, "A Definitive History of the National Council for Stream Improvement." $140,000 in 1946 was about $1.23 million in 2017.

53 Don. E. Bloodgood, "The Industrial Waste Problem II: Strawboard, Petroleum, and Distillery Wastes," *Sewage Works Journal* 19:4 (July 1947), 607. $140,000 in 1946 was approximately $1.38 million in 2017.

54 The NCSI is currently known as the National Council for Air and Stream Improvement (NCASI). See "About Us," http://www.ncasi.org/about/default.aspx, accessed Sept. 30, 2008; Winget, "A Definitive History of the National Council for Stream Improvement." For summaries of research funded by the NCSI in the 1940s and 1950s, see Isaiah Gellman, "Support of Water Pollution Research Projects at Research Institutions by the National Council for Stream Improvement and Other Industrial Associations," in Edward F. Eldridge, ed., *Proceedings of the Second Pacific Northwest Symposium on Water Pollution Research: Financing Water Pollution Research Projects, Jan. 20, 1958* (Portland, Oreg., US Department of Health, Education, & Welfare, Public Health Service, Region 9, 1958), 7–11.

55 OSSA Minutes Jan. 23, 1948, vol. 2, 86–107.

56 Guthrie, *The Economics of Pulp and Paper*, 168–175. There are other reasons why industry profit computations are difficult. The industry reported transportation costs that were often simplified and did not reflect administrative expenses and interest on long-term debt. Also, private industry generally—particularly until the 1970s—was under minimal regulatory requirements or social sanction to report this kind of information. Later research adds even more nuance to our understanding of behavior and decisions at the level of the individual firm with regard not only to the pursuit of profit but also to compliance with environmental regulations; see Guthrie, *The Economics of Pulp and Paper*, 165–166 and 174–175 and Neil Gunningham, Robert A. Kagan, and Dorothy Thornton, *Shades of Green: Business, Regulation, and Environment* (Stanford, Calif.: Stanford University Press, 2003).

57 Sources for this and subsequent graphs of state general fund and OSSA/DEQ funding provided in Bibliographic Essay. $21.3 million in 1939 was approximately

$291 million in 2017; $192.7 million in 1954 was approximately $1.38 billion in 2017.

58 US Dept. of Commerce Bureau of the Census, *US Census of Manufactures*, Washington, DC: US Government Printing Office, 1954. Other analyses show that during these years the national pulp and paper industry realized a rate-of-return comparable to the average realized in the broader national manufacturing sector, about 6 percent. Further, West Coast mills were relatively more profitable in the 1930s and 1940s than the national industry average; see Ohanian, *The American Pulp and Paper Industry*, 67–68; Guthrie, *The Economics of Pulp and Paper*, 165–166. $862.8 million in 1939 was approximately $1.18 billion in 2017, while $4.6 billion in 1954 was approximately $32.9 billion.

59 Another interesting comparison of the relative importance among different constituencies regarding water quality is found in revisiting the matter of the $12 million bond measure Portland voters approved in May 1944 to upgrade the city's sewer infrastructure and provide for primary sewage treatment. This amount equated to about half of the entire Oregon State General Fund during the 1943–1945 biennium, and voters approved it in the midst of World War II. This shows again that, in spite of the financial burden, Portland residents viewed the matter as a serious one and were willing to pay for it.

60 Oregon State Sanitary Authority, "Fifth Biennial Report for the Period July 1, 1946 to June 30, 1948"; Oregon State Sanitary Authority, "Stream Survey Report: Willamette River, 1953."

61 It may not be an accident that the Crown Zellerbach Company was one of the top six most profitable pulp and paper industry firms from the mid-1940s through the 1960s, as reported in Guthrie, *An Economic Analysis of the Pulp and Paper Industry*, Table 8-7, 197.

62 August S. Erspamer and William D. Rice, "Treatment of Pulp and Paper Mill Waste," *Industrial and Engineering Chemistry* 41:8 (Aug. 1949), 1806.

63 OSSA Minutes, May 7, 1940, vol. 1, 32–33 and 38–50.

64 OSSA Minutes, March 9, 1945, vol. 2, 10–11.

65 OSSA Minutes, June 22, 1945, vol. 2, 14–15.

66 "Report Asked on Sewage," *Oregonian*, Jan. 24, 1948, 7; see also OSSA Minutes, Jan. 23, 1948, vol. 2, 86–107.

67 OSSA Minutes, Jan. 23, 1948, vol. 2, 86–107.

68 OSSA Minutes, Jan. 23, 1948, vol. 2, 86–107.

69 "Willamette River Pollution," Dec. 4, 1948, Charlton Papers. For Charlton's materials to Sanitary Authority Chairman Harold Wendel, see OSSA Minutes, Aug. 20, 1948, vol. 2, 119–120; Weyerhaeuser constructed a pilot recovery plant in Longview in 1946. See "Pulp Wastes Solution Seen in Weyerhaeuser's Tests," *Oregonian*, March 24, 1950, sec. 2, 10.

70 "Stream Pollution Program Draws Walton League Rap," *Oregonian*, Dec. 21, 1948, 12.

71 OSSA Minutes, Dec. 2, 1948, vol. 2, 145.

72 "Session Open on Pollution," *Oregonian*, Feb. 18, 1949, sec. 1, 13. In addition to the 1948 IWLA report, Peter W. Welch, representing the Federation of Community Clubs, had recently criticized the OSSA in radio addresses and

public presentations; see OSSA Minutes, July 11, 1947, vol. 2, 62, April 11, 1947, vol. 2, 52–55, and Feb. 25, 1949, vol. 2, 162–163.

73 OSSA Minutes Feb. 25, 1949, vol. 2, 170–174.

74 "State Warns Pollution Must End," *Oregonian*, April 27, 1949, 1; see also OSSA Minutes April 26, 1949, vol. 2, 179–196.

75 "State Warns Pollution Must End."

76 OSSA Minutes, May 9, 1950, vol. 2, 230–234.

77 US Bureau of the Census, *Census of Population: 1950 Oregon—Volume II, Part 37,* 12–13.

78 "State Warns Pollution Must End"; see also OSSA Minutes April 26, 1949, vol. 2, 190–195, and Glenn D. Carter, "Pioneering Water Pollution Control in Oregon." *Oregon Historical Quarterly* 107:2 (Summer 2006), 263.

79 John T. Cumbler, "Conflict, Accommodation, and Compromise: Connecticut's Attempt to Control Industrial Wastes in the Progressive Era," *Environmental History* 5:3 (2000), 328.

80 OSSA Minutes, July 22, 1949, vol. 2, 199–200.

81 "Pulp Wastes Solution Seen in Weyerhaeuser's Tests," *Oregonian*, March 24, 1950, sec. 2, 10.

82 Terence Kehoe, *Cleaning Up the Great Lakes: From Cooperation to Confrontation* (DeKalb, Ill.: Northern Illinois University Press, 1997), particularly 5–11.

83 "Industry Said Eager to Find Solution to River Pollution," *Oregonian*, Oct. 21, 1949, 19.

84 Charlton to Wendel, Oct. 25, 1949, folder Correspondence re: Water Pollution 1949, box 18, Charlton Papers.

85 OSSA Minutes, Nov. 4, 1949, vol. 2, 212–217.

86 OSSA Minutes, Nov. 4, 1949.

87 "Crackdown on Pollution," *Oregonian*, Dec. 12, 1949, 12.

88 "Freeing the River," *Oregon Journal*, Feb. 19, 1950, 18A. See also Robbins, *Landscapes of Conflict*, 256–257.

89 "Pulp Concern Denies Threat," *Oregonian*, Feb. 18, 1950, sec. 3, 3. For Publishers' Pulp & Paper Company's earlier threats of possible mill closures the company attributed to OSSA policies, see OSSA Minutes, Nov. 4, 1949, vol. 2, 212–217.

90 For this analysis, see Robbins, *Landscapes of Conflict*, 257.

91 "Pollution Showdown," *Oregon Journal*, Feb. 17, 1950, 12.

92 For experiments in Oregon on the lagooning of sulfite wastes, see "Paper Mills to Investigate Lagooning Sulphite Waste," *Oregonian*, Feb. 17, 1950, 8.

93 "Sanitary Authority Issues Orders to Paper Mills on Waste Dumping," *Oregon Journal*, May 10, 1950, 2; "State Edict Orders End of Pollution," *Oregonian*, May 10, 1950, 1. For Attorney General Neuner's finding of fact, see OSSA Minutes, May 9, 1950, vol. 2, 230–236.

94 "River Waste Takes Toll," *Oregonian*, Sept. 21, 1950, sec. 1, 11.

95 George W. Gleeson, *The Return of a River: The Willamette River, Oregon* (Corvallis: Oregon State University Water Resources Research Institute, 1972), 24, 26.

96 "National Pollution Control Meetings Slated in Portland," *Oregonian*, May 23, 1950, sec. 1, 13; "Adviser Board to Tour Area," *Oregonian*, July 7, 1950, sec. 2, 7. See also "State's Water Draws Study," *Oregonian*, April 28, 1950, sec. 3, 8.

97 "Pollution Board May Ask Congress for Funds for Columbia Research,"
    Oregonian, July 12, 1950, sec. 2, 12; "Local, Federal Officials Discuss Northwest
    River Pollution Puzzle," *Oregon Journal*, July 11, 1950, sec. 1, 7; OSSA Minutes,
    May 9, 1950, vol. 2, 230–234.

98 OSSA Minutes, Dec. 7, 1951, vol. 2, 289–290.

99 OSSA Chairman Wendel, for example, "refused to indorse the lagoon methods,
    and emphasized that the mills must develop a process to rid the river of sulfite
    pollution." See "Praise Given Paper Plants," *Oregonian*, June 2, 1951, sec. 1, 7;
    "Mills Praised on Cleansing," *Oregonian*, June 7, 1951, sec. 1, 19. In spite of
    Wendel's misgivings, the OSSA considered this practice an adequate short-term
    solution; see OSSA Minutes, June 1, 1951, vol. 2, 272–273, and Sept. 7, 1951, vol.
    2, 280–281.

100 "Gain in War on Pollution," *Oregonian*, Aug. 29, 1951, sec. 1, 14. Clackamas
    County expanded its program to use sulfite wastes as a road binder in 1952 and
    again in 1953. See "Roads to Get Sulphite Acid," *Oregonian*, Feb. 19, 1952, sec. 1,
    9; "Roads Using Mill's Waste," *Oregonian*, July 17, 1953, sec. 1, 18.

101 "Plan Mapped for Sulphite," *Oregonian*, May 13, 1953, sec. 1, 19. OSSA Minutes,
    Feb. 13, 1953, vol. 2, 364.

102 "Paper Firm Seeks to Halt Sulphite Stream Pollution," *Oregonian*, July 22, 1953,
    sec. 1, 1.

103 Oregon counties stopped using sulfite wastes as a road binder some time later
    when experience showed that the practice kept down dust but tended to get
    washed off roads by rain and ended up in lakes and streams anyway. See Carter,
    "Pioneering Water Pollution Control," 263.

104 OSSA Minutes, March 21, 1952, vol. 2, 293–297.

105 "Fighting Stream Pollution," *Oregonian*, March 28, 1952, sec. 1, 16.

106 "Our Rivers Come Clean," *Oregonian*, July 18, 1952, sec. 1, 18.

CHAPTER 5: POLLUTED PARADISE

1 "Oregon's Cleaner Waters," *Oregonian*, March 23, 1954, sec. 1, 13.

2 Glen D. Carter, "Pioneering Water Pollution Control in Oregon," *Oregon
   Historical Quarterly* 107:2 (Summer 2006), 267–268.

3 "Cleaner But Not Clean," *Oregonian*, Aug. 20, 1956, sec. 1, 10; Herbert Lundy,
   "River Pollution Control on Treadmill," *Oregonian*, Aug. 15, 1957, sec. 1, 14.

4 "Cleaner But Not Clean."

5 William G. Robbins, *Landscapes of Conflict: The Oregon Story, 1940–2000* (Seattle:
   University of Washington Press, 2004), 10–11.

6 State of Oregon, *Progress Report of the Postwar Readjustment Commission*, Jan. 1,
   1947–Dec. 31, 1948 (Salem, Oreg.: 1949), 4.

7 The next three most populous cities and counties in 1950 included Salem
   (43,410) and Marion County (101,401), Eugene (35,879) and Lane County
   (125,776), and Corvallis (16,207) and Benton County (31,570). See US Bureau
   of the Census, *Census of Population: 1950 Oregon—Volume II, Part 37,
   Characteristics of the Population, Number of Inhabitants, Oregon*, 12–13, available at
   http://www.census.gov/prod/www/abs/decennial/1950.htm, accessed Oct. 6,
   2008; US Bureau of the Census, *Census of Population: 1960 Oregon—Volume I,
   Part 39, Characteristics of the Population*, 39-13 and 39-14, available at http://

www.census.gov/prod/www/abs/decennial/1960cenpopv1.html, accessed Dec. 31, 2011. Wartime and postwar expansion taxed sewerage infrastructures across the country.

8   For the example of Houston, Texas, see Martin V. Melosi, "Sanitary Services and Decision Making in Houston, 1876–1945," *Journal of Urban History* 20:3 (May 1994), 387, 393–395.

9   Lundy, "River Pollution Control on Treadmill."

10 Lundy, "River Pollution Control on Treadmill."

11 "Cleaner But Not Clean."

12 Ron Moxness, "Progress Report Made on $16,000,000 Project," *Sunday Oregonian*, July 31, 1949, sec. 2, 6.

13 "Sewer Lines Link in River Marks Disposal Plan Gain," *Oregonian*, Oct. 3, 1952, sec. 1, 14.

14 OSSA Minutes, May 9, 1950, vol. 2, 230–234.

15 Clarence J. Velz, "Report on Natural Purification Capacities: Willamette River" (Ann Arbor: University of Michigan School of Public Health, for the National Council for Stream Improvement of the Pulp, Paper and Paperboard Industries, Inc., May 1951); Velz, "Supplementary Report on Lower Willamette River Waste Assimilation Capacity" (Ann Arbor: University of Michigan School of Public Health, for the National Council for Stream Improvement of the Pulp, Paper and Paperboard Industries, Inc., March 1961).

16 A partial list would include: Velz, "Factors Influencing Self-Purification and Their Relation to Pollution Abatement," *Sewage Works Journal* 19:4 (July 1947), 629–644; Velz, "Efficient Utilization of Water Resources for Industrial Waste Disposal," paper presented at the meeting of the National Council for Stream Improvement of the Pulp, Paper and Paperboard Industries, Inc., White Sulphur Springs, W. Va., Aug. 7, 1958; Velz, "Report on Natural Purification Capacities: Willamette River"; Velz and J. J. Gannon, "Forecasting Heat Loss in Ponds and Streams," *Journal (Water Pollution Control Federation)* 32:4 (April 1960), 392–417; Velz, *Applied Stream Sanitation* (2nd ed.) (New York: John Wiley and Sons Inc., 1984). In 1947, Velz was head of the Civil Engineering Department and professor of Sanitary Engineering at Manhattan College in New York; by 1950, he was professor and chair of the Department of Public Health Statistics and the University of Michigan's School of Public Health.

17 Velz, "Report on Natural Purification Capacities: Willamette River," 1, 3.

18 OSSA Minutes, July 22, 1949, vol. 2, 199–200.

19 Melosi, *The Sanitary City: Urban Infrastructure in America from Colonial Times to the Present* (Baltimore, Md.: Johns Hopkins University Press, 2000), 255–257.

20 See, generally, Peter J. Bowler, *The Norton History of the Environmental Sciences* (New York: W. W. Norton & Company, 1993).

21 Harold W. Streeter, "Surveys for Stream Pollution Control," *Proceedings of the ASCE* 64 (Jan. 1938), 6, quoted in Melolsi, *The Sanitary City*, 256. On the process of applying and refining procedures and techniques established in the early twentieth century, see Joel A. Tarr, *The Search for the Ultimate Sink: Urban Pollution in Historical Perspective* (Akron, Ohio: University of Akron Press, 1996), 368–370.

22 Velz, "Factors Influencing Self-Purification," 629.

23 Arn Keeling describes this dynamic as it applied to the Fraser River in Vancouver, BC, during this same period; see Keeling, "Urban Waste Sinks as a Natural Resource: The Case of the Fraser River," *Urban History Review* 34:1 (2005), 58–70.

24 See, for example, "Proceedings of Local Associations," *Sewage Works Journal* 10:4 (July 1938), 794–800, and Board of Review (R. H. Corey, Wellington Donaldson, Carl E. Green, and Abel Wolman), "Report on the Collection and Disposal of Sewage," Portland, Oreg., Aug. 19, 1939, in folder 8402-01 Sewage Disposal Project—Sewage Collection and Disposal Report 1939, box 1, Public Works Administration-City Engineer's Historical/Subject Records, City of Portland Archives, Portland, Oreg.

25 "Proceedings of Local Associations."

26 Velz, "Report on Natural Purification Capacities: Willamette River," 13–17.

27 Melosi, *The Sanitary City*, 256–257.

28 Velz, "Report on Natural Purification Capacities: Willamette River," 41.

29 Table 5.1 data from OSSA, *Biennial Report, 1956–1958*, 30.

30 Lundy, "River Pollution Control on Treadmill."

31 "Interim Report on Status of Water Pollution Control in the Willamette River Basin" (Portland: Oregon State Sanitary Authority, 1957), 13[14].

32 "Interim Report on Status of Water Pollution," 11–18.

33 Jouni Paavola, "Interstate Water Pollution Problems and Elusive Federal Water Pollution Policy in the United States, 1900–1948," *Environment and History* 12:4 (2006), 442–446, 450.

34 For an overview of federal public health and water quality research at the Cincinnati facility, see James E. Smith, James F. Kreiss, Gary S. Logsdon, and Norbert Schomaker, "Environmental Research in Cincinnati: A Century of Federal Partnership," *Environmental Engineer* 13 (Summer 2011), available at www.aaee.net/DownloadCenter/EEJournalV13P1.pdf, accessed Dec. 30, 2011.

35 William L. Andreen, "The Evolution of Water Pollution Control in the United States—State, Local, and Federal Efforts, 1789–1972: Part II," *Stanford Environmental Law Journal* 22 (June 2003), 235–239.

36 Paul Charles Milazzo, *Unlikely Environmentalists: Congress and Clean Water, 1945–1972* (Lawrence: University Press of Kansas, 2006), 19–20.

37 Milazzo, *Unlikely Environmentalists*, 20–21.

38 John H. Hall, "Fourteenth Biennial Budget, State of Oregon, 1949–1951, Submitted to the Forty-Fifth Legislative Assembly" (Salem, Oreg.: Budget Division of the Executive Department, 1948), 81; Douglas McKay, "Fifteenth Biennial Budget, State of Oregon, 1951–1953, Submitted to the Forty-Sixth Legislative Assembly" (Salem, Oreg.: Budget Division of the Executive Department, 1950), 95; Paul L. Patterson, "Sixteenth Biennial Budget, State of Oregon, 1953–1955, Submitted to the Forty-Seventh Legislative Assembly" (Salem, Oreg.: Budget Division of the Executive Department, 1952), 124–127; Paul L. Patterson, "Seventeenth Biennial Budget, State of Oregon, 1955–1957, Submitted to the Forty-Eighth Legislative Assembly" (Salem, Oreg.: Budget Division of the Executive Department), 1954, 112–114.

39 Richard W. Judd and Christopher S. Beach, *Natural States: The Environmental Imagination in Maine, Oregon, and the Nation* (Washington, DC: Resources for the

Future, 2003), 38–39; Adam W. Rome. "'Give Earth a Chance': The Environmental Movement and the Sixties," *The Journal of American History* 89:2 (Sept. 2003), 525–554.

40 David B. Charlton to The Honorable John A. Blatnik, March 15, 1948, folder Correspondence re: Water Pollution 1947–1948, box 18, David B. Charlton Papers (MSS 1900) [hereafter Charlton Papers], Oregon Historical Society Research Library, Portland, Oreg.

41 Milazzo, *Unlikely Environmentalists*, 18.

42 Richard W. Judd, "The Coming of the Clean Waters Acts in Maine, 1941–1961," *Environmental History Review* 14:3 (Fall 1990), 61–63; Milazzo, *Unlikely Environmentalists*, 17–37.

43 For California, see Donald J. Pisani, "Fish Culture and the Dawn of Concern Over Water Pollution," *Environmental Review* 8:2 (1984), 117–131; for Oregon, see chapter 1 of this book. See also Virgil Smith, "Oregon Waters: Crystal Ball Shows Them Clean Again," *Oregonian*, Jan. 21, 1945, Magazine section, 5.

44 Milazzo, *Unlikely Environmentalists*, 17–37. Congress further strengthened the 1956 amendments in 1961. These modifications increased federal assistance for municipal wastewater treatment projects and modestly enhanced federal enforcement powers, particularly for interstate waters. See William L. Andreen, "The Evolution of Water Pollution Control in the United States—State, Local, and Federal Efforts, 1789–1972: Part II," 242–244.

45 Elmo Smith, "Eighteenth Biennial Budget, State of Oregon, 1957–1959, Submitted to the Forty-Ninth Legislative Assembly" (Salem, Oreg.: Budget Division of the Executive Department, 1956), 128–130; Robert D. Holmes, "Nineteenth Biennial Budget, State of Oregon, 1959–1961, Submitted to the Fiftieth Legislative Assembly" (Salem, Oreg.: Budget Division of the Executive Department, 1958), p. 179–185; Mark O. Hatfield, "Twentieth Biennial Budget, State of Oregon, 1961–1963, Submitted to the Fifty-First Legislative Assembly" (Salem, Oreg.: Budget Division of the Executive Department, Dec. 1, 1960), 63–69.

46 OSSA Biennial Reports. The following are approximated 2017 values: $21,785 in 1956 equals $149,000; $66,745 in 1958 equals $431,000; $71,032 in 1960 equals $446,000; and $647,125 in 1957 equals $4,270,000.

47 Curtiss M. Everts Jr. and Arve H. Dahl, "The Federal Water Pollution Control Act of 1956," *American Journal of Public Health* 47:3 (March 1957), 305–310.

48 Bob Boxberger, "Scientist Preaches Clean Stream Doctrine," *Oregon Journal*, May 25, 1960, 7; see also Boxberger, "Pollutants Gradually Stifling Many Forms of Aquatic Life," *Oregon Journal*, May 26, 1960, Part 1, 6.

49 Bob Boxberger, "63 Research Projects Seek Key to Cause-Effect of Pollution," *Oregon Journal*, May 17, 1960, Part 1, 6. For more on Eldridge's career, see "Eldridge, Edward Franklin; ASCE Fellow (1894–1964)," *Transactions of the American Society of Civil Engineers* (1965), 819.

50 US Public Health Service, *Symposium of Research on Problems Relating to Water Pollution in the Pacific Northwest (1st)* (Springfield, Va.: US Department of Commerce, National Technical Information Service, 1957).

51 For an overview of these and additional symposia topics, see Edward F. Eldridge, ed., *Proceedings of the Eleventh Pacific Northwest Symposium on Water Pollution Research: The Social and Economic Aspects of Water-Resource Quality Control,*

November 8 and 9, 1962 (Portland, Oreg.: US Department of Health, Education, & Welfare, Public Health Service, Region IX, 1962). For a complete collection of these proceedings, visit the US Environmental Protection Agency Region 10 library, Seattle, Wash.

52 William F. Willingham, *Army Engineers and the Development of Oregon: A History of the Portland District US Army Corps of Engineers* (Washington, DC: US Government Printing Office, 1983), 10–11.

53 Patricia A. Benner and James R. Sedell, "Upper Willamette River Landscape: A Historic Perspective," in Antonius Laenen and David A. Dunnette, eds., *River Quality: Dynamics and Restoration.* (Boca Raton, Fla.: Lewis Publishing, 1997), 23–48.

54 William G. Robbins, *Landscapes of Promise*, 283–295. Quote from US Army Corps of Engineers, "A Report on Willamette River and Tributaries, Oregon," *H. Doc. 544*, 75th Cong., 3rd sess. (1938), excerpted at http://www.ccrh.org/comm/cottage/primary/544.htm, accessed Dec. 31, 2011.

55 For completion dates, locations, storage capacity, and primary uses of each dam, see Appendix 2.

56 US Army Corps of Engineers, "Willamette Valley Project," http://www.nwp.usace.army.mil/locations/willamettevalley.asp, accessed Dec. 31, 2011.

57 George W. Gleeson, *The Return of a River: The Willamette River, Oregon* (Corvallis: Oregon State University Water Resources Research Institute, 1972), 62–63; Appendix 2.

58 Gleeson, *The Return of a River*, 62–64.

59 For a clear articulation and example of this approach to river systems and waste management, see Keeling, "Urban Waste Sinks as a Natural Resource," 58–70.

60 Gleeson, *The Return of a River*, 62–64.

61 Carter, "Pioneering Water Pollution Control in Oregon," 260.

62 Walter Mattila, "Millions of Crawfish Perishing; Pollution Hits Hardy Bullheads," *Oregon Journal*, Sept. 18, 1949, A9.

63 "State's Water Draws Study," *Oregonian*, April 28, 1950, sec. 3, 8.

64 "National Pollution Control Meetings Slated in Portland," *Oregonian*, May 23, 1950, sec. 1, 13; "Adviser Board to Tour Area," *Oregonian*, July 7, 1950, sec. 2, 7. See also "State's Water Draws Study," *Oregonian*, April 28, 1950, sec. 3, 8.

65 Oregon State Water Resources Board, "First Biennial Report" (Salem: Oregon State Water Resources Board, July 1956).

66 OSSA Minutes, April 7, 1960, vol. 3, 362.

67 John H. Lincoln and Richard F. Foster, *Report on Investigation of Pollution in the Lower Columbia River* (Salem, Oreg., Interstate Technical Advisory Committee, 1943).

68 Lawrence Barber, "Pulp Mills Strive to Whip Pollution Problem in Columbia," *Oregonian*, March 15, 1953, sec. 1, 28.

69 For reports on progress abating lower Columbia River slime pollution, see Barber, "Pulp Mills Strive to Whip Pollution Problem in Columbia"; "Survey Made on Pollution," *Oregonian*, Sept. 28, 1953, sec. 1, 19; and "Getting Together on Pollution," *Oregon Journal*, May 27, 1958, sec. 4, 4.

70 "Clean Rivers in Sight," *Oregonian*, Oct. 3, 1953, sec. 1, 10.

71 "Joint Pollution Control," *Oregonian*, May 28, 1958, 26. For the 1948 survey, see Frederic F. Fish and Robert R. Rucker, "Pollution in the Lower Columbia Basin in 1948–with Particular Reference to the Willamette River." *Special Scientific Report: Fisheries No. 30* (Washington, DC: US Department of the Interior, US Fish and Wildlife Service, June 1950).

72 For proceedings of this conference, see Edward F. Eldridge, ed., *Proceedings of the Third Pacific Northwest Symposium on Water Pollution Research: The* Sphaerolitus *(Slime) Problem, Sept. 10–11, 1958* (Portland, Oreg., US Department of Health, Education, & Welfare, Public Health Service, Region 9, 1958).

73 Lawrence Barber, "Columbia Flow Said Polluted," *Oregonian*, Sept. 11, 1958, sec. 1, 1, 10, and Barber, "Conferees Urge Two-State Action to Remove Pollution," *Oregonian*, Sept. 12, 1958, sec. 1, 17. See also Barber, "Savants to Plan Control of Columbia River Slime," *Oregonian*, Aug. 22, 1958, sec. 1, 23.

74 "Columbia Pollution Attacked," *Oregonian*, Aug. 6, 1959, sec. 1, 1, 9.

75 "New Attack on Pollution," *Sunday Oregonian*, Aug. 9, 1959, sec. 1, 30. See also Robbins, *Landscapes of Conflict*, 260–263.

76 Blaine Schulz, "The Southwest Wants Our Water: Columbia Siphoning Said Threat to Nature," *Oregonian*, Jan. 24, 1966, sec. 2, 10; "Lawyer Raps River Pollution," *Oregonian*, Sept. 8, 1965, sec. 1, 11; Carter, "Pioneering Water Pollution Control," 264.

77 See "Antipollution Deadlines Set," *Oregonian*, Sept. 4, 1959, sec. 1, 1, 19; see also "River Slime End Visioned," *Oregonian*, Sept. 22, 1959, sec. 1, 16.

78 For a brief history of the Columbia Slough, see Lawrence Barber, *Columbia Slough* (Portland, Oreg., Columbia Slough Development Corporation, Oct. 1977). For an interpretation of the environmental history of the slough that finds evidence of environmental racism, see Ellen Stroud, "Troubled Waters in Ecotopia: Environmental Racism in Portland, Oregon," *Radical History Review* 74 (1999), 65–95.

79 "Shingle Men Tell of Filth," *Oregonian*, March 2, 1944, sec. 2, 3.

80 Wolman Report, 16.

81 "State Brackets Slough Waste," *Oregonian*, April 22, 1954, sec. 1, 1.

82 OSSA Minutes, Oct. 23, 1953, vol. 3, 2–4.

83 "State Brackets Slough Waste," *Oregonian*.

84 OSSA Minutes, Oct. 23, 1953, vol. 3, 2–4.

85 "State Brackets Slough Waste," *Oregonian*.

86 OSSA Minutes, Jan. 29, 1954, vol. 3, 5–19; OSSA Minutes, Nov. 5, 1954, vol. 3, 58–59; "Here's Chance to Review Editors' Recommendations," *Oregonian*, Nov. 1, 1954, sec. 1, 18.

87 See OSSA Minutes Sept. 2, 1955, vol. 3, 90–104, and Dec. 2, 1955, vol. 3, 114–115.

88 OSSA Minutes, Sept. 7, 1956, vol. 3, 163; "Supreme Court Affirms Ban on Dumping Waste," *Oregonian*, March 23, 1961, sec. 2, 7.

89 Lundy, "River Pollution Control on Treadmill."

90 "Oregon Ahead in Stream Purity," *Oregon Journal*, Oct. 9, 1955, A22.

91 "We Want Clean River," *Oregonian*, Feb. 20, 1959, 30.

92 OSSA Minutes, Sept. 22, 1960, vol. 3, 396.

93 "Sewage Treatment Unit Operating; Completion of System Due by 1953," *Oregonian*, Oct. 7, 1951, sec. 1, 1. For 1950 population figure, see US Bureau of the Census, *Census of Population: 1950 Oregon—Volume II, Part 37*, 12.

94 "East Side Sewage Put into Columbia," *Oregonian*, Oct. 6, 1953, sec. 1, 1; "Pump Helps Clean River, Move Wastes to Facility," *Oregonian*, Oct. 24, 1953, sec. 1, 7.

95 "Sewage Treatment Unit Operating." For the dedication ceremony, see "Rites Dedicate Sewage Plant," *Oregonian*, Oct. 27, 1952, sec. 1, 12, sec. 2, 4, and "Visiting Experts to Inspect City's Sewage Plant," *Oregonian*, Oct. 22, 1952, sec. 1, 1.

96 OSSA Minutes, Sept. 22, 1960, vol. 3, 394. Former OSSA fisheries biologist Glenn D. Carter is not entirely accurate in citing George Gleeson when he writes that "all communities along the main stem of the Willamette except for Portland had primary treatment facilities" by 1957. The OSSA had to bring suit against Portland city officials for not expanding its sewage infrastructure to the Tryon Creek and Linnton areas, not connecting other outfalls, and because primary treatment was found not to be adequate for discharge into the lower Columbia River. However, most of the city's sewage *was* receiving primary treatment as of October 1952. Gleeson was asserting that not *all* of Portland's sewage was receiving primary treatment by 1957. See Carter, "Pioneering Water Pollution Control," 260, and Gleeson, *The Return of a River*, 59.

97 OSSA Minutes, July 14, 1960, vol. 3, 381–383.

98 "Master Sewage Plan," *Oregonian*, Nov. 11, 1955, sec. 1, 30.

99 "Tri-County Sewer Program Brings Flurry of Queries from Residents," *Oregonian*, Feb. 9, 1958, sec. 1, 38. $27 million in 1958 was approximately $174 million in 2017.

100 "Right from the Start," *Oregonian*, May 31, 1958, sec. 1, 10.

101 "Sewer Emergency," *Oregonian*, Oct. 20, 1956, sec. 1, 10. For the inability of Portland city officials to establish effective long-term urban infrastructure plans through the 1950s, see Carl Abbott, *Portland: Planning, Politics, and Growth in a Twentieth-Century City* (Lincoln: University of Nebraska Press, 1983), 165. $5 million in 1956 was approximately $34.1 million in 2017.

102 OSSA Minutes, July 14, 1960, vol. 3, 381.

103 "Water Clean-Up Order Aimed at Cities, Firms," *Oregonian*, Jan. 25, 1958, sec. 1, 4. See also OSSA Minutes, July 14, 1960, vol. 3, 381. For number of sewer outfalls, see "City Maps Pollution Campaign," *Oregonian*, Jan. 15, 1959, sec. 1, 1.

104 "Clean River Goal Far Off," *Oregon Journal*, Jan. 30, 1958, sec. 4, 4.

105 "City Club Asks Action on Problem of Sewage," *Oregonian*, March 30, 1958, 21.

106 "River Waste Gets Study," *Oregonian*, July 23, 1958, sec. 1, 11.

107 "River Waste Gets Study"; OSSA Minutes vol. 3, Aug. 8, 1958, 252. $39.555 million in 1958 was approximately $255 million in 2017.

108 OSSA Minutes, Feb. 6, 1959, vol. 3, 263–279.

109 OSSA Minutes, Oct. 17, 1958, vol. 3, 256–260. See also "City Faces State Demand to Explain Pollution," *Oregonian*, Sept. 30, 1958, sec. 1, 1; "Pollution Ultimatum Given City," *Oregonian*, Oct. 18, 1958, sec. 1, 1, 6.

110 "City Maps Pollution Campaign."

111 OSSA Minutes, Feb. 6, 1959, vol. 3, 267–269. See also "City Faces Pollution
　　　Court Suit," *Oregonian*, Feb. 7, 1959, sec. 1, 1, 6; "State Suit Asks City Stop 'Filth,'"
　　　Oregonian, March 3, 1959, sec. 1, 1.

112 "We Want Clean River."

113 For a detailed analysis of the explicit and implicit morality and religiosity in
　　　postwar environmentalism, see Thomas R. Dunlap, *Faith in Nature:
　　　Environmentalism as Religious Quest* (Seattle: University of Washington Press,
　　　2004). For an interpretation of changes in Americans' conception of the
　　　environment in the postwar era based on the experiences of suburban residents,
　　　see Adam W. Rome, *The Bulldozer in the Countryside: Suburban Sprawl and the Rise
　　　of American Environmentalism* (Cambridge, UK: Cambridge University Press,
　　　2001).

114 "Mayor Snaps at Oregon Air, Sanitary Units," *Oregonian*, March 4, 1959, sec. 1, 1.
　　　The state air pollution authority was also pressing the city.

115 "Pollution Law to Be Tested by City," *Oregonian*, Aug. 7, 1959, sec. 1, 27.

116 "Sanitary Authority Delays River Pollution Hearing," *Oregonian*, Nov. 15, 1959,
　　　15; see also OSSA Minutes, Nov. 24, 1959, vol. 3, 315–316.

117 "Pollution Problems to Be Considered," *The League Alert*, Dec. 1959. Washburn
　　　had worked in radio and television broadcasting in Portland since the 1940s, and
　　　would, in the 1960s and 1970s, work for the Oregon State Board of Health. He
　　　followed his water pollution film of 1959 with a television documentary focused
　　　on air pollution two years later, *How Now Brown Cloud*; see "Guide to the
　　　Sherman A. Washburn Photographs, 1962," http://nwda-db.orbiscascade.org/
　　　findaid/ark:/80444/xv92688 and "Sherman A. Washburn Obituary," *Oregonian*
　　　July 2, 2014, http://obits.oregonlive.com/obituaries/oregon/obituary.
　　　aspx?pid=171591391, both accessed Feb. 28, 2017.

118 OSSA Minutes, Dec. 15, 1960, vol. 4, 33–35.

119 For creation of the air pollution authority in 1951 and absorption of this agency
　　　within the OSSA in 1959, see "Oregon Department of Environmental Quality,
　　　Administrative Overview" (Salem: State of Oregon Office of the Secretary of
　　　State, Nov. 2009), arcweb.sos.state.or.us/doc/recmgmt/sched/special/state/
　　　overview/20080009deqadov.pdf, accessed Sept. 9, 2012; for reference to the
　　　authority's revised rules of procedure based on this change, see OSSA Minutes,
　　　Nov. 24, 1959, vol. 3, 311–312. Attention is measured in terms of relative
　　　coverage of water and air pollution issues in OSSA minutes from November 24,
　　　1959, through January 4, 1963.

120 See Watford Reed, "Dirty River: Big Swallow Dangerous," *Oregon Journal*, May 24,
　　　1959, 2C; Reed, "Dirty River Once Stank, Killed Fish," *Oregon Journal*, May 25,
　　　1959, 2B; and Reed, "Dirty River Cleanup Job Partly Done," *Oregon Journal*, May
　　　26, 1959, 18A.

121 For contacts with Charlton, see Tom Lawson McCall to Charlton [handwritten
　　　"1954"], folder Correspondence re: Water Pollution 1951–58, box 18, Charlton
　　　Papers; for award, see "League Pays Honor to 4," *Oregon Journal*, Nov. 20, 1959 /
　　　Oregonian, Nov. 21, 1959, 4, and Brent Walth, *Fire at Eden's Gate: Tom McCall and
　　　the Oregon Story* (Portland, Oreg.: Oregon Historical Society Press, 1994), 121–
　　　124.

122 Walth, *Fire at Eden's Gate*, 134–137, for details of McCall's work on this
　　　documentary.

123 Thomas L. McCall, producer, *Pollution in Paradise*, television documentary (Portland, Oreg.: KGW-TV, 1962).

124 For more analysis of this documentary and the significant role it played in McCall's career, see Robbins, *Landscapes of Conflict*, 266–280, and Walth, *Fire at Eden's Gate*, especially beginning on p. 134 and following.

125 OSSA Minutes, Jan. 4, 1963, vol. 4, 272–273.

126 Donald Hillebrand to Tom McCall, n.d. [handwritten "April 3, 1963"], folder KGW 1956–1966, box 32, sub-series 10, series C, Thomas Lawson McCall Papers (MSS 625-1), Oregon Historical Society Research Library, Portland. Although no further identifying information is provided, Mr. Hillebrand likely lived near The Dalles, as he references ongoing struggles to get the Harvey Aluminum plant to abate its smoke pollution.

CHAPTER 6: AT LONG LAST?

1 "Proposed Water Quality Management Plan for Willamette River Basin," Portland: Oregon Department of Environmental Quality, 1976.

2 Earl Deane, "Willamette Pollution Called Health Hazard," *Oregonian*, Aug. 10, 1965, sec. 1, 1.

3 Thomas L. McCall, producer, *Pollution in Paradise*, television documentary. (Portland, Oreg.: KGW-TV, 1962).

4 Deane, "Willamette Pollution Called Health Hazard."

5 "Pollution Unlimited?" *Oregonian*, Aug. 11, 1965, sec. 1, 22; "Red Flag on Pollution," *Oregonian*, Aug. 16, 1965, sec. 1, 26.

6 Harold Hughes, "Grip Given Sanitary Authority," *Oregonian*, March 27, 1963, sec. 1, 1; "Pollution Bill Gains Approval," *Oregonian*, April 19, 1963, sec. 1, 17.

7 Harold Hughes, "Grip Given Sanitary Authority"; "Pollution Bill Gains Approval." For text of this law, see chapter 171, *General Laws of Oregon, 1963*, 235–239.

8 William G. Robbins, *Landscapes of Conflict: The Oregon Story, 1940–2000* (Seattle: University of Washington Press, 2004), 266–280.

9 Matthew Klingle, *Emerald City: An Environmental History of Seattle* (New Haven, Conn.: Yale University Press, 2007). There is much literature on this topic, but see, for example, Samuel P. Hays, *Beauty, Health, and Permanence: Environmental Politics in the United States, 1955–1985* (Pittsburgh: University of Pittsburgh Press, 1987).

10 See, for example, Adam W. Rome, *The Bulldozer in the Countryside: Suburban Sprawl and the Rise of American Environmentalism* (Cambridge, UK: Cambridge University Press, 2001), and Gregory Summers, *Consuming Nature: Environmentalism in the Fox River Valley, 1850–1950* (Lawrence: University Press of Kansas, 2006). Rachel Carson herself asserted that the 1959 "cranberry scare," in combination with radiological contamination of milk and fear of birth defects caused by the drug thalidomide, contributed to the strong popular reception of her serialized articles and book; see Priscilla Coit Murphy, *What a Book Can Do: The Publication and Reception of Silent Spring*, (Amherst, Mass.: University of Massachusetts Press, 2005), 37.

11 See Peter J. Bowler, *The Norton History of the Environmental Sciences* (New York, W. W. Norton & Company: 1993), and Paul Charles Milazzo, *Unlikely*

Environmentalists: Congress and Clean Water, 1945–1972 (Lawrence: University Press of Kansas, 2006).

12 Ben A. Minteer, *The Landscape of Reform: Civic Pragmatism and Environmental Thought in America* (Cambridge, Mass.; The MIT Press, 2006); Samuel P. Hays, *Explorations in Environmental History: Essays* (Pittsburgh: University of Pittsburgh Press, 1998); Richard W. Judd and Christopher S. Beach, *Natural States: The Environmental Imagination in Maine, Oregon, and the Nation* (Washington, DC: Resources for the Future, 2003); G. Ray Funkhouser, "The Issues of the Sixties: An Exploratory Study in the Dynamics of Public Opinion," *The Public Opinion Quarterly* 37:1 (Spring 1973), 62–75.

13 Murphy, *What a Book Can Do*, 11–14. Terence Kehoe also finds television and print media reports in the early 1960s critical in bringing about a shift in Americans' approach to the environment; see Kehoe, *Cleaning Up the Great Lakes: From Cooperation to Confrontation* (DeKalb, Ill.: Northern Illinois University Press, 1997), 8.

14 Murphy, *What a Book Can Do*, 3.

15 Laura Jane Gifford, *Crisis in the Klamath Basin* (documentary film), *Oregon Encyclopedia*, http://oregonencyclopedia.org/articles/crisis_in_the_klamath_basin_documentary_film/#.VziqRr4YTkE, accessed May 14, 2016.

16 For these aspects of Rachel Carson's *Silent Spring*, see Murphy, *What a Book Can Do*.

17 For more on this concept from Gifford Pinchot and Progressive Era conservationism generally, see Samuel P. Hays, *Conservation and the Gospel of Efficiency: The Progressive Conservation Movement, 1890–1920* (Cambridge, Mass.: Harvard University Press, 1959).

18 Murphy, *What a Book Can Do*, 199–221.

19 McCall, *Pollution in Paradise*.

20 "State Sanitary Unit Places River Pollution Blame," *Oregonian*, June 21, 1964, sec 1, 30.

21 Kenneth H. Spies to Harold Wendel, April 13, 1966, box 1, folder Sanitary Authority 1967 (part 1) (1/8), Harold F. Wendel Papers (Coll. 196), Oregon Historical Society Research Library, Portland, Oregon.

22 "Pollution Draws Fire at Wildlife Hearing," *Oregonian*, May 27, 1964, sec. 1, 11.

23 Peter Tugman, "Sanitary Board Chairman Looks Back on Work After 25 Years of Service," *Oregonian*, March 31, 1964, sec. 2, 8.

24 Harold E. Hughes, "Oregon Votes Funds to Cover Cost," *Oregonian*, March 5, 1965, sec. 1, 16. $1 million in 1965 was approximately $5.87 million in 2015.

25 Deane, "Willamette Pollution Called Health Hazard"; "Dams to Aid Fouled River," *Oregonian*, Aug. 13, 1965, sec. 3, 8.

26 Deane, "Willamette Pollution Called Health Hazard."

27 "Sanitary Authority Tells Pulp Mills to Slash Willamette River Pollution," *Oregonian*, Aug. 17, 1965, sec. 1, 15.

28 "Pollution Reading Dips," *Oregonian*, Aug. 28, 1965, sec. 1, 13.

29 "State Sanitary Authority Pledges Halt to Pollution of Rivers," *Oregonian*, Aug. 27, 1965, sec. 1, 27.

30 "Red Flag on Pollution," *Oregonian*, Aug. 16, 1965, sec. 1, 26.

31 "Foul River Rap Refuted," *Oregonian*, Aug. 16, 1965, 17.

32 Harold Hughes, "McCall, Straub Agree State Moved Too Slow On Pollution," *Oregonian*, Oct. 9, 1965, sec. 1, 8.

33 See chapter 3 for this overview. Also note that in 1941 the state changed its biennial calendar from the period January 1 through December 31 to the period July 1 through June 30.

34 In approximate 2017 amounts, $50,000 equals $646,000; $23,000 equals $297,000; $15,000 equals $194,000; and $10,000 equals $129,000.

35 For more on the USS *Oregon*, see Ken Lomax, "A Chronicle of the Battleship Oregon," *Oregon Historical Quarterly* 106:1 (Spring 2005), 132–145.

36 Charles A. Sprague, "Tenth Biennial Budget, State of Oregon, 1941–1943, Submitted to the Forty-First Legislative Assembly," 213. $37,500 in 1941 was approximately $510,000 in 2017.

37 State of Washington State Pollution Commission, *Progress Report on Field Investigations and Research* (Seattle: State of Washington State Pollution Commission, December 1938).

38 Washington State Pollution Control Commission, "The Seattle Sewage Treatment Problem" (Olympia: The Commission, 1948), http://www.washington.edu/ uwired/outreach/cspn/Website/Classroom%20Materials/Curriculum%20 Packets/Building%20Nature/Documents/27.html, accessed May 14, 2011. Washington State would not have a state water quality agency equivalent to the OSSA until 1945; see Washington State Pollution Control Commission, "Minimum Requirements for the Control of Industrial Wastes," *Sewage and Industrial Wastes* 22:4 (April 1950), 514–520.

39 L. A. Powe Jr., "Water Pollution Control in Washington," *Washington Law Review* 43 (1967), 427.

40 OSSA Meeting Minutes, Sept. 15, 1939, vol. 1, 14–15.

41 OSSA Meeting Minutes, May 7, 1940, vol. 1, 36–37; OSSA Meeting Minutes, June 21, 1940, vol. 1, 57.

42 The data in this and subsequent graphs is drawn mostly from biennial governor's budget proposals and budget analyses. Where these reports have been not been forthcoming, I have also drawn data from the Harold F. Wendel Papers at the Oregon Historical Society Research Library and, as a last resort, the *Oregonian*. Full discussion of these sources is to be found in the bibliographic note.

43 $27,895 in 1947 was approximately $238,000 in 2017, and $33,840 was approximately $288,000.

44 $65,045 in 1949 was approximately $526,000 in 2017, and $110,599 was approximately $894,000.

45 Oregon State Sanitary Authority, "Thirteenth Biennial Report on Water Pollution Control, for the Period July 1, 1962 to June 30, 1964," 18. $700,000 in 1957 was approximately $4.62 million in 2017, and $1,387,983 in 1964 was approximately $8.3 million.

46 Legislative Fiscal Committee, "Analysis of the Governor's 1967–1969 Budget Report to the 54th Legislative Assembly," Jan. 1967, 38.1–38.4.

47 $159,012 in 1957 was approximately $1.05 million in 2017, and $635,627 in 1967 was approximately $3.53 million. The dashed lines in this figure connect reliable data points.

48 The rate of inflation from the 1941–1943 biennium to the 1943–1945 biennium is figured by comparing the year 1943 (end of one biennium) with 1945 (end of the following biennium). For the inflation rate calculator, see http://www.dollartimes.com/calculators/inflation.htm.

49 The literature on this is extensive, but see, for example, Martin V. Melosi, *The Sanitary City: Urban Infrastructure in America from Colonial Times to the Present* (Baltimore, Md.: Johns Hopkins University Press, 2000); Terence Kehoe, *Cleaning Up the Great Lakes*; Hugh S. Gorman, "Laying the Foundation for the Control of Industrial Pollution, 1930–1970: Two Canals, A Refinery, and Clarence W. Klassen," *Journal of Illinois History* 8:3 (2005), 182–208; and Arn Keeling, "Sink or Swim: Water Pollution and Environmental Politics in Vancouver, 1889–1975." *BC Studies* 142/143 (Summer/Autumn 2004), 69–101.

50 Brent Walth, *Fire at Eden's Gate: Tom McCall and the Oregon Story* (Portland, Oreg.: Oregon Historical Society Press, 1994), 183.

51 David B. Charlton to Dr. Frank Terraglio, Portland State University, May 9, 1975, 1–2, folder Water Pollution: Portland Area, box 18, David B. Charlton Papers (MSS 1900), Oregon Historical Society Research Library, Portland, Oreg.

52 "Campaign Echoes," *Oregonian*, Jan. 14, 1967, sec. 1, 20.

53 "H. F. Wendel Dies at 74," *Oregonian*, April 18, 1967, sec. 1, 1, 16.

54 "Pollution Fighter," *Oregonian*, April 19, 1967, sec. 1, 18.

55 Glen D. Carter, "Pioneering Water Pollution Control in Oregon," *Oregon Historical Quarterly* 107:2 (Summer 2006), 257–258.

56 "Governor Takes Control of Sanitary Authority," *Oregonian*, April 27, 1967, sec. 1, 1, 27. See also Walth, *Fire at Eden's Gate*, 183.

57 Barney McPhillips's quote from William G. Robbins, *Landscapes of Conflict*.

58 "Air, Water Pollution Bills Get Senate Nod," *Oregonian*, May 26, 1967, sec. 1, 19.

59 "Job Unfinished," *Oregonian*, June 15, 1967, sec. 1, 50.

60 "McCall Resigns Sanitary Authority; Mosser, Chemist Named to Posts," *Oregonian*, July 11, 1967, sec. 1, 1. McCall's biographer Brent Walth's interpretation of his decision is that "he had found the authority's work too tedious and detail-driven for his tastes"; see Walth, *Fire at Eden's Gate*, 198.

61 "McCall Resigns Sanitary Authority," 1.

62 "New Look," *Oregonian*, July 15, 1967, sec. 1, 14.

63 Matt Kramer, "Big Shakeup Becomes Effective in State Government," *Oregonian*, July 1, 1969, sec. 1, 13.

64 "Senate Bills Introduced," *Oregonian*, March 4, 1969, sec. 3, 9.

65 Oregon Department of Environmental Quality, "About the Environmental Quality Commission," http://www.deq.state.or.us/about/eqc/eqc.htm, accessed Jan. 28, 2012.

66 Hughes, "McCall, Straub Agree State Moved Too Slow On Pollution."

67 Legislative Fiscal Committee, "Analysis of the Governor's 1969–1971 Budget Report to the 55th Legislative Assembly," Jan. 1969, 42.1–42.7; "McCall Signs Last of Bills," *Oregonian*, June 17, 1969, sec. 1, 13.

68 Walth, *Fire at Eden's Gate*.

69 Hughes, "McCall, Straub Agree State Moved Too Slow On Pollution."

70 Hughes, "McCall, Straub Agree State Moved Too Slow On Pollution."

71 Robert Olmos, "Willamette River Now Said Reasonably Safe for Portland Area Water Sport Enthusiasts," *Oregonian,* July 7, 1969, sec. 1, 27; "Pollution Gains Made," *Oregonian,* July 26, 1969, sec. 1, 12.

72 Doug Seymour, "End of Technical Requirements Brings New Faces to State Jobs," *Oregonian,* Sept. 19, 1971, F6.

73 Carter, "Pioneering Water Pollution Control in Oregon," 270–271.

74 Robbins, *Landscapes of Conflict,* 268–272.

75 John Painter, "Slaughterhouse to Use Sewers for Disposal of Plant Wastes," *Oregonian,* Feb. 25, 1971, sec. 1, 11; "County Demands State Hearing on Columbia Slough Pollution by Meatpacking Firm," *Oregonian,* Nov. 20, 1970, sec. 1, 18.

76 Painter, "Slaughterhouse to Use Sewers for Disposal of Plant Wastes."

77 Painter, "Slaughterhouse to Use Sewers for Disposal of Plant Wastes"; Jim Kadera, "Polluted Columbia Slough Faces Clean-Up Problems," *Oregonian,* Aug. 1, 1971, sec. 1, 41. $150,000 in 1971 was approximately $688,000 in 2017.

78 Walth, *Fire at Eden's Gate,* 198–199.

79 "Court Order Sought to Prevent Pollution," *Oregonian,* July 25, 1972, sec. 1, 23; Jim Kadera, "Boise Cascade Shuts Salem Plant." *Oregonian,* July 26, 1972, sec. 1, 1.

80 "Court Order Sought to Prevent Pollution"; Kadera, "Boise Cascade Shuts Salem Plant." $6.5 million in 1975 was approximately $28.6 million in 2017.

81 "Court Order Sought to Prevent Pollution"; Kadera, "Boise Cascade Shuts Salem Plant."

82 "Court Order Sought to Prevent Pollution"; Kadera, "Boise Cascade Shuts Salem Plant."

83 "Boise Cascade Plant Reopens," *Oregonian,* Aug. 1, 1972, sec. 3, 5.

84 Tom McCall and Steve Neal, *Tom McCall: Maverick* (Portland, Oreg.: Binford & Mort, 1977), 182–183; Walth, *Fire at Eden's Gate,* 327–331.

85 Don Billingsley, "Heel for McCall?" *Oregonian,* July 31, 1972, sec. 1, 16.

86 Walth, *Fire at Eden's Gate,* 317.

87 See, generally, Walth, *Fire at Eden's Gate.*

88 Ethel Starbird, "A River Restored: Oregon's Willamette," *National Geographic* 141:6 (June 1972), 816–834.

89 Mark O. Hatfield, "Willamette Cleanup," *New York Times,* Jan. 6, 1970, 40.

90 Phil Cogswell, "State, Federal Agency Attorneys Jockey for Jurisdiction Over Pollution," *Oregonian,* July 5, 1970, sec. 1, 57.

91 E. W. Kenworthy, "Tough Rules Saving a Dying Oregon River," *New York Times,* Sept. 8, 1970, 50.

92 Charlton to Terraglio, May 9, 1975.

CHAPTER 7: HYDRA-HEADED POLLUTION

1 Don Holm, "Willamette Crawdads Many, Tasty," *Oregonian,* Sept. 29, 1976, 3M.

2 "Proposed Water Quality Management Plan for Willamette River Basin" (Portland: Oregon Department of Environmental Quality, 1976); William G. Robbins, *Landscapes of Conflict: The Oregon Story, 1940–2000* (Seattle: University of Washington Press, 2004), 269–270.

3 Jack Lewis, "The Birth of the EPA." *EPA Journal,* Nov. 1985. http://www.epa.gov/aboutepa/history/topics/epa/15c.html, accessed Feb. 4, 2012.

4 Laurance S. Rockefeller, Chair. "Report to the President and the Council on Environmental Quality," Washington, DC: Citizens' Advisory Committee on Environmental Quality, April 1971; Brent Walth, *Fire at Eden's Gate: Tom McCall and the Oregon Story* (Portland, Oreg.: Oregon Historical Society Press, 1994), 266.

5 Walth, *Fire at Eden's Gate*, 369–370.

6 Paul Charles Milazzo, *Unlikely Environmentalists: Congress and Clean Water, 1945–1972* (Lawrence: University Press of Kansas, 2006), 129–138; Lewis, "The Birth of the EPA."

7 Lewis, "The Birth of the EPA."

8 William L. Andreen, "The Evolution of Water Pollution Control in the United States—State, Local, and Federal Efforts, 1789–1972: Part I," *Stanford Environmental Law Journal* 22 (Jan. 2003), 145–200, 215–294; Neil S. Shifrin, "Pollution Management in the Twentieth Century." *Journal of Environmental Engineering* 131:5 (May 2005), 676–691.

9 Andreen, "The Evolution of Water Pollution Control in the United States," 215–294.

10 Lester Edelman, "Congressional Intent of the 1972 Federal Water Pollution Control Act," *Water Pollution Control Federation Journal* 45:1 (Jan. 1973), 1, quoted in Harvey Lieber and Bruce Rosinoff, *Federalism and Clean Waters: The 1972 Water Pollution Control Act* (Lexington, Mass.: Lexington Books [D.C. Heath and Co.], 1975), 7.

11 Lieber and Rosinoff, *Federalism and Clean Waters*, 9–10.

12 Lieber and Rosinoff, *Federalism and Clean Waters*, 10–11. $20 billion in 1972 was approximately $87.9 billion in 2017.

13 Lieber and Rosinoff, *Federalism and Clean Waters*, 11; Andrew Karvonen, *Politics of Urban Runoff: Nature, Technology, and the Sustainable City* (Cambridge, Mass.: The MIT Press, 2011), 9.

14 Lieber and Rosinoff, *Federalism and Clean Waters*, 14–19; David Stradling and Richard Stradling, "Perceptions of the Burning River: Deindustrialization and Cleveland's Cuyahoga River," *Environmental History* 13:3 (July 2008), 515–535.

15 Lieber and Rosinoff, *Federalism and Clean Waters*, 14–19.

16 Dominick's comment cited in Mark O. Hatfield, "Willamette Cleanup," *New York Times*, Jan. 6, 1970, 40.

17 Andy Schaedel, "The Three 'O's,' or, the Evolution of Assessments and the Origins of 303(d) Listings and TMDLs [draft]," March 31, 2008 [unpublished presentation, available at DEQ Headquarters, Portland, Oreg.].

18 Schaedel, "The Three 'O's.'"

19 Oregon Department of Environmental Quality, Water Quality Division, "Oregon's 1994 Water Quality Status Assessment Report (305[b])," Portland: Oregon Department of Environmental Quality, April 1994.

20 Schaedel, "The Three 'O's.'"

21 Schaedel, "The Three 'O's.'"

22 "Proposed Water Quality Management Plan for Willamette River Basin," 16.

23 "Proposed Water Quality Management Plan for Willamette River Basin," 14–16, 30, 34.

24 "State Opens New Attack on Pollution," *Oregonian*, Aug. 25, 1976, B3.

25 "State Opens New Attack on Pollution."

26 Schaedel, "The Three 'O's.'"

27 "Proposed Water Quality Management Plan for Willamette River Basin," 41.

28 Schaedel, "The Three 'O's'"; Oregon Department of Environmental Quality, Water Quality Division, "Oregon's 1994 Water Quality Status Assessment Report (305[b])."

29 US Environmental Protection Agency, "Impaired Waters and Total Maximum Daily Loads," http://water.epa.gov/lawsregs/lawsguidance/cwa/tmdl/, accessed Feb. 4, 2012.

30 Oregon Department of Environmental Quality, "Oregon's 1994 Water Quality Status Assessment Report," 1–3.

31 Schaedel, "The Three 'O's.'" Professors, students, and alumni of the law school established the nonprofit NEDC in 1969 to provide legal support and expertise in the service of environmental protection and resource conservation; see "Northwest Environmental Defense Center," http://law.lclark.edu/centers/ northwest_environmental_defense_center/, accessed Feb. 4, 2012.

32 "Lawsuit Seeks to Force Cleanup of Tualatin River," *Oregonian*, Dec. 13, 1986, A19.

33 "Lawsuit Seeks to Force Cleanup of Tualatin River."

34 Joan Laatz, "Water-Quality Management Faces Change," *Oregonian*, June 5, 1987, B2.

35 Schaedel, "The Three 'O's.'"

36 Laatz, "Water-Quality Management Faces Change."

37 Schaedel, "The Three 'O's,'" 14. For broader context on EPA's shift, see Edmund P. Russell III, "Lost Among the Parts Per Billion: Ecological Protection at the United States Environmental Protection Agency, 1970–1993," *Environmental History* 2:1 (Jan. 1997), 29–51.

38 Schaedel, "The Three 'O's'"; Oregon Department of Environmental Quality, "Oregon's 1994 Water Quality Status Assessment Report," 2–8.

39 Roberta J. Lindberg, "Non-Regulatory Efforts to Combat Nonpoint Source Pollution in Oregon" (MA internship report, University of Oregon, 1995), 56–57.

40 Andrew Karvonen, *Politics of Urban Runoff: Nature, Technology, and the Sustainable City* (Cambridge, Mass.: The MIT Press, 2011), 9.

41 Oregon Department of Environmental Quality, "Oregon's 1994 Water Quality Status Assessment Report," 1–6.

42 Oregon State Legislature, Minutes of the House Committee on Natural Resources, June 28, 1993, Tapes 67–73, available at http://arcweb.sos.state.or.us/ doc/records/legislative/legislativeminutes/1993/house/natural_resources/ hNR062893.txt, accessed Feb. 18, 2012.

43 Oregon Department of Environmental Quality, "Oregon's 1994 Water Quality Status Assessment Report," 2–11.

44 Hal Bernton, "Environmental Groups Sue EPA Over Pace of Water Cleanup," *Oregonian*, Sept. 13, 1996, B7.

45 Bernton, "Environmental Groups Sue EPA Over Pace of Water Cleanup."

46 Bernton, "Environmental Groups Sue EPA Over Pace of Water Cleanup."

47 Schaedel, "The Three 'O's,'" 15.

48 Hal Bernton, "Kitzhaber Unveils Plan to Clean Up Oregon Streams," *Oregonian*, Nov. 19, 1996, A1; "Healthy Streams Partnership," *Oregonian*, Nov. 20, 1996, B8.

49 "Oregon Plan for Salmon and Watersheds: About the Oregon Plan," http://www. oregon-plan.org/OPSW/about_us.shtml#Origins_of_the_Oregon_Plan_for_ Salmon_and_Watersheds, accessed Feb. 4, 2012.

50 Schaedel, "The Three 'O's,'" 2–3.

51 Oregon Department of Environmental Quality, "Oregon's 1994 Water Quality Status Assessment Report," 1–7.

52 Oregon Department of Environmental Quality, "Oregon's 1994 Water Quality Status Assessment Report," 1–7.

53 Portland Bureau of Environmental Services, "Combined Sewer Overflow (CSO) Program," http://www.portlandonline.com/bes/index.cfm?c=31030&a=47260, accessed Sept. 30, 2008. Many people refer to this project as the "Big Pipe."

54 Martin V. Melosi, *The Sanitary City: Urban Infrastructure in America from Colonial Times to the Present* (Baltimore, Md.: Johns Hopkins University Press, 2000).

55 Portland Bureau of Environmental Services, "Combined Sewer Overflow (CSO) Program."

56 Gordon Oliver, "City Council Adopts Plan to Cut Pollution in Metro-Area Rivers," *Oregonian*, March 29, 1990, B3.

57 "Northwest Environmental Advocates—Our History," http:// northwestenvironmentaladvocates.org/about_us/, accessed Aug. 21, 2016.

58 John Painter, "Environmental Suit Seeks to Halt Flushing of City Sewage in Rivers," *Oregonian*, April 17, 1991, D8.

59 Cathy Kiyomura, "Council OKs 20-Year Plan for Sewers," *Oregonian*, Aug. 2, 1991, C4.

60 US Bureau of the Census, *Census of Population: 1950 Oregon—Volume II, Part 37, Characteristics of the Population, Number of Inhabitants, Oregon*, 12, available at http://www.census.gov/prod/www/abs/decennial/1950.htm, accessed Oct. 6, 2008; "Portland: Population Profile," City-Data.com, http://www.city-data.com/ us-cities/The-West/Portland-Population-Profile.html, accessed April 23, 2016.

61 Barnes C. Ellis, "Judge Upholds Portland's Plan to Fix Antiquated Sewer System," *Oregonian*, Dec. 19, 1991, C4; Kiyomura, "Council OKs 20-Year Plan For Sewers."

62 Ellis, "Judge Upholds Portland's Plan."

63 Ellis, "Judge Upholds Portland's Plan."

64 Portland Bureau of Environmental Services, "Combined Sewer Overflow (CSO) Program."

65 Portland Bureau of Environmental Services, "Sustainable Stormwater Management," http://www.portlandonline.com/bes/index. cfm?c=31030&a=201839, accessed Sept. 30, 2008. See also commentary on Portland City Commissioner Sam Adams's "Gray to Green" initiative involving these and other elements; Bob Sallinger and Mike Houck, "A New Approach to Portland Watersheds," *Oregonian*, April 16, 2008.

66 "Combined Sewer Overflow Projects: The Big Pipe," http://www.portlandonline. com/cso/, accessed Feb. 4, 2012.

67 Scott Learn, "Accidental Sewer Overflow May Affect Triathlon Swim," *Oregonian* Aug. 31, 2007. Quote from "Officials: Don't Swim, Fish in the Willamette River for 48 Hours," *Oregonian*, May 20, 2008.

68 Oregon Department of Environmental Quality, "News Release: DEQ Issues Penalties Totaling Nearly $450,000 to City of Portland for Sewage Discharges to Willamette and Other Waterways," http://www.deq.state.or.us/news/prDisplay.asp?docID=1976, accessed Jan. 1, 2012.

69 Beth Slovic, "Portland's $1.4 Billion Big Pipe Project Comes to an End after 20 Years," Oregonian, Nov. 25, 2011.

70 US Environmental Protection Agency, "Superfund's 30th Anniversary: 30 Years of Protecting Communities and the Environment," http://www.epa.gov/superfund/30years/, accessed Feb. 4, 2012; US Environmental Protection Agency, "Administration Proposes Hazardous Waste Cleanup Fund," press release, June 13, 1979, http://www.epa.gov/aboutepa/history/topics/cercla/01.html, accessed Feb. 4, 2012.

71 US Environmental Protection Agency, "Superfund: Basic Information," http://www.epa.gov/superfund/about.htm, accessed Feb. 4, 2012.

72 US Environmental Protection Agency, "Superfund's 30th Anniversary"; US Environmental Protection Agency, "Administration Proposes Hazardous Waste Cleanup Fund."

73 Brent Hunsberger, "Portland Harbor Joins Superfund Today," Oregonian, Dec. 1, 2000, A1.

74 EPA Office of Solid Waste and Emergency Response (OSWER) authorized CAG formation with its Directive 9230.0-28; see EPA, "Community Advisory Groups (CAGs) at Superfund Sites," Publication 9230.0-28AFS, Aug. 1996.

75 "Citizen Advisory Committee Meeting Information Meeting Portland Harbor Superfund Site Tuesday, March 25, 2002," http://yosemite.epa.gov/r10/CLEANUP.NSF/PH/Portland+Harbor+Community+Advisory I Group I Archiv e!OpenDocument#_Section7, accessed Feb. 22, 2012; Willamette Riverkeeper, "Our History," http://www.willamette-riverkeeper.org/WRK/about.html, accessed April 24, 2016.

76 Portland Harbor Community Advisory Group, "What's Already Happened?" June 2, 2011, http://portlandharborcag.info/node/186, and "Introduction to the CAG and the Portland Harbor," Oct. 5, 2008, http://www.portlandharborcag.info/node/4, both accessed Feb. 19, 2012.

77 US Environmental Protection Agency, "Finding Potentially Responsible Parties (PRP)," Nov. 22, 2015, https://www.epa.gov/enforcement/finding-potentially-responsible-parties-prp, accessed April 24, 2016.

78 Lower Willamette Group, "Lower Willamette Group," http://lwgportlandharbor.org/intro/lwg.htm, n.d., accessed April 24, 2016.

79 Brent Hunsberger, "A River of Risk: 'It Wasn't a Healthy Place To Work': The Decades-Long Story of Two Chemical Plants in Northwest Portland Includes Deception, Cooperation and Abiding Threats to the Willamette River Cleanup," Oregonian, Dec. 19, 2000, A1.

80 EPA Persistent Bioaccumulative and Toxic (PBT) Chemical Program, "Dioxins and Furans," April 18, 2011, http://www.epa.gov/pbt/pubs/dioxins.htm, accessed Feb. 19, 2012.

81 EPA PBT Chemical Program, "Dioxins and Furans."

82 EPA PBT Chemical Program, "Dioxins and Furans."

83 See, for example, Samuel M. Peck, "Dermatitis from Cutting Oils, Solvents, and Dialectrics, Including Chloracne," *Journal of the American Medical Association* 125:3 (1944), 190–196; George E. Morris and Irving R. Tabershaw, "'Cable Rash'—A Note on a New Cleansing Mixture," *Journal of the American Medical Association* 121:3 (1943), 192–193; and Alice Hamilton, "The Toxicity of the Chlorinated Hydrocarbons," *Yale Journal of Biology and Medicine* 15:6 (July 1943). 787–801.

84 "2,4-D" is 2,4-Dichlorophenoxyacetic acid; "2,4,5-T" is 2,4,5-Trichlorophenoxy-acetic acid; and "2,3,7,8-TCDD" is 2,3,7,8-Tetrachlorodibenzo-p-Dioxin. For more on this, see Carol Van Strum and Paul Merrell, *No Margin of Safety: A Preliminary Report on Dioxin Pollution and the Need for Emergency Action in the Pulp and Paper Industry* (Greenpeace USA, 1987), III-1-III-18, and the sources cited in this work.

85 Victor Cohn, "Federal Report Blasted: Use of Herbicide Called Threat," *Oregonian*, July 16, 1971, 5; Jerry M. Flint, "Dow Aides Deny Herbicides Risk," *New York Times*, Mar. 18, 1970, 96; Les Ledbetter, "Science: A Weed Killer Banned in US and Vietnam," *New York Times*, April 19, 1970, 176.

86 For more on this, see Van Strum and Merrell, *No Margin of Safety.*

87 "Nursing Mother's Milk Found to Contain Dioxin," *Oregonian*, Feb. 18, 1977, D16.

88 "Scientists Score Testing for Dioxin," *Oregonian*, March 6, 1977, B8. This and related discoveries centered on the human health effects of herbicide spraying, and is covered more fully in Carol Van Strum, *A Bitter Fog: Herbicides and Human Rights* (San Francisco: Sierra Club Books, 1983), and in William G. Robbins, *Landscapes of Conflict*, 194–205.

89 Dana Tims, "EPA Reply Ordered on Study of Dioxin," *Oregonian*, Feb. 4, 1987, B3.

90 Cohn, "Federal Report Blasted," *Oregonian.*

91 Carmel Finley, "Dioxin in Mill Emissions Will Prompt New EPA Rules," *Oregonian*, Sept. 6, 1987, B6.

92 Finley, "Dioxin in Mill Emissions"; Keith R. Solomon, "Chlorine in the Bleaching of Pulp and Paper," *Pure and Applied Chemistry* 68:9 (1996), 1721–1730.

93 Finley, "Dioxin in Mill Emissions."

94 Van Strum and Merrell, *No Margin of Safety.*

95 Philip Shabecoff, "Traces of Dioxin Found in Range of Paper Goods," *New York Times*, A1, A22.

96 Kathryn Harrison, "The Regulator's Dilemma: Regulation of Pulp Mill Effluents in the Canadian Federal State," *Canadian Journal of Political Science* 29:3 (Sept. 1996), 469–496; Mark R. Powell, "Control of Dioxins (and Other Organochlorines) from the Pulp and Paper Industry under the Clean Water Act and Lead in Soil at Superfund Mining Sites: Two Case Studies in EPA's Use of Science," Discussion Paper 97-08 (Washington, DC: Resources for the Future, March 1997).

97 "Hydra-Headed Pollution," *Morning Oregonian*, July 30, 1925, 10.

CONCLUSION: SPEAKING FOR OURSELVES

1 For more on the continued importance of Portland Harbor as part of this "working river," see Carl Abbott, "Rivers Still Run Through It," *Oregonian*, Jan. 13, 2008.

2 Brent Walth, *Fire at Eden's Gate: Tom McCall and the Oregon Story* (Portland, Oreg.: Oregon Historical Society Press, 1994),141.

3 William G. Robbins, *Landscapes of Conflict: The Oregon Story, 1940–2000* (Seattle: University of Washington Press, 2004), 260.

4 Robbins, *Landscapes of Conflict: The Oregon Story, 1940–2000*; Richard White, *Organic Machine: The Remaking of the Columbia River* (New York: Hill and Wang, 1995); James C. Scott, *Seeing Like a State: How Certain Schemes to Improve the Human Condition Have Failed* (New Haven, Conn.: Yale University Press, 1998).

BIBLIOGRAPHIC ESSAY

1 Legislative Fiscal Committee, "Analysis of the Governor's 1967–1969 Budget Report to the 54th Legislative Assembly," Jan. 1967, 38.1–38.4.

2 For more information about this method and a comparison calculator, see "Seven Ways to Compute the Relative Value of a US Dollar Amount, 1774 to Present," http://www.measuringworth.com/uscompare/, accessed Aug. 20, 2016.

3 Ellen Stroud, "Troubled Waters in Ecotopia: Environmental Racism in Portland, Oregon," *Radical History Review* 74 (1999), 65–95, quote p. 69.

4 Brent Walth, *Fire at Eden's Gate: Tom McCall and the Oregon Story* (Portland, Oreg.: Oregon Historical Society Press, 1994), 139.

5 Walth, *Fire at Eden's Gate*, 141.

6 E. Kimbark MacColl, *The Growth of a City: Power and Politics in Portland, Oregon, 1915–1950* (Portland, Oreg.: The Georgian Press, 1979).

7 George W. Gleeson, *The Return of a River: The Willamette River, Oregon* (Corvallis: Oregon State University Water Resources Research Institute, 1972).

8 Richard W. Judd and Christopher S. Beach, *Natural States: The Environmental Imagination in Maine, Oregon, and the Nation* (Washington, DC: Resources for the Future, 2003).

9 Judd and Beach, *Natural States*, x–xi.

Bibliographic Essay

This section provides a brief overview and discussion of some of the sources used in *Speaking for the River*.

OREGON STATE BIENNIAL BUDGET APPROPRIATIONS

Figures for biennial Oregon State General Fund appropriations and allocations to the Oregon State Sanitary Authority, Department of Environmental Quality, and other state agencies mostly come from biennial governor's budget requests. These bound volumes allow for unambiguous budget comparisons from the late 1930s into the early 1960s. Sanitary authority funding is difficult to track, however, beginning with the 1965–1967 biennia for three reasons.

First, beginning with the 1965–1967 period, biennial governor's budget requests do not contain the same information, nor are they organized in the same way. Sanitary authority funding is subsumed within the broader board of health. Second, beginning in the mid-1960s Oregon's newspaper-of-record, the *Oregonian*, seems not to have provided the kind of detailed coverage of agency budgets it had in the past, making that source no longer useful for cross-referencing state records. Third, OSSA funding was complicated both by the integration of its budget with the Oregon Air Pollution Authority and, beginning in 1967, with the creation of regional air pollution authorities that received general fund appropriations through the OSSA. State officials in the late 1960s identified additional difficulties in comparing state budget appropriations. The legislative fiscal committee's January 1967 analysis of Governor Tom McCall's 1967–1969 budget states that "because of the complex organization and lack of accepted standards of performance, legislative review of this budget should be on a 'line-by-line' basis." This complexity comes from the fact that:

The Board's activities are carried out by many separate units within the agency. The programs of the Board are financed by five separate sources of moneys and many of the programs are financed by several different sources. . . . There are no objective workload indices that can be coupled with accepted standards of service by which the over-all program of the Board of Health can be reviewed, analyzed and approved at any given level of service. This problem has confronted legislative committees in the past and will be a problem for the 1967 Legislature in its consideration of this budget.[1]

INFLATION CALCULATOR

One of my goals in this book has been to provide, as much as possible, an apples-to-apples comparison between historical and current dollar amounts so readers can more readily understand the scale of relative costs. For example, I wanted to help readers answer such questions as the following: In 1944 Portland's voters approved a $24 million infrastructure funding measure for postwar sewers, highways, and other projects, but how much would this amount be circa 2014? The answer is approximately $260 million, and to provide such comparisons I have relied on Lawrence Officer and Samuel Williamson's MeasuringWorth calculator. No method of inflation comparison is without its strengths and weaknesses, but following Officer and Williamson's advice I have used their Gross Domestic Product (GDP) deflator formula. They recommend this formula to compare opportunity costs associated with investment and government projects because the measure of average prices is akin to the Consumer Price Index (CPI), but includes in its calculation *all* things produced in the economy, not merely consumer goods and services.[2] In the end, translating historical dollar amounts into comparable present-day figures will, I hope, shed some light on the significance of the amounts involved.

COMMENTS ON LITERATURE PERTINENT TO THE WILLAMETTE RIVER POLLUTION ABATEMENT STORY

In addition to the previous scholarship on Willamette River pollution identified in the Introduction, a few other sources warrant discussion. The first is Ellen Stroud's 1999 article on the social repercussions of the highly polluted Columbia Slough in North Portland. This article contains some

valuable material and sheds light on Portland's lamentable history of insti-
tutionalized racism. However, her thesis is not entirely accurate. She writes
that Portland officials failed to prioritize cleanup of the slough because this
area was home to a concentrated population of people of color, particularly
after World War II. She asserts that "Portlanders in power thought the North
Portland Peninsula was a disaster, and so it was."[3] Many sources in *Speaking
for the River* contradict this. Portland city officials were acutely aware of the
severely polluted Columbia Slough in North Portland at least as early as the
1930s. City commissioners, engineers, and the experts they hired recognized this and accordingly prioritized sewer construction in this area. The
Columbia Slough Interceptor line was the first segment of the city's postwar sewer program begun in summer 1947, and was also the first interceptor line connected to the Columbia Boulevard Treatment Plant a few years
later. Stroud provides evidence that actions of some city officials—in this
case the Housing Authority of Portland—were implicitly or explicitly rac-
ist; the housing authority was not, however, involved in sewer system planning. Whereas it certainly took extensive effort and many years of debate and
planning to improve water conditions in the slough, *Speaking for the River*
disproves the assertion that *all* city officials were complicit in perpetuating environmental racism by sacrificing North Portland and the Columbia
Slough to unmitigated pollution.

Journalist Brent Walth devotes a chapter in his biography of Tom McCall
to the issue of Willamette River pollution. In an otherwise commendable
book on a very important Oregonian, Walth's study unnecessarily denigrates
the sanitary authority's work for the sake of elevating McCall. Walth asserts
the sanitary authority's foremost abatement measure was reliance upon dilution that would result from increased summer and fall water flow on the main
stem of the river from the thirteen-dam Willamette Valley Project.[4] *Speaking
for the River* provides evidence that the sanitary authority pursued the opposite approach until the early 1960s, when it became apparent that water
quality could not be improved without both abating point- and nonpoint
sources and providing augmented flow from the Willamette Valley Project
dams. Walth also insinuates the sanitary authority was engaged in the "job
of convincing Oregonians that the Willamette was in fact growing cleaner
all the time," as if authority members and staff were not working diligently
on a complex topic, often with significantly constrained resources. That the
authority's achievements from the 1940s through the 1960s might fall short

of the desired results of some 1990s observers does not alter the authority's tangible and hard-won accomplishments. Walth writes that in the early 1960s, "Tom McCall was one of the few journalists who did not buy the [A] uthority's story," as if the authority were actively engaged in subterfuge and trying to bamboozle people.[5] This is simply not supported by the facts. The OSSA was *not* trying to put a "job" over on Oregonians.

The sanitary authority is not the only group to be unfairly diminished in Walth's elevation of McCall. Walth criticizes eight prior governors who served since creation of the sanitary authority, claiming they did not make the Willamette River cleanup a "true priority." Based on the primary sources in this book and the work of historians, geographers, and other scholars cited herein, I have found no evidence indicating *any* governor in the United States prior to the 1960s prioritized water pollution abatement. Further, the evidence I found for the sanitary authority's biennial budgets between the late 1930s and early 1970s (see chapter 6) shows they grew significantly in real and relative terms. This indicates that both governors and legislators were gradually realizing the state needed to address its water pollution problems, and they were committing funds to do so. To his credit McCall did step out ahead of his predecessors in his advocacy for cleaning up the Willamette, and during his gubernatorial tenure the successful Willamette River cleanup became national news. These achievements, however, came as a result of forty years of efforts from a diverse array of abatement advocates, consistent pressure from the sanitary authority, and incrementally increased funding requested by previous Oregon governors and authorized by legislators.

These and other examples illuminate two patterns in Walth's chapter on Willamette River pollution. The first is that Walth comes to his conclusions through reliance on a handful of sources. Rather than consulting primary sources—such as newspaper articles, meeting minutes, and correspondence—he makes extensive use of two particular secondary sources for his interpretation. One of these is E. Kimbark MacColl's 1979 book, *The Growth of a City: Power and Politics in Portland, Oregon, 1915–1950*, which addresses Willamette River pollution only in a cursory way.[6] The other is George W. Gleason's 1972 report *The Return of a River*. This is a solid work that provides an excellent summary narrative, but is focused primarily on science and technology, with only general background on the complex political and economic factors involved. As Gleeson himself wrote in his foreward, his work "attempts to provide a semi-historical description" but it was "impossible to give proper credit to persons, agencies, and organizations whose collective efforts effected river improvement."[7]

The second pattern is a consequence of the first: Walth makes sweeping statements about the accomplishments and motivations of previous Oregon governors and the sanitary authority that turn out to be inaccurate. This is because he has not delved very deeply into news accounts from the *Oregonian* and *Oregon Journal*, where the issues were being debated in a public forum, and he does not consult the meeting minutes of the sanitary authority itself, which contain the most thorough extant record of conversations, arguments, and debates between authority members and polluters. Without a doubt, Tom McCall is a central figure in the Willamette River pollution story. His accomplishments in this and many other areas are of such lasting significance that the necessary contributions of others working toward the same goals need not be denigrated for his further elevation. Doing so risks perpetuating anachronistic interpretations that inaccurately convey the complexity of historical events and misrepresent the actions and motivations of the people and organizations involved. This book offers extensive evidence from a broad range of primary and secondary sources that complement and—in some cases—update Walth.

Speaking for the River focuses on the more tangible and pragmatic aspects of water pollution abatement, but other scholars have approached the subject by emphasizing "forces" or "impulses." Historians Richard W. Judd and Christopher S. Beach frame their analysis of Oregon's and Maine's environmental history in terms of an "environmental imagination" with two components, the "pastoral" and the "wilderness." The pastoral ideal is traced to Thomas Jefferson and centers on America as a largely rural and agricultural society. The wilderness ideal is based on the romantic view of untrammeled nature John Muir espoused in the late nineteenth century.[8] Judd and Beach provide some important insights in their comparative analysis. Because they focus on water pollution abatement within states outside the heavily industrialized Northeast and Great Lakes regions, they help balance literature on this topic. However, their assertion that Oregonians' push to abate water pollution "emerged out of a nostalgic post–World War II literature of place" that combined elements of the "pastoral" and "wilderness" traditions is only partly accurate.[9] *Speaking for the River* identifies a citizen-led movement to abate Willamette River pollution that coalesced in the mid-1920s—well before Judd and Beach pick up their narrative. One finds a diverse range of motivations among these abatement advocates and allied groups. Some of these motivations contained romanticized "pastoral" and "wilderness" views of untrammeled nature, but many others were framed in a conservationist "wise use" perspective that sought to address the many conflicting uses of the river so that economic, recreational, and aesthetic values could be balanced. Finally, in focusing on one strand of motivations among abatement advocates, questions inevitably arise about how words and ideas get translated into tangible improvements in environmental conditions. Pastoral and wilderness ideals do not themselves produce empirical data, actionable legislation, or enforceable regulations. It was through the long process of gathering scientific data and pressuring polluters to change that Oregon's clean stream advocates achieved their results. Further, if the nexus of cultural values and motivations that effect environmental change are simplified into leading national figures such as Jefferson and Muir, one is left without any real understanding of what motivated the vast majority of people involved in environmental issues who were not among this elite.

Bibliography

LOCAL AND STATE GOVERNMENT AND GOVERNMENT-CONTRACTED
REPORTS

Amberg, Herman R. *Bacterial Fermentation of Spent Sulphite Liquor for the Production of Protein Concentrate Animal Feed Supplement, Bulletin Series, No. 38.* Corvallis: Oregon State Agricultural College Engineering Experiment Station, [Oct.] 1956.

Board of Consulting Engineers on Sewage Disposal (Ray E. Koon, John W. Cunningham, and Robert G. Dieck). *Report on General Survey of the Problems of Sewage Treatment and Disposal in Willamette River Valley Oregon.* Salem: State of Oregon Reconstruction Advisory Board, [Aug.] 1933.

Board of Review (R. H. Corey, Wellington Donaldson, Carl E. Green, and Abel Wolman), "Report on the Collection and Disposal of Sewage," Portland, Oreg.: Aug. 19, 1939.

Dimick, R. E., and Fred Merryfield. *The Fishes of the Willamette River System in Relation to Pollution, Bulletin Series, No. 20.* Corvallis: Oregon State Agricultural College Engineering Experiment Station, [June] 1945.

Gibbs, Charles V. *Metropolitan Seattle Sewerage and Drainage Survey: A Report for the City of Seattle, King County and the State of Washington on the Collection, Treatment and Disposal of Sewage and the Collection and Disposal of Storm Water in the Metropolitan Seattle Area.* Seattle: Brown and Caldwell Civil and Chemical Engineers, March 1958.

Gleeson, George W. *A Sanitary Survey of the Willamette River from Sellwood Bridge to the Columbia River, Bulletin Series, No. 6.* Corvallis: Oregon State Agricultural College Engineering Experiment Station, [April] 1936.

Gleeson, George W., and Fred Merryfield, *Industrial and Domestic Wastes of the Willamette Valley, Bulletin Series, No. 7* (parts 1, 2, and 3). Corvallis: Oregon State Agricultural College Engineering Experiment Station, [May] 1936.

Langton, C. V., and H. S. Rogers. *Preliminary Report on the Control of Stream Pollution in Oregon, Bulletin Series, No. 1.* Corvallis: Oregon State Agricultural College Engineering Experiment Station, [March] 1929.

Lincoln, John H., and Richard F. Foster. *Report on Investigation of Pollution in the Lower Columbia River.* Salem, Oreg.: Interstate Technical Advisory Committee, 1943.

Merryfield, Fred. *A Preliminary Survey of Industrial Pollution of Oregon Streams.* Portland: Oregon State Planning Board, [June] 1937.

Merryfield, Fred. "Stream Pollution Studies—1944: Preliminary Report Submitted to Governor's Special Committee, Oregon State Sanitary Authority." Corvallis: Oregon State Agricultural College Engineering Experiment Station, [January] 1945.

Merryfield, Fred, W. B. Bollen, and F. C. Kachelhoffer. *Industrial and City Wastes, Bulletin Series, No. 22.* Corvallis: Oregon State Agricultural College Engineering Experiment Station, [March] 1947.

Merryfield, Fred, and W. G. Wilmot. *1945 Progress Report on Pollution of Oregon Streams, Bulletin Series, No. 19.* Corvallis: Oregon State Agricultural College Engineering Experiment Station, [June] 1945.

Moses, Robert. *Portland Improvement.* New York: William E. Rudge's Sons, Nov. 10, 1943.

Northcraft, Martin, and Warren C. Westgarth. *Oregon State Water Resources Board, Bulletin No. 2: Water Quality Data Inventory Supplement.* Corvallis: Oregon State College Engineering Experiment Station, [June] 1957.

Oregon Department of Environmental Quality, Water Quality Division. "Oregon's 1994 Water Quality Status Assessment Report (305[b])." Portland: Oregon Department of Environmental Quality, April 1994.

Oregon Department of Environmental Quality. "Proposed Water Quality Management Plan for Willamette River Basin." Portland: Oregon Department of Environmental Quality, 1976.

Oregon State Planning Board Advisory Committee on Stream Purification. *Efforts to Reduce Stream Pollution in Oregon.* Portland, Oreg.: Oregon State Planning Board, [Feb.] 1937.

Oregon State Sanitary Authority. "Implementation and Enforcement Plan for the Public Waters of the State of Oregon." Portland, Oreg.: Oregon State Sanitary Authority, May 1967.

Oregon State Sanitary Authority. "Interim Report on Status of Water Pollution Control in the Willamette River Basin." Portland, Oreg.: Oregon State Sanitary Authority, 1957.

Oregon State Water Resources Board. *First Biennial Report.* Salem, Oreg.: Oregon State Water Resources Board, [Jan.] 1957.

Oregon State Water Resources Board. *Middle Willamette River Basin.* Salem, Oreg.: Oregon State Water Resources Board, [June] 1963.

Oregon State Water Resources Board. *Willamette River Basin.* Salem, Oreg.: Oregon State Water Resources Board, [June] 1967.

Rodgers, H. S., C. A. Mockmore, and C. D. Adams. *A Sanitary Survey of the Willamette Valley, Bulletin Series, No. 2.* Corvallis: Oregon State Agricultural College Engineering Experiment Station, [June] 1930.

Ronchetto, John C. *An Analytical Digest of Existing Legislation of Oregon and Other States Relating to Stream Pollution.* Salem, Oreg.: Oregon State Planning Board Advisory Committee on Stream Purification, Aug. 1936.

Ronchetto, John C. *Recommended Principles for Stream Purification Legislation in Oregon.* Salem, Oreg.: Oregon State Planning Board Advisory Committee on Stream Purification, June 1937.

State of Oregon. *Progress Report of the Postwar Readjustment Commission* [biennial]. Salem, Oreg., 1943–1948.

FEDERAL GOVERNMENT AND GOVERNMENT-CONTRACTED REPORTS

Bingham, Tayler H., Timothy R. Bondelid, Brooks M. Depro, Ruth C. Figueroa, A. Brett Hauber, Suzanne J. Unger, and George L. Van Houtven. "A Benefits Assessment of Water Pollution Control Programs Since 1972: Part 1, The Benefits of Point Source Controls for Conventional Pollutants in Rivers and Streams, Final Report," Research Triangle Park, N.C.: Research Triangle Park Institute [Prepared for the US Environmental Protection Agency Office of Water Office of Policy, Economics, and Innovation], Jan. 2000.

Eldridge, Edward F., ed. *Proceedings of the Second Pacific Northwest Symposium on Water Pollution Research: Financing Water Pollution Research Projects, Jan. 20, 1958.* Portland, Oreg.: US Department of Health, Education, & Welfare, Public Health Service, Region 9, 1958.

Eldridge, Edward F., ed. *Proceedings of the Third Pacific Northwest Symposium on Water Pollution Research: The* Sphaerolitus *(Slime) Problem, Sept. 10–11, 1958.* Portland, Oreg.: US Department of Health, Education, & Welfare, Public Health Service, Region 9, 1958.

Eldridge, Edward F., ed. *Proceedings of the Eleventh Pacific Northwest Symposium on Water Pollution Research: The Social and Economic Aspects of Water-Resource Quality Control, November 8 and 9, 1962.* Portland, Oreg.: US Department of Health, Education, & Welfare, Public Health Service, Region 9, 1962.

Fish, Frederic F., and Robert R. Rucker. *Pollution in the Lower Columbia Basin in 1948— with Particular Reference to the Willamette River. Special Scientific Report: Fisheries No. 30.* Washington, DC: US Department of the Interior US Fish and Wildlife Service, June 1950.

National Resources Committee. *Progress Report with Statements of Coordinating Committees.* Washington, DC: US Government Printing Office, June 15, 1936.

National Resources Committee. *Regional Planning Part I—Pacific Northwest.* Washington, DC: US Government Printing Office, May 1936.

National Resources Committee Special Advisory Committee on Water Pollution. *Report on Water Pollution.* Washington, DC: US Government Printing Office, July 1935; reprint Feb. 1936.

National Resources Committee Special Advisory Committee on Water Pollution. *Water Pollution in the United States.* Washington, DC: US Government Printing Office, 1939.

Rockefeller, Laurence S., Chair. "Report to The President and The Council on Environmental Quality." Washington, DC: Citizens' Advisory Committee on Environmental Quality, April 1971.

US Environmental Protection Agency. "Community Advisory Groups (CAGs) at Superfund Sites." Publication 9230.0-28AFS, Aug. 1996.

US Department of Health, Education, and Welfare, Public Health Service. *Proceedings: The National Conference on Water Pollution, December 12–14, 1960, Sheraton-Park Hotel, Washington, DC.* Washington, DC: US Government Printing Office, 1961.

US Department of the Interior, Census Office. *Abstract of the Twelfth Census of the United States, 1900.* Washington, DC: US Government Printing Office, 1904.

US Department of the Interior, Bureau of the Census, *Census of Population: 1950.* Washington, DC: US Government Printing Office, 1953.

US Department of the Interior, Bureau of the Census, *Census of Population: 1960.* Washington, DC: US Government Printing Office, 1963.

US Federal Security Agency, Public Health Service, Division of Water Pollution Control. *Pacific Northwest Drainage Basins Summary Report on Water Pollution.* Washington, DC: US Government Printing Office, 1951.

BIOGRAPHY

Clements, Kendrick. *Hoover, Conservation, and Consumerism: Engineering the Good Life.* Lawrence: University Press of Kansas, 2000.

Clucas, Richard A. "The Political Legacy of Robert W. Straub." *Oregon Historical Quarterly* 104:4 (Winter 2003), 462–477.

Mathewson, Worth. *William L. Finley: Pioneer Wildlife Photographer.* Corvallis: Oregon State University Press, 1986.

McCall, Tom, and Steve Neal. *Tom McCall: Maverick.* Portland, Oreg.: Binford & Mort, 1977.

McKay, Floyd J. *An Editor for Oregon: Charles A. Sprague and the Politics of Change.* Corvallis: Oregon State University Press, 1998.

Melosi, Martin V. "Lyndon Johnson and Environmental Policy," in Robert A. Divine, ed., *The Johnson Years, Volume 2: Vietnam, the Environment, and Science* (Lawrence: University Press of Kansas, 1987), 113–149.

Murphy, Priscilla Coit. *What a Book Can Do: The Publication and Reception of* Silent Spring. Amherst: University of Massachusetts Press, 2005.

Murrell, Gary. *Ironpants: Oregon's Anti-New Deal Governor, Charles Henry Martin.* Pullman: Washington State University Press, 2000.

Walth, Brent. *Fire at Eden's Gate: Tom McCall and the Oregon Story.* Portland, Oreg.: Oregon Historical Society Press, 1994.

THE NEW DEAL ERA

Beito, David T. *Taxpayers in Revolt: Tax Resistance During the Great Depression.* Chapel Hill, N.C.: The University of North Carolina Press, 1989.

Bessey, Roy F. *Pacific Northwest Regional Planning: A Review.* Olympia: State of Washington Printing Plant, 1963.

Clawson, Marion. *New Deal Planning: The National Resources Planning Board.* Baltimore: The Johns Hopkins University Press (for Resources for the Future), 1981.

Lowitt, Richard. *The New Deal and the West.* Bloomington: Indiana University Press, 1984.

McKinley, Charles. *Uncle Sam in the Pacific Northwest: Federal Management of Natural Resources in the Columbia River Valley.* Berkeley: University of California Press, 1952.

Patterson, James T. *The New Deal and the States: Federalism in Transition.* Princeton, N.J.: Princeton University Press, 1969.

Shanley, Robert A. "Franklin D. Roosevelt and Water Pollution Control Policy."
   Presidential Studies Quarterly 18:2 (Spring 1988), 319–330.

Smith, Jason Scott. *Building New Deal Liberalism: The Political Economy of Public Works,
   1933–1956.* Cambridge: Cambridge University Press, 2006.

Warken, Philip W. *A History of the National Resources Planning Board, 1933–1943.* New
   York: Garland Publishing, Inc., 1979.

ENVIRONMENTAL SCIENCE AND PUBLIC HEALTH

Bowler, Peter J. *The Norton History of the Environmental Sciences.* New York: W. W.
   Norton & Company, 1993.

Cumbler, John T. *Reasonable Use: The People, the Environment, and the State, New England
   1790–1930.* New York: Oxford University Press, 2001.

Duffy, John. *The Sanitarians: A History of American Public Health.* Urbana, Ill.: University
   of Illinois Press, 1990.

Lindberg, Roberta J. "Non-Regulatory Efforts to Combat Nonpoint Source Pollution in
   Oregon." MA internship report, University of Oregon, 1995.

Gorman, Hugh S. *Redefining Efficiency: Pollution Concerns, Regulatory Mechanisms, and
   Technological Change in the US Petroleum Industry.* Akron, Ohio: University of Akron
   Press, 2001.

Russell, Edmund P., III. "Lost Among the Parts Per Billion: Ecological Protection at the
   United States Environmental Protection Agency, 1970–1993." *Environmental History*
   2:1 (Jan. 1997), 29–51.

Schaedel, Andy. "The Three 'O's,' or, the Evolution of Assessments and the Origins of
   303(d) Listings and TMDLs [draft]." March 31, 2008 [Unpublished presentation,
   available at DEQ Headquarters, Portland, Oreg.].

Smith, James E., James F. Kreiss, Gary S. Logsdon, and Norbert Schomaker.
   "Environmental Research in Cincinnati: A Century of Federal Partnership."
   Environmental Engineer 13 (Summer 2011).

WATER POLLUTION POLICY, ADMINISTRATION, AND POLITICS

Adler, Jonathan H. "Fables of the Cuyahoga: Reconstructing a History of Environmental
   Protection." *Fordham Environmental Law Journal* 14 (Fall 2002), 89–146.

Andreen, William L. "The Evolution of Water Pollution Control in the United States—
   State, Local, and Federal Efforts, 1789–1972: Part I." *Stanford Environmental Law
   Journal* 22 (Jan. 2003), 145–200.

Andreen, William L. "The Evolution of Water Pollution Control in the United States—
   State, Local, and Federal Efforts, 1789–1972: Part II." *Stanford Environmental Law
   Journal* 22 (June 2003), 215–294.

Benedickson, Jamie. *The Culture of Flushing: A Social and Legal History of Sewage.*
   Vancouver, BC: University of British Columbia Press, 2007.

Casner, Nicholas. "Polluter versus Polluter: The Pennsylvania Railroad and the
   Manufacturing of Pollution Policies in the 1920s." *Journal of Policy History* 11:2
   (1999), 179–200.

Cowdrey, Albert E. "Pioneering Environmental Law: The Army Corps of Engineers and
   the Refuse Act." *Pacific Historical Review* 44:3 (Aug. 1975), 331–349.

Gnoss, George H., Jr. "Special Problems of Water Pollution: The Private Sector." *University of California, Davis, Law Review* 1 (1969), 105–139.

Judd, Richard W. "The Coming of the Clean Waters Acts in Maine, 1941–1961." *Environmental History Review* 14:3 (Fall 1990), 51–74.

Kehoe, Terence. *Cleaning Up the Great Lakes: From Cooperation to Confrontation.* DeKalb, Ill.: Northern Illinois University Press, 1997.

Lieber, Harvey, and Bruce Rosinoff, *Federalism and Clean Waters: The 1972 Water Pollution Control Act.* Lexington, Mass.: Lexington Books (D. C. Heath and Co.), 1975.

Milazzo, Paul Charles. *Unlikely Environmentalists: Congress and Clean Water, 1945–1972.* Lawrence: University Press of Kansas, 2006.

Murphy, Earl Finbar. *Water Purity: A Study in Legal Control of Natural Resources.* Madison: University of Wisconsin Press, 1961.

Paavola, Jouni. "Interstate Water Pollution Problems and Elusive Federal Water Pollution Policy in the United States, 1900–1948." *Environment and History* 12:4 (2006), 435–465.

Rome, Adam W. "Coming to Terms with Pollution: The Language of Environmental Reform, 1865–1915." *Environmental History* 1:3 (1996), 6–28.

ENVIRONMENTAL POLICY, ADMINISTRATION, AND POLITICS

Adler, Jonathan H. "The Fable of Federal Environmental Regulation: Reconsidering the Federal Role in Environmental Protection." *Case Western Reserve Law School Review* 55:1 (Sept. 2004), 93–113.

Brooks, Karl Boyd. *Before Earth Day: The Origins of American Environmental Law, 1945–1970.* Lawrence: University Press of Kansas, 2009.

Hays, Samuel P. *Beauty, Health, and Permanence: Environmental Politics in the United States, 1955–1985.* Pittsburgh: University of Pittsburgh Press, 1987.

Hays, Samuel P. *Conservation and the Gospel of Efficiency: The Progressive Conservation Movement, 1890–1920.* Cambridge, Mass.: Harvard University Press, 1959.

Hays, Samuel P. *Explorations in Environmental History: Essays.* Pittsburgh: University of Pittsburgh Press, 1998.

Huffman, Thomas R. "Defining the Origins of Environmentalism in Wisconsin: A Study in Politics and Culture." *Environmental History Review* 16:3 (Autumn 1992), 47–69.

Lewis, Jack. "The Birth of the EPA." *EPA Journal*, Nov. 1985. http://www.epa.gov/aboutepa/history/topics/epa/15c.html, accessed Feb. 4, 2012.

Markowitz, Gerald, and David Rosner. *Deceit and Denial: The Deadly Politics of Industrial Pollution.* Berkeley: University of California Press, 2003.

McEvoy, Arthur F. *The Fisherman's Problem: Ecology and Law in the California Fisheries, 1850–1980.* New York: Cambridge University Press, 1986.

McEvoy, Arthur F. "Toward an Interactive Theory of Nature and Culture: Ecology, Production, and Cognition in the California Fishing Industry." *Environmental Review (ER)* 11:4 [Special Issue: Theories of Environmental History] (Winter 1987), 289–305.

Neuzil, Mark. *The Environment and the Press: From Adventure Writing to Advocacy.* Evanston, Ill.: Northwestern University Press, 2008.

Revesz, Richard L. "Federalism and Environmental Regulation: A Public Choice Analysis." *Harvard Law Review* 115:2 (Dec. 2001), 553–641.

Robertson, Thomas. "'This is the American Earth': American Empire, the Cold War, and American Environmentalism." *Diplomatic History* 32:4 (Sept. 2008), 561–584.

Scott, James C. *Seeing Like a State: How Certain Schemes to Improve the Human Condition Have Failed.* New Haven, Conn.: Yale University Press, 1998.

Sellers, Christopher. "Body, Place and the State: The Makings of an 'Environmentalist' Imaginary in the Post–World War II US." *Radical History Review* 74 (Spring 1999), 31–64.

Steinberg, Ted. "Down to Earth: Nature, Agency, and Power in History." *The American Historical Review* 107:3 (June 2002), 798–820.

Switzer, Jacqueline Vaughn. *Green Backlash: The History and Politics of Environmental Opposition in the US.* Boulder, Colo.: Lynne Rienner Publishers, 1997.

Rosen, Christine M. "Differing Perceptions of the Value of Pollution Abatement Across Time and Place: Balancing Doctrine in Pollution Nuisance Law, 1840–1906." *Law and History Review* 11:2 (Fall 1993), 303–381.

Shifrin, Neil S. "Pollution Management in the Twentieth Century." *Journal of Environmental Engineering* 131:5 (May 2005), 676–691.

PULP AND PAPER AND OTHER INDUSTRIAL POLLUTERS

Adams, W. Claude. "History of Papermaking in the Pacific Northwest: I." *Oregon Historical Quarterly* 52:1 (March 1951), 21–37.

Adams, W. Claude. "History of Papermaking in the Pacific Northwest: II." *Oregon Historical Quarterly* 52:2 (June 1951), 83–100.

Adams, W. Claude. "History of Papermaking in the Pacific Northwest: III." *Oregon Historical Quarterly* 52:3 (Sept. 1951), 154–185.

Bartlett, Richard A. *Troubled Waters: Champion International and the Pigeon River Controversy.* Knoxville: University of Tennessee Press, 1995.

Bourhill, Bob. *History of Oregon's Timber Harvests and/or Lumber Production: State Data, 1849 to 1992, County Data, 1925 to 1992.* Salem: Oregon Dept. of Forestry, 1994.

Burke, John G. "Wood Pulp, Water Pollution, and Advertising." *Technology and Culture* 20:1 (Jan. 1979), 175–195.

Erspamer, August S., and William D. Rice. "Treatment of Pulp and Paper Mill Waste." *Industrial and Engineering Chemistry* 41:8 (Aug. 1949), 1806–1809.

Gorman, Hugh S. "Laying the Foundation for the Control of Industrial Pollution, 1930–1970: Two Canals, A Refinery, and Clarence W. Klassen." *Journal of Illinois History* 8:3 (2005), 182–208.

Gunningham, Neil, Robert A. Kagan, and Dorothy Thornton. *Shades of Green: Business, Regulation, and Environment.* Stanford, Calif.: Stanford University Press, 2003.

Guthrie, John A. *An Economic Analysis of the Pulp and Paper Industry.* Pullman: Washington State University Press, 1972.

Guthrie, John A. *The Economics of Pulp and Paper.* Pullman: State College of Washington Press, 1950.

Harrison, Kathryn. "Between Science and Politics: Assessing the Risks of Dioxins in Canada and the United States." *Policy Sciences* 24:4 (Nov. 1991), 367–388.

Harrison, Kathryn. "The Regulator's Dilemma: Regulation of Pulp Mill Effluents in the Canadian Federal State." *Canadian Journal of Political Science* 29:3 (Sept. 1996), 469–496.

Hurst, W. M., E. G. Nelson, Jesse E. Harmond, Leonard M. Klein, and D. W. Fishler. *The Fiber Flax Industry in Oregon: Its History, Present Status, and Future Possibilities.* Corvallis: Oregon State College Agricultural Experiment Station, 1953.

Lucas, William. *Canning in the Valley: Canneries of the Salem District.* Salem, Ore.: William Lucas, 1998.

Norberg-Bohm, Vicki, and Mark Rossi. "The Power of Incrementalism: Environmental Regulation and Technological Change in Pulp and Paper Bleaching in the US." *Technology Analysis & Strategic Management* 10:2 (1998), 225–245.

Ohanian, Nancy Kane. *The American Pulp and Paper Industry, 1900–1940: Mill Survival, Firm Structure, and Industry Relocation.* Westport, Conn.: Greenwood Press, 1993.

Powell, Mark R. "Control of Dioxins (and other Organochlorines) from the Pulp and Paper Industry under the Clean Water Act and Lead in Soil at Superfund Mining Sites: Two Case Studies in EPA's Use of Science." Discussion Paper 97–08, Washington, DC: Resources for the Future, March 1997.

Ross, Benjamin, and Steven Amter. *The Polluters: The Making of Our Chemically Altered Environment.* New York: Oxford University Press, 2010.

Solomon, Keith R. "Chlorine in the Bleaching of Pulp and Paper." *Pure and Applied Chemistry* 68:9 (1996), 1721–1730.

Van Strum, Carol, and Paul Merrell. *No Margin of Safety: A Preliminary Report on Dioxin Pollution and the Need for Emergency Action in the Pulp and Paper Industry.* Greenpeace USA, 1987..

Winget, Russell L. "A Definitive History of the National Council for Stream Improvement (of the Pulp, Paper, and Paperboard Industries), Inc., 1943–1966." National Council for Stream Improvement, 1966.

CITIZEN ACTIVISM

Caldwell, Lynton K., Lynton R. Hayes, and Isabel M. MacWhirter. *Citizens and the Environment: Case Studies in Popular Action.* Bloomington, Ind.: Indiana University Press, 1976.

Casner, Nicholas. "Angler Activist: Kenneth Reid, the Izaak Walton League, and the Crusade for Federal Water Pollution Control." *Pennsylvania History* 66:4 (1999), 535–553.

Dunlap, Thomas R. *Faith in Nature: Environmentalism as Religious Quest.* Seattle: University of Washington Press, 2004.

Funkhouser, G. Ray. "The Issues of the Sixties: An Exploratory Study in the Dynamics of Public Opinion." *The Public Opinion Quarterly* 37:1 (Spring 1973), 62–75.

Hayden, Thomas Howard (Paul E. Toren, ed.). *Citizen Leadership in Conservation: The Minnesota Izaak Walton League, 1922–1973.* St. Paul, Minn., Izaak Walton League Minnesota Division, 2001.

Izaak Walton League of America. "Dynamic Action: A Record of Goals and Accomplishments by Izaak Walton League Divisions and Chapters." Chicago, Ill., Izaak Walton League of America, 1946.

The League of Women Voters Education Fund. *The Big Water Fight: Trials and Triumphs in Citizen Action on Problems of Supply, Pollution, Floods, and Planning across the U.S.A.* Brattleboro, Vt.: The Stephen Greene Press, 1966.

Merritt, Dawn. "From the Fast-moving Fifties to the 'Sensible' Sixties." *Outdoor America,* Summer 2012, 12–23.

Merritt, Dawn. "From the Jazz Age to World War II." *Outdoor America,* Spring 2012, 12–20.

Merritt, Dawn. "The Roaring 20s: A Call to Action." *Outdoor America,* Winter 2012, 24–33.

Minteer, Ben A. *The Landscape of Reform: Civic Pragmatism and Environmental Thought in America.* Cambridge, Mass.: The MIT Press, 2006.

Reiger, John F. *American Sportsmen and the Origins of Conservation* (3rd ed.). Corvallis: Oregon State University Press, 2001.

Rome, Adam W. "'Give Earth a Chance': The Environmental Movement and the Sixties." *The Journal of American History* 89:2 (Sept. 2003), 525–554.

Searle, R. Newell. "Autos or Canoes? Wilderness Controversy in the Superior National Forest." *Journal of Forest History* 22:2 (1978), 68–77.

URBAN ADMINISTRATION AND SANITATION

Buttenweiser, Ann L. *Manhattan Water-Bound: Planning and Developing Manhattan's Waterfront from the Seventeenth Century to the Present.* New York: New York University Press, 1987.

Dolin, Eric Jay. *Political Waters: The Long, Dirty, Contentious, Incredibly Expensive but Eventually Triumphant History of Boston Harbor—A Unique Environmental Success Story.* Amherst: University of Massachusetts Press, 2004.

Elkind, Sarah S. *Bay Cities and Water Politics: The Battle for Resources in Boston and Oakland.* Lawrence: University Press of Kansas, 1998.

Gandy, Matthew. *Concrete and Clay: Reworking Nature in New York City.* Cambridge, Mass.: The MIT Press, 2002.

Goldman, Joanne Able. *Building New York's Sewers: Developing Mechanisms of Urban Management.* West Lafayette, Ind.: Purdue University Press, 1997.

Karvonen, Andrew. *Politics of Urban Runoff: Nature, Technology, and the Sustainable City.* Cambridge, Mass.: The MIT Press, 2011.

Keeling, Arn. "Sink or Swim: Water Pollution and Environmental Politics in Vancouver, 1889–1975." *BC Studies* 142/143 (Summer/Autumn 2004), 69–101.

Keeling, Arn. "Urban Waste Sinks as a Natural Resource: The Case of the Fraser River." *Urban History Review* 34:1 (2005), 58–70.

Laurgaard, Olaf. *Treatise on the Design, Test, & Construction of the Front St. Intercepting Sewer and Drainage System in Portland, Oregon, Including Intercepting Sewer, Pumping Plant, & Concrete Bulkhead-Wall on Gravel Filled Timber Cribs.* Corvallis: Oregon State Agricultural College, 1933.

Melosi, Martin V. *The Sanitary City: Urban Infrastructure in America from Colonial Times to the Present*. Baltimore, Md.: Johns Hopkins University Press, 2000.

Melosi, Martin V. "Sanitary Services and Decision Making in Houston, 1876–1945." *Journal of Urban History* 20:3 (May 1994), 365–406.

Olson, Sherry. "Downwind, Downstream, Downtown: The Environmental Legacy in Baltimore and Montreal." *Environmental History* 12:4 (2007), 845–866.

Schneider, Daniel. *Hybrid Nature: Sewage Treatment and the Contradictions of the Industrial Ecosystem*. Cambridge, Mass.: The MIT Press, 2011.

Skowronek, Stephen. *Building a New American State: The Expansion of National Administrative Capacities, 1877–1920*. Cambridge, UK: Cambridge University Press, 1982.

Tarr, Joel A. "Industrial Waste Disposal in the United States as a Historical Problem." *Ambix* 49:1 (March 2002), 4–20.

Tarr, Joel A. "The Metabolism of the Industrial City: The Case of Pittsburgh." *Journal of Urban History* 28:5 (2002), 511–545.

Tarr, Joel A. *The Search for the Ultimate Sink: Urban Pollution in Historical Perspective*. Akron, Ohio: University of Akron Press, 1996.

Tierno, Mark J. "The Search for Pure Water in Pittsburgh: The Urban Response to Water Pollution, 1893–1914." *The Western Pennsylvania Historical Magazine* 60:1 (1977), 23–36.

URBAN HISTORIES

Klingle, Matthew. *Emerald City: An Environmental History of Seattle*. New Haven, Conn.: Yale University Press, 2007.

Rawson, Michael. *Eden on the Charles: The Making of Boston*. Cambridge, Mass.: Harvard University Press, 2010.

Rome, Adam W. *The Bulldozer in the Countryside: Suburban Sprawl and the Rise of American Environmentalism*. Cambridge, UK: Cambridge University Press, 2001.

Stradling, David. *Smokestacks and Progressives: Environmentalists, Engineers and Air Quality in America, 1881–1951*. Baltimore, Md.: Johns Hopkins University Press, 1999.

Thrush, Coll. *Native Seattle: Histories from the Crossing-Over Place*. Seattle: University of Washington Press, 2007.

Voyer, Richard A., Carol Pesch, Jonathan Garber, Jane Copeland, and Randy Comeleo. "New Bedford, Massachusetts: A Story of Urbanization and Ecological Connections." *Environmental History* 5:3 (2000), 352–376.

RIVERS, LAKES, AND ESTUARIES

Colten, Craig E. "Illinois River Pollution Control, 1900–1970," in Lary M. Dilsaver and Craig E. Colten, eds. *The American Environment: Interpretations of Past Geographies*, (Lanham, Md.: Rowman & Littlefield, 1992), 254–272.

Colten, Craig E. "Contesting Pollution in Dixie: The Case of Corney Creek." *Journal of Southern History* 72:3 (Aug. 2006), 605–634.

Cumbler, John T. "Conflict, Accommodation, and Compromise: Connecticut's Attempt to Control Industrial Wastes in the Progressive Era." *Environmental History* 5:3 (2000), 314–335.

Dunwell, Frances F. *The Hudson: America's River*. New York: Columbia University Press, 2008.

Egerton, Frank N. "Pollution and Aquatic Life in Lake Erie: Early Scientific Studies." *Environmental Review* 11:3 (Fall 1987), 189–206.

Harris, Glenn, and Seth Wilson. "Water Pollution in the Adirondack Mountains: Scientific Research and Governmental Response, 1890–1930." *Environmental History Review* 17:4 (Winter 1993), 47–72.

Kehoe, Terence. "The Persistence of Cooperation: Government Regulation of Great Lakes Water Pollution, 1945–1978." *Business and Economic History* 24:1 (Fall 1995), 147–154.

Kehoe, Terence. "'You Alone Have the Answer': Lake Erie and Federal Water Pollution Control Policy, 1960–1972." *Journal of Policy History* 8:4 (1996), 440–469.

McFarlane, Wallace Scot. "The Limits of Progress: Walter Lawrance and the Shifting Terrain of Science, Pollution, and Environmental Politics on Maine's Androscoggin River, 1941–1977." Senior thesis, Bowdoin College, 2009.

Mallea, Amahia. "Downstreamers: Public Health and Relationships on the Missouri River." *Agricultural History* 76:2 (Spring 2002), 393–404.

Nash, Linda. "The Changing Experience of Nature: Historical Encounters with a Northwest River." *Journal of American History* 86:4 (Mar. 2000), 1600–1629.

Pisani, Donald J. "Fish Culture and the Dawn of Concern Over Water Pollution." *Environmental Review* 8:2 (1984), 117–131.

Pisani, Donald J. "The Polluted Truckee: A Study in Interstate Water Quality, 1870–1934." *Nevada Historical Society Quarterly* 20:3 (1977), 151–166.

Read, Jennifer. "'A Sort of Destiny': The Multi-Jurisdictional Response to Sewage Pollution in the Great Lakes, 1900–1930." *Scientia Canadensis* 22–23:51 (1998–1999), 103–129.

Scarpino, Philip V. *Great River: An Environmental History of the Upper Mississippi, 1890–1950*. Columbia: University of Missouri Press, 1985.

Steinberg, Theodore. *Nature Incorporated: Industrialization and the Waters of New England*. Amherst; University of Massachusetts Press, 1994.

Stradling, David, and Richard Stradling. "Perceptions of the Burning River: Deindustrialization and Cleveland's Cuyahoga River." *Environmental History* 13:3 (July 2008), 515–535.

Summers, Gregory. *Consuming Nature: Environmentalism in the Fox River Valley, 1850–1950*. Lawrence: University Press of Kansas, 2006.

White, Richard. *Organic Machine: The Remaking of the Columbia River*. New York: Hill and Wang, 1995.

THE WILLAMETTE RIVER AND ITS ENVIRONMENT

Allen, John Eliot, Marjorie Burns, and Scott Burns. *Cataclysms on the Columbia: The Great Missoula Floods* (2nd ed.). Portland, Oreg: Oooligan Press, 2009.

Alt, David, and Donald W. Hyndman. *Northwest Exposures: A Geologic Story of the Northwest*. Missoula, Mont.: Mountain Press Publishing Company, 1995.

Barber, Lawrence. *Columbia Slough*. Portland, Oreg.: Columbia Slough Development Corporation, Oct. 1977.

Benner, Patricia A., and James R. Sedell. "Upper Willamette River Landscape: A Historic Perspective," in Antonius Laenen and David A. Dunnette, eds., *River Quality: Dynamics and Restoration*, (Boca Raton, Fla.: Lewis Publishing, 1997), 23–48.

Carter, Glenn D. "Pioneering Water Pollution Control in Oregon." *Oregon Historical Quarterly* 107:2 (Summer 2006), 254–272.

Everts, Curtiss M. "Willamette Clean-up." *Oregon State Game Commission Bulletin* 8:7 (July 1953), 3–5.

Gleeson, George W. *The Return of a River: The Willamette River, Oregon*. Corvallis: Oregon State University Water Resources Research Institute, 1972.

Hillegas, James V. "Working for the 'Working River'": Willamette River Pollution, 1926–1962." MA thesis, Portland State University, 2009.

Honey, William D. Jr. *The Willamette River Greenway: Cultural and Environmental Interplay*. Corvallis: Oregon State University Water Research Institute, 1975.

Lang, William L. "From Where We Are Standing: The Sense of Place and Environmental History," in David D. Goble and Paul W. Hirt, *Northwest Lands, Northwest Peoples: Readings in Environmental History*, (Seattle: University of Washington Press, 1999), 79–94.

Lang, William L. "One City, Two Rivers: Columbia and Willamette Rivers in the Environmental History of Twentieth-Century Portland, Oregon," in Char Miller, ed., *Cities and Nature in the American West*, (Reno: University of Nevada Press, 2010), 96–111.

Larson, Douglas W. "Reservoir Limnology in the Pacific Northwest: Willamette River Basin, OR." *Lakeline* 21:4 (Winter 201/2002), 11–16.

Mahaffey, Juanita. "Pollution: The Worst Poacher." *Oregon State Game Commission Bulletin* 15:6 (June 1960), 3–4.

Mullane, Neil. "The Willamette River of Oregon: A River Restored?" in Antonius Laenen and David A. Dunnette, eds., *River Quality: Dynamics and Restoration*, (Boca Raton, Fla.: Lewis Publishing, 1997), 65–76.

Noble, Richard Earle. "The Willamette River Fishes as Biological Indicators of Pollution." MS thesis, Oregon State College, 1952.

Orr, Elizabeth, and William Orr. *Oregon Water: An Environmental History*. Portland, Oreg.: Inkwater Press, 2005.

Robbins, William G. "The Willamette Valley Project of Oregon: A Study in the Political Economy of Water Resource Development." *Pacific Historical Review* 47:4 (1978), 585–605.

Robbins, William G. "Willamette Eden: The Ambiguous Legacy." *Oregon Historical Quarterly* 99:2 (Summer 1998), 189–218.

Starbird, Ethel. "A River Restored: Oregon's Willamette." *National Geographic* 141:6 (June 1972), 816–834.

Stroud, Ellen. "Troubled Waters in Ecotopia: Environmental Racism in Portland, Oregon." *Radical History Review* 74 (1999), 65–95.

Taylor, Joseph. *Making Salmon: An Environmental History of the Northwest Fisheries Crisis.* Seattle: University of Washington Press, 1999.

Van Strum, Carol. *A Bitter Fog: Herbicides and Human Rights.* San Francisco: Sierra Club Books, 1983.

Velz, Clarence J. "Report on Natural Purification Capacities: Willamette River." Ann Arbor: University of Michigan School of Public Health, for the National Council for Stream Improvement of the Pulp, Paper and Paperboard Industries, Inc., May 1951.

Velz, Clarence J. "Supplementary Report on Lower Willamette River Waste Assimilation Capacity." Ann Arbor: University of Michigan School of Public Health, for the National Council for Stream Improvement of the Pulp, Paper and Paperboard Industries, Inc., March 1961.

Willingham, William F. *Army Engineers and the Development of Oregon: A History of the Portland District US Army Corps of Engineers.* Washington, DC: US Government Printing Office, 1983.

OREGON AND THE PACIFIC NORTHWEST

Abbott, Carl. *Boosters and Businessmen: Popular Economic Thought and Urban Growth in the Antebellum Middle West.* Westport, Conn.: Greenwood, 1981.

Beckham, Stephen Dow. "History of Western Oregon Since 1846," in William C. Sturtevant, ed., *Handbook of North American Indians, vol. 7: Northwest Coast* (Washington, DC: Smithsonian Institution, 1990), 180–188.

Boag, Peter. *Environment and Experience: Settlement Culture in Nineteenth-Century Oregon.* Berkeley: University of California Press, 1992.

Boyd, Robert T. "Demographic History," in William C. Sturtevant, ed., *Handbook of North American Indians, vol. 7: Northwest Coast* (Washington, DC: Smithsonian Institution, 1990), 135–148.

Boyd, Robert T. "Strategies of Indian Burning in the Willamette Valley." *Canadian Journal of Anthropology* 5 (1986), 65–86.

Bunting, Robert. *The Pacific Raincoast: Environment and Culture in an American Eden, 1778–1900.* Lawrence: University Press of Kansas, 1997.

Burney, W. T., ed. *The Gold-Gated West: Songs and Poems by Samuel L. Simpson.* Philadelphia & London: J. B. Lippincott Company, 1910.

Burton, Robert E. *Democrats of Oregon: The Pattern of Minority Politics, 1900–1956.* Eugene: University of Oregon Books, 1970.

Fidler, W. W. "Personal Reminiscences of Samuel L. Simpson." *Oregon Historical Society Quarterly* 15:4 (Dec. 1914), 264–276.

Herzig, Jill Hopkins. "The Oregon Commonwealth Federation: The Rise and Decline of a Reform Organization." MA thesis, University of Oregon, 1963.

Jetté, Melinda Marie. "'At the hearth of the crossed races': Intercultural Relations and Social Change in French Prairie, Oregon, 1812–1843." PhD diss., University of British Columbia, 2004.

Jetté, Melinda Marie. "'we have allmost Every Religion but our own': French–Indian Community Initiatives and Social Relations in French Prairie, Oregon, 1834–1837," *Oregon Historical Quarterly* 108:2 (Summer 2007), 222–245.

Judd, Richard W., and Christopher S. Beach. *Natural States: The Environmental Imagination in Maine, Oregon, and the Nation.* Washington, DC: Resources for the Future, 2003.

Lipin, Lawrence M. *Workers and the Wild: Conservation, Consumerism, and Labor in Oregon, 1910–30.* Urbana, Ill.: University of Illinois Press, 2007.

Lomax, Ken. "Research Files: A Chronicle of the Battleship 'Oregon.'" *Oregon Historical Quarterly* 106:1 (Spring 2005), 32–145.

Mauldin, Frank. *Sweet Mountain Water: The Story of Salem, Oregon's Struggle to Tap Mt. Jefferson Water and Protect the North Santiam River.* Salem, Oreg.: Oak Savanna Publishing, 2004.

Mockford, Jim. "Before Lewis and Clark, Lt. Broughton's River of Names: The Columbia River Exploration of 1792." *Oregon Historical Quarterly* 106:4 (Winter 2005), 542–567.

Moulton, Gary E., ed., *The Journals of the Lewis and Clark Expedition.* Lincoln: University of Nebraska Press, 1983–2001.

Robbins, William G. *Landscapes of Conflict: The Oregon Story, 1940–2000.* Seattle: University of Washington Press, 2004.

Robbins, William G. *Landscapes of Promise: The Oregon Story, 1800–1940,* Seattle: University of Washington Press, 1997.

Robbins, William G. "Narrative Form and Great River Myths: The Power of Columbia River Stories." *Environmental History Review* 17:2 (Summer 1993), 1–22.

Robbins, William G., and Katrine Barber. *Nature's Northwest: The North Pacific Slope in the Twentieth Century.* Tucson: University of Arizona Press, 2011.

Seligman, Lester G. "Political Change: Legislative Elites and Parties in Oregon." *The Western Political Quarterly* 17: 2 (June 1964), 177–187.

Silverstein, Michael. "Chinooks of the Lower Columbia," in William C. Sturtevant, ed., *Handbook of North American Indians, vol. 7: Northwest Coast* (Washington, DC: Smithsonian Institution, 1990), 533–546.

Vogel, Eve. "The Columbia River's Region: Politics, Place, and Environment in the Pacific Northwest, 1933–Present." PhD dissertation, University of Oregon, 2007.

Wilkinson, Charles. *The People Are Dancing Again: The History of the Siletz Tribe of Western Oregon.* Seattle: University of Washington Press, 2010.

Zenk, Henry B. "Kalapuyans," in William C. Sturtevant, ed., *Handbook of North American Indians, vol. 7: Northwest Coast* (Washington, DC: Smithsonian Institution, 1990), 547–553.

Zenk, Henry B. "Notes on Native American Place-Names of the Willamette Valley Region." *Oregon Historical Quarterly* 109:1 (Spring 2008), 6–33.

PORTLAND, OREGON

Abbott, Carl. *Portland: Planning, Politics, and Growth in a Twentieth-Century City.* Lincoln: University of Nebraska Press, 1983.

Abbott, Carl. *The Metropolitan Frontier: Cities in the Modern American West.* Tucson: University of Arizona Press, 1993.

Ballestrem, Val C. "'In the Shadow of a Concrete Forest': Transportation Politics in Portland, Oregon, and the Revolt Against the Mount Hood Freeway, 1955–1976." MA thesis, Portland State University, 2009.

Lang, William L. "One City, Two Rivers: Columbia and Willamette Rivers in the Environmental History of Twentieth-Century Portland, Oregon," in Char Miller, ed., *Cities and Nature in the American West* (Reno: University of Nevada Press, 2010), 96–111.

Lucia, Ellis. *The Conscience of a City: Fifty Years of City Club Service in Portland.* Portland, Oreg.: The City Club of Portland, 1966.

MacColl, E. Kimbark. *The Growth of a City: Power and Politics in Portland, Oregon, 1915–1950.* Portland, Oreg.: The Georgian Press, 1979.

Harmon, Rick. "The Bull Run Watershed: Portland's Enduring Jewel." *Oregon Historical Quarterly* 96:2/3 (Summer/Fall 1995), 242–270.

SEWAGE AS FERTILIZER

Barles, Sabine, and Laurence Lestel. "The Nitrogen Question: Urbanization, Industrialization, and Water Quality in Paris, 1830–1939." *Journal of Urban History* 33:5 (July 2007), 794–812.

Goddard, Nicholas. "'A mine of wealth'? The Victorians and the Agricultural Value of Sewage." *Journal of Historical Geography* 22:3 (1996), 274–290.

Oleszkiewicz, J. A., and D. S. Mavinic. "Wastewater Biosolids: An Overview of Processing, Treatment, and Management." *Journal of Environmental Engineering Science* 1 (2002), 75–88.

Sheail, John. "Town Wastes, Agricultural Sustainability, and Victorian Sewage." *Urban History* 23:2 (Aug. 1996), 189–210.

Tarr, Joel A. "From City to Farm: Urban Wastes and the American Farmer." *Agricultural History* 49:4 (Oct. 1975), 598–612.

VISUAL MEDIA

McCall, Thomas L., producer. *Pollution in Paradise*, television documentary. Portland, Oreg.: KGW-TV, 1962.

Smith, William Joyce, producer. *Pollution in the Willamette*, film. Portland, Oreg., William Joyce Smith, 1940.

NEWS PERIODICALS AND NEWSLETTERS

Commonwealth Review [Eugene, Oregon]

The League Alert [Portland, Oregon, Chapter, League of Women Voters]

League of Oregon Cities Newsletter [Eugene]

The New York Times

Oregon Business & Investors, Inc., Bulletin [Portland]

The Oregon Journal [Portland]

Oregon State Game Commission Bulletin [Portland]

The Oregon Voter [Portland]
The Oregonian [Portland]
Outdoor America
The Portland Telegram [Oregon]
Scientific American
The Scientific Monthly
The Spectator [Portland, Oregon]

PROFESSIONAL JOURNALS

American Journal of Public Health
Chemical & Metallurgical Engineering
Chemical Engineering Progress
Chemtech
Civil Engineering
Engineering News-Record
Economic Botany
The George Washington Law Review
Industrial and Engineering Chemistry
Journal of the American Medical Association
Journal of Forestry
The Journal of Politics
Proceedings: American Society of Civil Engineers
Sewage Works Journal / Sewage and Industrial Wastes / Journal (Water Pollution Control Federation)
Science
Stanford Law Review
Tappi [Technical Association of the Pulp and Paper Industry]
University of California, Davis, Law Review
University of Pennsylvania Law Review
Washington Law Review
Wastes Engineering
Wisconsin Law Review
Yale Journal of Biology and Medicine

Index

Note: Page numbers in italics refer to figures or tables.

Abbott, Carl, 105
activated sludge sewage treatment process, 42–43, 63, 127
Advisory Committee on Stream Purification (ACSP), 69–71, 74, 82
    established, 69
    in relation to other groups, 77, 234
    legislative proposal of, 72
    members of, 101, 120
    publications by, 71–72
Agent Orange, 228
agriculture
    water pollution caused by, 6, 216, 220
    white settlers and, 4, 20, 42
Albany, OR, 49, 52, 168
    pulp mill in, 123, 180, 241
American Can Co. mill (Halsey, OR), 202, 230, 241
American Petroleum Institute, 129
American Pulp and Paper Association, 129
And On to the Sea (film by Sherman Washburn), 172
Androscoggin River, ME, 129
Androscoggin River Technical Committee, 129
Anti-Stream Pollution League (A-SPL)
    established, 50
    in relation to other groups, 63, 74, 77, 234
    legislative proposal of, 51–53, 78
    membership, 51

role of in raising public awareness, 53–54
Apartment House Owners Association, 107
assimilative capacity (of wastes in water), 43, 55, 153, 164, 215–17
    Columbia Slough and, 164
    as standard approach to water quality management, 96, 235
    NCSI Willamette River model of, 151–53, 217
Atiyeh, Victor, 199
Audubon Society (of Oregon), 38, 49
Averill, Edgar F.
    biography of, 48–49
    investigation of fish kill by, 47
    IWLA and, 48, 71
    pollution abatement advocacy of, 49–51, 98, 205, 211, 231, 234
    Portland sewers and, 66, 97–98, 167

bacteria, 31–35, 38, 43, 163
    fecal coliform, 127, 148, 231
    decrease in, 150, 214, 231
    in Willamette River, 33, 41, 54, 148
    nonpoint sources of, 215
    sewage treatment and, 168
    typhoid, 31, 32, 33, 34–35, 38
bacteriological theory of disease etiology, 41
Baer, Walter E., Portland sewer planning and, 63–65, 95
Baker, George, 41
    A-SPL and, 50–52
    opposition to water quality bill, 51–52, 55, 60–65, 172

Beach, Christopher, 299
Bean, Ormond
    advocate of home rule, 81
    OSPL and, 78
    Portland sewer planning and, 63–66,
        79–81, 104–105, 234
Bell, Nina, 220, 224
Bennett, J. E. "Jake"
    biography of, 103–104
    Portland sewer planning and, 79–81,
        97, 103–105
"Beautiful Willamette" poem (Simpson),
    4, 13, 45, 221
    parody of, 82
    publication of, 19
    references to in calls for pollution
        abatement, 40, 82, 122
    reflecting white romantic ideals, 19–
        20, 26
Benton, Thomas Hart, 26
Big Cliff Dam, OR, 159, 243
"Big Pipe" project. See Portland sewers
bioaccumulation, 8, 207, 228
biochemical oxygen demand (BOD)
    applied in characterizing water
        quality, 124, 127, 152, 164,
        215–16
    definition of, 91
    population equivalent, 140, 164, 180,
        180
    in relation to dissolved oxygen in
        water, 91, 127, 152, 215
    of industrial wastes, 125, 127–28, 152,
        164, 180, 230
    reduction of, 113, 168, 140
Black, William M., 91
Blatnik, John A., 155–56
Blue Heron Paper mill (Oregon City,
    OR). See Publishers' Paper mill
    (Oregon City, OR)
Blue River Dam, OR, 159, 244
Blumenauer, Earl, 222, 224
Board of Equalization. See Portland
    sewers
Boise Cascade mill (Salem, OR), 201–
    204, 203, 240

Boise Cascade mill (St. Helens, OR). See
    Crown Zellerbach Paper mill (St.
    Helens, OR)
Boise Cascade mill (Vancouver, WA), 44,
    240
Bonneville Power Administration, 73,
    183
Boston, MA, 6, 30, 64, 93
Bowes, William A.
    biography of, 104
    Portland sewer planning and, 11, 99,
        103–108, 110, 116, 167, 170,
        172, 225
    postwar planning and, 101
Brooklyn, NY, 6, 30, 32
Broughton, William R., 21
Bullivant, R. R., 68
Bull Run watershed, OR, 31–32
Burch, Albert, 87
Bureau of Environmental Services (of the
    City of Portland), 222, 225

Calapooia River, OR, 22, 23
California, 22, 28, 92, 125, 181–82
    water quality in, 156, 246
Camas, WA, pulp mill in, 44, 138, 239
Canada, 34, 43, 118
carbohydrates, as component of wood
    pulping wastes, 162
Carney, Byron G.,
    biography of, 72–73
    drafting of 1938 citizen's initiative by,
        78, 98
    OSPL and, 77
    political views of, 72–73
    water quality legislation, sponsor of,
        73–75
Carson, Joseph K., 80, 104
    support for 1938 clean streams vote,
        80–81, 233
Carson, Rachel, comparison to Tom
    McCall, 177–79
Carter, Glen D., 147–48, 197, 200
Cascade Mountains, 3, 20, 22, 23, 69
Cascade Pacific Pulp, LLC mill (Halsey,
    OR). See American Can Co. mill
    (Halsey, OR)

Cascades Tissue Group mill (St. Helens, OR). *See* Crown Zellerbach Paper Company mill (St. Helens, OR)

Chapman, C. C.
    characterization of Byron Carney, 72
    legislator, 86
    Oregon Voter, editor of, 52, 72

Charlton, David B., 109, *121*, 149, 155, 167, 170, 180, 195, 211, 231, 234
    ACSP and, 71
    biography of, 120
    Charlton Laboratories, 120
    IWLA and, 71, 117–22, 128, 135, 138–39, 181, 235
    OSSA criticism from, 117–22, 138, 140
    Postwar Readjustment and Development Commission and, 101
    pulp industry criticism from, 117–22, 130, 135, 138–39, 142
    Tom McCall and, 173–74, 205
    water pollution narrative told by, 9–10, 206
    William L. Finley and, 120

Chicago, IL, 36–38, 39, 64, 93, 94

Chinook jargon. *See* Chinuk Wawa

Chinookans, 1, 3, 22, 25, 27, *27*, 28, 238

Chinuk Wawa, 25

chlorine
    water pollution from, 230–31
    water purification from, 34–35, 172

Cincinnati, OH, 34, 91, 94, 154

Citizens' Advisory Committee on Environmental Quality, 211

Citizens' Committee for Wartime Tax Saving, 107, 108

Citizens Sewage Disposal Committee, 109–111, *111*

Civic Emergency Committee (of Oregon), 63

Civic Emergency Federation (of Portland, OR), 63–64

Civil Works Administration (CWA), 62

Civilian Conservation Corps (CCC), 62, 75

Clackamas County, OR, 145, 149, 169

Clackamas River, OR, 22, 23

Clanton, R. E., 50

Clark, William, 1, 21–22, 25

Clean River Program (of the City of Portland), 222–24

clean streams movement. *See* pollution abatement

Clean Water Act. *See* Federal Water Pollution Control Act Amendments of 1972 (PL 92-500)

Clements, Kendrick, 36

Clyde, Ralph C., 97

Coast Fork Willamette River, OR, 22, 23, 219

Coast Range, OR, 3, 20, 22, 23, 28, 33, 229

Columbia Boulevard Wastewater Treatment Plant, 140, 150, *168*, 181, 224
    completion of, 167, *167*, 188
    funding for, 11, 97–98, 100, 102–13, 188
    groundbreaking for, 115–16
    inadequacy of, 165, 166, 201,
    planning for, 12, 95–98, *112*, 113–14
    precursors to, 51–52, 64, 88, 104
    wastes reduced by, 150, 223

Columbia River, 3, 25, 230
    assimilative capacity of, 96, 97, 162, 164
    backflow into Willamette River of, 41, 58, 59, 167
    development of, 13–14, 29, 39, 75, 158
    dioxins detected in, 230
    Missoula Floods shaping of, 23–24
    Portland sewage released to, 113, 172, 225
    pulp and paper making wastes discharged to, 118, 122, 123, 138, 145–46, 161–64, 230
    pulp and paper mills along, 44, 138, 230, 239–40
    slime (sphaerotilus) problems in, 148, 157, 161–64, 233
    water quality and, 7, 33, 43, 92, 93, 149, 161–64, 172, 181

Columbia River Basin Comprehensive Project for Water Supply and Pollution Control, 182

Columbia River Fisherman's Protective Union, 69, 100
Columbia River Gorge, 21, 109, 174
Columbia River Paper mill (Camas, WA). *See* Crown Zellerbach mill (Camas, WA)
Columbia River Paper mill (Vancouver, WA). *See* Boise Cascade mill (Vancouver, WA)
Columbia Slough, OR, *165*
  environmental racism and, 296–297
  meat processing industry along, 148, 161, 164–66, 201
  sewer infrastructure and, 41, 60, 94, 113, 116, 164–66, 222–23
  sewer outfalls to, 60, 94, 95, 113, 116, 201
  water pollution and, 76, 95, 110, 149, 164–66, 225, 296–97
combined sewer overflows (CSOs), 215, 218, 222, 231, 234
Commoner, Barry, 179, 214
Commonwealth Conference (at the University of Oregon), 68, 56
Comprehensive Environmental Response, Compensation and Liability Act (CERCLA) of 1980, 16, 225–27
Confederated Tribes of Grand Ronde, 227
Congress of Industrial Organizations (CIO), 108
conservation organizations. *See* Audubon Society (of Oregon); Izaak Walton League of America (IWLA); Multnomah Angler's Club; National Wildlife Federation; Oregon State Sportsmen's Association; Salem Hunters' and Anglers' Club; Salmon Protective League; Troutdale Rod and Gun Club
conservationism
  approach to natural resource management, 11, 35, 69, 72, 96, 109, 122, 152, 300
  Herbert Hoover and, 36–37, 61
  in relation to environmentalism, 178–79
  Rachel Carson and Tom McCall in relation to, 178–79

conservationists, 36–38, 51, 77, 109, 120, 140, 173, 206, 234
Coolidge, Calvin, 35, 37
Cooper, Kenneth, 106–108
cooperative pragmatic approach to water pollution abatement, 88, 141, 197, 235
Corps of Discovery Expedition, 21, 25
Corvallis, OR, 30, 52, 56, 83, 84
Cottage Grove Dam, OR, 159, *159*, 161, 244
Cottage Grove, OR, 33, 144
Cougar Dam, OR, *159*, 244
crawfish, 76, 82, 101, 117, 160, 209, 210, 211, 233
Crisis in the Klamath Basin (film by Tom McCall), 173, 178
Crown Columbia Paper mill (Camas, WA). *See* Crown Zellerbach mill (Camas, WA)
Crown Columbia Paper mill (West Linn, OR). *See* Crown Zellerbach Paper mill (West Linn, OR)
Crown Paper mill (West Linn, OR). *See* Crown Zellerbach Paper mill (West Linn, OR)
Crown Willamette Paper mill (Camas, WA). *See* Crown Zellerbach mill (Camas, WA)
Crown Willamette Paper mill (Lebanon, OR). *See* Crown Zellerbach Paper mill (Lebanon, OR)
Crown Willamette Paper mill (West Linn, OR). *See* Crown Zellerbach Paper mill (West Linn, OR)
Crown Zellerbach mill (Camas, WA), 44, 138, 239
Crown Zellerbach Paper mill (Lebanon, OR), 29, 44, 123, 135, 143, 181, 240
Crown Zellerbach Paper mill (St. Helens, OR), 44, 76, 123, 230, 240
Crown Zellerbach Paper mill (West Linn, OR), 1–2, 29, 44, 123, 135, 136, *137*, 143, 145, 180, 240
Cumbler, John T., 141
Cummins, Mrs. C. D., 109
Cuyahoga River, OH, 93, 214

Dana, Marshall N., 109
Dargan, Tom, 176–77
Darling, Jay N. "Ding," 38
Day, L. B., 200, 204
DDT (Dichlorodiphenyltrichloro-
ethane), 178
demographics as contributor to
pollution, 2, 19, 29, 30, 33, 41, 45, 68,
74, 88, 90, 92, 146, 148, 149–50, 154,
161, 167, 170
Detroit Dam, OR, 149, 159, 160, 168,
243
Dexter Dam, OR, 159, 244
Dillehunt, Richard B., 81
dioxins, 8, 15, 128, 207, 211
cancer cluster in western OR, 229–31
description of, 227–28
Portland Harbor and, 231
pulp and paper making wastes and,
128
disease etiology. See miasma theory of
disease etiology; bacteriological
theory of disease etiology
dissolved oxygen (DO)
definition of, 91
water flow regulation and, 160, 221,
237
water quality metric, 54–55, 91, 127,
152, 160
Willamette River and, 67, 76, 148,
150, 153, 170, 175, 210, 210,
215, 231
Dorena Dam, OR, 159, 243

Earth Day, 212, 214
ecology, science of, 152, 157, 177, 178,
179
economics
conservationism and, 35, 55, 122
considered with other factors in
pollution abatement, 11–12,
13, 37, 300
disease etiology and, 41
federal water pollution legislation and,
155–56
fishing industry and, 38, 136
industrial waste treatment decisions
based on, 51, 52, 53, 119
law and, 12

NCSI research and, 152
pollution abatement stances based on,
32, 51, 52, 61, 75, 81, 115, 122,
141, 156, 142, 174, 197–98,
202, 205, 234, 236, 237
pollution as threat to, 55, 115
private sector influence based on, 51,
pulp and paper industry and, 126,
130, 134, 136, 138–39
sewage infrastructure decisions based
on, 31, 52, 101–102, 110–11,
114
white cultural values based on, 2, 3,
12, 16, 19, 21, 32
ecoroofs, 224
Eddy, Harrison P., 64–65, 81, 93, 96, 97
Eldridge, Edward F., 157
Endangered Species Act of 1973 (PL
93-205), 221
Environmental Action, 214
Environmental Impact Statements, 212
environmentalism, 178–179
environmental racism, 297
Eugene, OR, 24, 30, 49, 50, 69, 76, 229
Commonwealth Conference in, 54
water pollution in, 33, 47–49, 82, 83,
101
European Americans
cultural values of, 1, 2, 19, 20, 28, 40
emigration of, 3–4, 8–9, 20–21, 26,
28, 30, 45, 92, 223
water quality and, 7–8, 13, 29–32
Willamette Valley exploration by, 21–
22, 25–28
Everts, Curtiss M., 116, 168, 170

Fall Creek Dam, OR, 159, 244
Federal Oil Pollution Control Act of
1924, 36,
Federal Water Pollution
Control Act Amendments of 1972
(PL 92-500), 6, 7, 211–20, 223
section 305(b) reports, 215
Federal Water Pollution Control Act of
1948 (PL 845), 5, 145, 154, 188, 189,
195, 212
Federal Water Pollution Control Act of
1956 (PL 660), 5, 189, 195, 212

Federal Water Pollution Control
Administration, 213–14
Federated American Engineering
Societies, 129
Federation of Women's Clubs, 100
Fern Ridge Dam, OR, 159, *159*, 161, 243
films and television
And On to the Sea (Washburn), 172
Crisis in the Klamath Basin (McCall),
173, 178
Pollution in Paradise (McCall), 173–
77, 179, 195, 196, 197, 202,
233
Pollution in the Willamette (Smith),
83-85
Finley, Carmel, 230
Finley, Irene, 38, 120
Finley, William Lovell
biography of, 38
conservationism and, 38, 84, 109
David Charlton and, 120
Edgar Averill and, 49
IWLA and, 38
OSPL and, 77, 78
Portland sewers and, 43–44, 60
water pollution abatement and, 50,
77, 78, 81, 205, 231, 234
fish kills, 38, 47, 49, 50, 69, 144, 147, 176
fish survival experiments, 80–81, 83–84
forest management contributing to water
pollution, 48, 89, 201, 215, 216
Fort James Corporation mill (Camas,
WA). *See* Crown Zellerbach mill
(Camas, WA)
Fox River, WI, 121, 129
Franciscovich, Frank, 77
French Prairie, OR, 26
Friends of the Earth, 214

Gabrielson, Ira N., 38
geologic processes, 22–24
Georgia-Pacific Corporation mill
(Camas, WA). *See* Crown Zellerbach
Corporation mill (Camas, WA)
Giesy, I. B., 53
Glacial Lake Missoula, 24
Gleeson, George W., 14–15, 101, 128,
289

Grand Ronde Reservation, 28
Grand Ronde River, 218
Great Depression, 61–62, 81, 111, 185,
188, 234, 236
Great Lakes region, 26, 125
pollution compared to Willamette
River, 92–93, 229
water quality management in, 7, 34–
35, 141
"great man" myth, 297–299
Great Migration of 1843, 26
Green, Carl E., 66, 87, 139
Green Peter Dam, OR, 159, 244
Guthrie, John A., 124

Hallock, Blaine, 87
Hallock, Stephanie, 225
Hallock, Ted, 176–177
Halsey, OR, pulp mill in, 123, 202, 230,
241
Harding, William G., 35
Hatfield, Mark O., 176, 180, 181, 200,
206
water quality funding as governor,
184, 190–93
Hawley Pulp and Paper mill (Oregon
City, OR). *See* Publishers' Paper mill
(Oregon City, OR)
Hayden Island, OR, 113, 164
Healthy Stream Partnership, 220
heavy metals as a pollutant, 6, 8, 207,
211, 219
herbicides, 177, 219, 227–28
Hills Creek Dam, OR, 159, 244
Hilton, Frank H., 107–108
Holcomb, Blair, 109
Holm, Don, 209, 210, 231
Holman, Rufus C., 77
home rule, principle of, 172
Hommon, Harry B., 39, 43, 54, 55, 60
Hoover, Herbert, 35–38, 61, 62, 69, 70,
92
Hoyt, Palmer, 109
Hudson River Valley, NY, 4, 26

Illinois River, IL, 36, 39,
Illinois Sanitary Water Board, 121
Indians. *See* Native Americans

industrial wastes
    flax retting, 4, 29, 30, 44, 89, 115, 201
    logging, 48, 89, 125, 201, 215
    meat processing, 44, 83, 84, 89, 110,
        148, 161, 164–66, 201, 203
    mining, 89, 128, 201, 215
    tanning, 201
    vegetable processing, 29, 30, 44, 93,
        115, 136, 201
    See also pulp and paper making wastes
Institute of Paper Chemistry, 130
International Joint Commission (IJC),
    34–35
International Paper mill (Albany, OR).
    See Weyerhaeuser mill (Albany, OR)
International Paper mill (Springfield,
    OR), 123, 181, 241
Izaak Walton League of America (IWLA)
    David Charlton and, 71, 117–22, 128,
        135, 138–39, 181, 235
    Edgar Averill and, 48, 71
    founding of, 35–36
    Herbert Hoover and, 35–37
    Oregon IWLA established, 38
    OSSA criticism from, 117–22, 138,
        140
    pulp industry criticism from, 117–22,
        130, 135, 138–39, 142
    William Finley and, 38

James River mill (West Linn, OR). See
    Crown Zellerbach Paper mill (West
    Linn, OR)
James River Paper mill (Camas, WA). See
    Crown Zellerbach mill (Camas, WA)
James River Paper mill (Wauna, OR),
    230
Jefferson-Smurfit mill (Oregon City,
    OR). See Publishers' Paper mill
    (Oregon City, OR)
Judd, Richard W., 299

Kaiser, Edgar, 102, 113
Kalapuyans, 3, 4, 28–29, 238
Kehoe, Terence, 141
Kelley, Hall Jackson, 26
Kenworthy, E. W., 206
KGW-TV, 173, 174, 176, 178–79

Kimmel, Rex, 138
Kitzhaber, John, 220, 226
Klamath Tribe, 28, 173, 178
Klassen, Clarence W., 121, 122
Klingle, Matthew, 14
kraft wood pulp making process. See
    sulfate wood pulp making process,
Kramer, Loren, 216

La Grande, OR, 94
labor and pollution abatement, 62–65,
    69, 80, 97–98, 100, 104, 108, 109
Langton, Clair V., 56
Latourette, Lyman, 97
Laurgaard, Olaf,
    A-SPL and, 51
    Portland harbor wall and, 11, 39, 60,
        84
    sewage infrastructure and, 11, 39, 44,
        54–55, 63, 84
Lawrence Experiment Station, 6, 30, 91,
    92
League of Women Voters, 155, 172, 195,
    221, 227
Lebanon Paper Mill (Lebanon, OR). See
    Crown Zellerbach Paper mill
    (Lebanon, OR)
Lebanon, OR, pulp mill in, 29, 44, 123,
    135, 143, 181, 240
Lewis, Meriwether, 1, 21, 25
Lewis & Clark (explorers). See Corps of
    Discovery Expedition
Lewis & Clark Law School, 218
lignins as component of wood pulping
    wastes, 123, 124, 162, 230
"live box" fish survival experiments. See
    Fish survival experiments
Lindbergh, Charles, 211
Linn County, OR, dioxin in fish from,
    229
Linnton neighborhood, 170, 209
London, England, 30, 31, 32
Lonergan, Frank, 172
Long Tom River, OR, 22, 23
Longview, WA, pulp mill in, 135, 139,
    140
Lookout Point Dam, OR, 149, 159, 160,
    168, 243

Love Canal incident, 226
Lower Willamette Group, 227
Luckiamute River, OR, 22, 23
Lundy, Herbert, 149

máånumaå, 22
MacColl, E. Kimbark, 298
Maine, 129, 211, 299
Manifest Destiny, 26
Manufacturing Chemists' Association, 36
Martin, Charles Henry, 77, 78, 86
    biography of, 75
    lack of funding for OSSA, 86, 185
    political views of, 75
    primary defeat of, 86
    water quality legislation veto by, 76
Marys River, OR, 22, 23
McCall, Thomas (Tom) L.,
    Boise Cascade Salem mill and, 202–
        204
    DEQ and, 199–200, 214
    conservationism and, 179, 198
    environmentalism and, 179
    environmental legacy of, 7, 191, 195,
        202, 204–207, 211, 231, 233
    citizens' advisory committee on
        environmental quality, 211–12
    industrialization and, 174, 179, 202
    IWLA and, 173, 174, 195
    KGW-TV and, 173, 176, 179
    legislative achievements of, 199–200
    mythologization of, 7, 9–10, 14–15,
        195–196, 206, 297–99
    OSSA chair of, 197–198
    OSSA/DEQ funding as governor,
        191–95, 200, 235
    Pollution in Paradise, 173–74, 175,
        176–79, 195, 196, 198, 202,
        213
    Robert Straub and, 184, 200
    Rachel Carson and, 178–79
McCullough Lee, Dorothy, 11, 106–108,
    167
McKay, Douglas, 161
McKay, Floyd, 73
McKenzie River, OR, 22, 23
McLoughlin, John, 1, 26

McPhillips, Barney A., 142, 182, 183,
    197, 200
Meier, Julius, 56, 63–64, 107
mercury as a pollutant, 217
Merryfield, Fred, 66, 128
Métis, 26
Metropolitan Tri-Counties Health and
    Sanitation Committee, 169
Meunch, Virgil J., 121–22
miasma theory of disease etiology, 41–42
Middle Fork Willamette River, OR, 22,
    23, 160
Milazzo, Paul C., 155
Milwaukee, WI, 43, 64, 94
Milwaukie, OR, 48, 49, 72,
Missoula Floods, 24
mixing zone, 83, 217
Molalla River, OR, 22, 23
morals, pollution abatement arguments
    based on, 11, 16, 37, 90, 122, 171–72,
    234
Morning Oregonian. See Oregonian
Moses, Robert, 102, 106–107, 110, 111,
    120, 188
"Moses Report." See Portland
    Improvement (Moses)
Mosser, John, 199, 200, 202
Mullane, Neil, 7–8, 234
Multnomah (Anglicized Chinookan
    name applied to Willamette River),
    22, 25
Multnomah Anglers and Hunters Club,
    10, 69, 142
Multnomah Channel, 25, 59, 67, 69, 209,
    230
Multnomah County Medical Society,
    109
Multnomah County, OR, 107, 127, 141,
    149, 166, 169
Municipal League of Eugene, 50
municipal wastes. See sewage
Murphy, Priscilla Coit, 177
Murrell, Gary, 75
Muskie, Edmund, 211

National Council for Stream
Improvement of the Pulp, Paper, and
Paperboard Industries, Inc. (NCSI)
assimilative capacity model developed
by, 144, 151–53, 160
budget of, 130, 131–35
David Charlton critical of, 122, 130,
135, 139
founding of, 129–30
pollution abatement research funding
by, 129–30, 138, 140, 142, 151
National Environmental Policy Act
(NEPA) of 1970 (PL 91-190), 6, 212,
226
National Geographic, 7, 9, 14, 38, 196,
205, 233
National Recovery Act (NRA), 62
National Resources Board (NRB). *See*
National Resources Planning Board
(NRPB)
National Resources Committee (NRC).
See National Resources Planning
Board (NRPB)
National Resources Planning Board
(NRPB), 70–71, 78
National Water Pollution Control
Advisory Board, 145, 161
National Wildlife Federation, 84
Native Americans, 8–9, 20–21, 24–29,
27, 173, 238
See also Chinookans; Confederated
Tribes of Grand Ronde; Grand
Ronde Reservation;
Kalapuyans; Klamath Tribe;
Rogue River Wars
Native Seattle (Thrush), 8
nature as agent, 14
Neuner, George, 105–106, 144, 161
New Deal
Byron Carney as exemplar of, 73
Charles Martin opposition to, 75, 86
Oregon sewage treatment funding
during, 62, 94
Portland inability to secure sewage
treatment funding during, 64–
65, 94, 96, 105, 236
public works funding during, 5, 60,
62, 94, 97, 105
regional planning during, 5, 68–71

water quality research during, 5, 7, 10,
70–71
New York, state of, 4, 26, 30, 86, 91, 202,
226
New York City, 26, 93, 102
New York Harbor Act of 1888, 36
New York Times, 206, 228, 231
Newberg Pool, OR, 152, 160
Newberg, OR, 26, 30, 168
pulp mill in, 44, 123, 145, 180, 241
nitrogen, 43, 162, 218
Nixon, Richard, 211–12
No Margin of Safety (Van Strum &
Merrell), 230
Northwest Environmental Advocates
(NEA), 223–24
Northwest Environmental Defense
Center (NEDC), 218–22
Northwest Pulp and Paper
Manufacturers Association, 182

Office of Industrial Hygiene and
Sanitation. *See* US Public Health
Service (USPHS)
oil as a pollutant. *See* petroleum products
as pollutants
O'Neil's and Callaghan's Mill (Lebanon,
OR). *See* Crown Zellerbach Paper
mill (Lebanon, OR)
Oregon Business & Investors, Inc., 77,
81, 87, 107, 108
Oregon City, OR, 1–2, 26, 29, 29, 30,
101, 144, 238
pulp mill in, 1–2, 44, 50, 123, 137,
145–46, 163, 180, 239
Oregon Commonwealth Federation
(OCF), 86, 108
Byron Carney and, 73, 98
Portland sewer funding, opposition
to, 97–98
Oregon Country, 3, 20, 26
Oregon Department of Environmental
Quality (DEQ)
EPA in relation to, 212, 214
established, 190, 199
funding of, 190–5, 199
industrial polluters and, 201–204
lawsuit against, 218–21

Oregon Department of Environmental
Quality (DEQ), *continued*
leadership of, 200
Portland sewers and, 224–26
river basin planning by, 217–18
work of, 200–201, 212, 214–21
Oregon Environmental Quality
Commission (EQC), 199, 200
Oregon Grange, 79
Oregon Journal, 68, 76, 101, 117, 136,
166, 299
editorial support for pollution
abatement by, 40, 49, 75, 79,
103, 143, 170, 172, 234
Marshall Dana and, 109
Oregon Land Conservation and
Development Commission, 219–20
Oregon Plan for Salmon and Watersheds,
221
Oregon Pulp and Paper Company mill
(Salem, OR). *See* Boise Cascade
Corporation mill (Salem, OR)
Oregon State Agricultural College. *See*
Oregon State University
Oregon State Air Pollution Authority
(OSAPA), 173, 295
Oregon State Board of Health
established, 32
OSSA as part of, 85, 184, 191, 199–
200
pollution abatement advocacy of, 9,
38, 44, 49, 51, 55, 78, 206, 234
water quality purview of, 33–34, 39,
48, 50, 85, 174, 225
Oregon State College. *See* Oregon State
University
Oregon State Fish Commission
established, 32
OSSA and, 87
pollution abatement advocacy of, 9,
44, 49–51, 71, 101, 206
water quality purview of, 32–33, 38,
47
Oregon State Game Commission
established, 32
OSSA and, 186

pollution abatement advocacy of, 9,
38, 44, 51, 55, 234
water quality purview of, 32–33, 38,
47–48, 50, 72, 74, 168
Oregon State Industrial Union, 109
Oregon state water quality laws, bills, and
citizen initiatives
Senate Bill 392 (1937), 72–74, 78
Senate Bill 414 (1937), 73–76
Senate Bill 259 (1963), 176–77
Senate Bill 396 (1969), 199
Water Purification and Prevention of
Pollution Bill (1938), 78, 82,
186
Oregon State Legislature
OSSA funding and, 87, 131–34, 185–
95, 200, 296
water quality and, 53, 69, 72–73, 77,
106, 158, 180, 195, 196, 198,
199
See also Oregon state water quality
laws, bills, and citizen initiatives
Oregon State Planning Board (OSPB),
69–70
Oregon State Planning Board Advisory
Committee on Stream Purification
(ACSP). *See* Advisory Committee on
Stream Purification (ACSP)
Oregon State Sanitary Authority (OSSA)
budgets of, 87, 131–34, 185–95, 200,
296
cooperative pragmatic approach of,
88, 141, 197, 235
citizen initiative establishing, 78, 82,
186
critiques of, 117–22, 138, 140, 235,
297–98
pollution defined by, 89–90, 93
disbursement of federal funds by,
156–57, 189–91, *190*
established, 87
water classification by, 91–93, 136
Harold Wendel as chair of, 87
initial members of, 87
Tom McCall as chair of, 197–98
pulp and paper industry and, 93, 118–
19, 122, 135–46

comparison to other state water
quality agencies, 186,
Portland sewers and, 85, 88, 93–110,
116, 141, 166–73
DEQ supersedes, 190, 199
Oregon State Sportsmen's Association,
49
Oregon State University, 55, 66, 70, 71,
101, 109, 120, 130, 138
Engineering Experiment Station, 56,
76, 88, 145, 151, 219
Oregon State Water Resources Board,
161, 175
Oregon Stream Purification League
(OSPL)
established, 77
members of, 77, 98, 234
work of, 77–82, 98, 107
Oregon Supreme Court, 65, 166
Oregon Territory, 20, 26, 30
Oregon Voter, 52, 72, 73, 79, 121
Oregon Wildlife Federation, 49, 66, 84,
85, 100
Oregon, State of, 3, 20, 30
Oregonian,
dioxin reporting in, 228, 230
editorial support for pollution
abatement by, 6, 33, 39, 79,
100, 143, 146, 147, 162–63,
166, 169, 171–72, 171, 176,
196–97, 198, 234
letters to the editor in, 101, 204
Marshall Dana and, 109
Portland sewer system reporting in,
95, 113–14, 114, 116
Tom McCall, reporting about, 197–
98, 199
water quality reporting in, 47–48, 66,
66, 76, 80–81, 80, 149–50, 181,
182, 199, 209, 216
"organic machine" metaphor, 13–14, 221,
237
Outdoor America, 36, 37

Pacific Coast Association of Pulp and
Paper Manufacturers, 71
Pacific Meat Company, 166, 201–203

Pacific Northwest Regional Planning
Commission, 70, 109
Pacific Ocean, 1, 3, 22, 58–59, 158, 160,
162, 229
tidal influence on Willamette River,
23, 58, 153
pay-as-you-go sewer funding for
Portland, 65, 79, 81–82, 99, 104–105
replacement of by bond measure, 106,
110–12
See also Portland sewers
Pennoyer, Sylvester, 32
Pennsylvania, 6, 37, 72, 92, 121, 130
Sanitary Water Board, 6, 37, 51, 85,
128, 136
Peterson, Fred, 106–108, 167
petroleum products as pollutants, 36–37,
47, 61, 147, 155, 209, 211, 214, 219,
226, 227
persistent toxic pollutants, 128, 207, 211
See also bioaccumulation; DDT,
dioxins; heavy metals;
herbicides; mercury as a
pollutant; petroleum products
as pollutants; pesticides;
polychlorinated biphenyls
(PCBs)
pesticides, 6, 177–79, 217, 228
Phelps, Earle B., 91
phosphorous as a pollutant, 162
Pinchot, Gifford, 35
Pioneer Paper Manufacturing mill
(Oregon City, OR). See Publishers'
Paper mill (Oregon City, OR),
Pittsburgh, PA, 64, 93, 94
pollution
evolving definitions of, 74, 76, 90, 218
industrial: See industrial wastes; pulp
and paper making wastes
metrics to characterize, 55, 89, 91,
127, 152, 218. See also
biochemical oxygen demand
(BOD); dissolved oxygen
(DO); population equivalent
municipal. See sewage
nonpoint source, 213–14, 218–19,
221–22, 224, 297
OSSA definition of, 89–90, 93

point source, 5, 6, 8, 15, 141, 148, 150,
  201, 207, 210–11, 215–18,
  221-22, 225, 233, 297
pollution abatement
  economic arguments in support of,
    32, 51, 52, 61, 75, 81, 115, 122,
    141, 156, 142, 174, 197–198,
    202, 205, 234, 236, 237
  moral arguments in support of, 11, 16,
    37, 90, 122, 171–72, 234
  "on a treadmill," 6, 148, 153, 166, 173,
    176, 188, 190, 193, 194, 211,
    233
  women and, 77–78, 100, 109, 155,
    172, 195, 221, 227
pollution abatement campaigns
  to compel Portland officials to fund
    sewers 1940s, 83–85, 93–115
  to compel Portland officials to
    upgrade sewers1960s, 166–73
  to compel pulp and paper industry
    1950s, 119–122, 135–46
  to create Oregon State Sanitary
    Authority 1930s, 76–82
pollution abatement organizations
  See Advisory Committee on Stream
    Purification (ACSP); Anti-
    Stream Pollution League
    (A-SPL); League of Women
    Voters; Northwest
    Environmental Defense Center
    (NEDC); Oregon Department
    of Environmental Quality
    (DEQ); Oregon State Sanitary
    Authority (OSSA); Oregon
    Stream Purification League
    (OSPL); Portland Anti-
    Pollution Council; Portland
    Harbor Community Advisory
    Group (PHCAG)
Pollution in Paradise (film by Tom
  McCall), 173–74, 175, 176–79, 195,
  196, 198, 202, 213
Pollution in the Willamette (film by
  William J. Smith), 83–85
polychlorinated biphenyls (PCBs), 211,
  226, 227
Pope & Talbot mill (Halsey, OR). See
  American Can Co. mill (Halsey, OR)

population equivalent, 127–28, 140, 164,
  181
population growth. See demographics as
  contributor to pollution
Port of Portland, 227
Portland Anti-Pollution Council, 63, 66,
  98
Portland Basin, OR, 25, 27, 28–29
Portland Bureau of Health, 120
Portland Central Labor Council, 100,
  109
Portland Chamber of Commerce, 79, 88,
  100, 109, 110, 120, 140, 234
Portland City Club, 15, 49, 53, 71, 79,
  120, 170, 172
Portland City Council
  DEQ and, 224–26
  New Deal public works funding
    sought by, 64–65, 94, 96, 105,
    236
  OSSA and, 85, 88, 93–110, 116, 141,
    166–73, 224–26
  pay-as-you-go sewer funding plan of,
    65, 79, 81–82, 99, 104–105
  sewer planning by, 52, 60–68, 69, 79–
    82, 93–110
  See also mayors and individual council
    members
Portland General Electric, 74, 75, 182
Portland Harbor
  combined sewer overflows (CSOs) to,
    222–24
  dredging of, 59, 158
  economic importance of, 11, 40, 58
  harbor wall, 58, 59, 60, 61
  sewer outfalls to, 40, 58, 60, 92, 113
  tidal reach in, 23, 58–59, 153
  tours of, 66–67, 80–81, 227
  US Navy ships soiled by wastes in,
    147
  USS Oregon battleship a floating
    museum in, 185–86
  waterfront, 40, 40, 60, 75, 113
  water quality in, 66–67, 67, 76, 82, 84,
    144, 147, 160, 175, 182, 210,
    223, 227, 229
  See also Portland Harbor Superfund
    Site

Portland Harbor Community Advisory Group (PHCAG), 226–27
Portland Harbor Superfund Site, 8, 226–31
  designation of, 226
  citizen involvement in decisions about, 226–27
  Potentially Responsible Parties, 227
  See also Comprehensive Environmental Response, Compensation and Liability Act (CERCLA) of 1980
Portland Improvement (Moses), 102, 106, 110, 120, 188
Portland sewer equalization board. See Board of Equalization,
Portland sewers
  "Big Pipe" project, 224–25, 231
  DEQ and, 224–26
  New Deal funding for, 64–65, 94, 96, 105, 236
  OSSA and, 85, 88, 93–110, 116, 141, 166–73, 224–26
  pay-as-you-go funding plan, 65, 79, 81–82, 99, 104–105
  relative to other cities, 94, 168
  sewer equalization board, 97, 108
  unemployment relief measure, 63–65, 69, 80, 86, 97–98, 104
Portland State University, 206, 218
Portland Telegram, 42
Portland, OR
  politics of sewer funding in, 65, 79, 81–82, 99, 104–105
  politics of sewer planning in, 52, 60–68, 69, 79–82, 93–110
  population of, 30, 41, 68, 74, 94, 149, 167, 223
  shipbuilding in during World War II, 102, 113, 231
  urban planning, 102, 105–107, 110, 111, 120, 188
  See also Portland Harbor; Portland sewers
Postwar Readjustment and Development Commission, 101, 113, 120
Potomac River, 93, 166
polychlorinated biphenyls (PCBs), 227

President's Council on Environmental Quality, 212
primary waste treatment, 42, 96, 113, 127, 163, 168, 172, 214, 223
Progressive Era, 151
Public Employment Bureau (of Portland, OR), 63
public health
  balancing with other factors, 115, 119, 152, 162, 174
  laws and policies addressed to, 32, 37, 39, 60, 72, 156, 200, 220, 225
  profession, 16, 41, 77, 90, 151, 164, 198
  reducing threats to, 31, 32, 91, 171
  threats to from pollution, 4, 6, 10, 30, 52, 55, 68, 90, 152, 182, 236
  See also US Public Health Service; Oregon State Board of Health
Public Works Administration (PWA), 5, 62, 75
Publishers' Paper mill (Newberg, OR), 44, 123, 145, 180, 241
Publishers' Paper mill (Oregon City, OR), 1–2, 44, 50, 123, 137, 145–46, 163, 180, 239
Puget Sound, WA, 92, 94
Pulp and paper industry
  assimilative capacity model developed by, 144, 151–53, 160
  DEQ and, 201–204
  economic importance of, 126, 130, 134, 136, 138–139
  denial of pollution by, 49–50, 52, 136
  OSSA and, 93, 118–19, 122, 135–46
  pollution abatement research funding by, 129–130, 138, 140, 142, 151
  See also National Council for Stream Improvement of the Pulp, Paper, and Paperboard Industries, Inc. (NCSI); pulp and paper making; pulp and paper making wastes
pulp and paper making
  history of, 123–26
  water use and, 124–25

See also soda wood pulping process;
   sulfate wood pulping process;
   sulfite wood pulping process
pulp and paper making wastes
   adhesives made from, 4, 130, 138
   barging of for disposal, 145–46, *146*,
      163
   cattle fodder made from, 130, 138
   dioxins in, 128
   effects on receiving waters,
   fungicides and insecticides from, 130,
      138
   industry assertions of no effects on
      fish life from, 49–50, 52, 136
   lagooning of, 8, 130, 138, 140, 145,
      202–204
   population equivalent of, 127–28,
      140, 164, 181
   research into making saleable
      products from, 4, 52, 126, 129–
      31, 138–39, 145, 149
   road binder made from, 140
   vanillin created from, 4, 130, 138

Rawn, A. M., 113–14, 217
Reconstruction Finance Corporation
   (RFC), 62, 63, 69
*Report on the Collection and Disposal of
   Sewage* (Wolman), 95–96, 113
Return of a River (Gleeson), 14–15, 298
Riley, Earl, 101, 107, 116
River Thames, 31
Rivers and Harbors Act of 1899, 34, 36,
   155
rivers as components of waste treatment
   systems, 31, 42, 151, 152, 222, 235
road construction as pollution source,
   216
Roberts, Barbara, 219,
Robbins, William G., 177, 205, 235
Robinson, David, 66
Rockefeller, Laurance, 211
Rockwell, Cleveland, *24*, *26*
Rogers, Emmett, 170
Rogers, H. S., 56
Rogue River Wars, 28
Ronchetto, John C., 71

Roosevelt, Franklin D., 5, 60–62, 69–70,
   75, 86

Safe Drinking Water Act of 1974 (PL
   93-523), 221
Salem Capital Journal, 48, 101
Salem Chamber of Commerce, 50
Salem Hunters' and Anglers' Club, 85
Salem, OR, 26, 30, 49, 50, 53, 69, 73, 75,
   76, 84, 101, 170, 200, 203–204
   pulp mill in, 201–204, *203*, 240
   Willamette River gauge at, 59, 175,
      182, 210
salmon
   commercial fishing of, 29, 68–69
   dams and, 237
   economic value of, 10–11, 238
   recreational fishing of, 33, 49, 68–69
   water quality and, 33, 38–39, 50, 67,
      76, 91, 101, 117, 175–76, 182–
      84, 216
Salmon Protective League, 49
sanitary engineering profession, 89, 90,
   160, 164, 235
Santa Barbara oil spill of 1969, 214
Santiam River, OR, 22, 23, 84, 144, 160
Sauvie Island, OR, 22, 25, 27, 67, 226
Schrunk, Terry, 170, 172
Seattle, WA, 8–9, 14, 58, 94
secondary waste treatment
   description of, 42, 127, 168
   Portland construction of, 172
   requirement for, 215
   Willamette Valley need for, 150, 168,
      170, 181, 201, 214, 216
Security Owners' Association, 73
sediment as a form of water pollution,
   215, 217
sediments, pollution in, 8, 15–16, 211,
   226–27, 229
Sellwood Bridge, OR, 120
Sellwood neighborhood (Portland, OR),
   59
sewage. *See* sewage treatment; sewer
   outfalls; sewers
sewage sludge as fertilizer, 4, 43, 114
sewage treatment

activated sludge process, 42–43, 63, 127
compared to beer brewing, 43
See also primary waste treatment; secondary waste treatment; sewers; tertiary waste treatment
sewage treatment plants in Oregon, 50, 55, 94, 116, 144, 149, 150, 168, 170, 181, 189–90, 218
See also Columbia Boulevard Wastewater Treatment Plant
sewer
separated compared to combined, 41–42, 60, 222, 224
storm, 41, 215, 219, 222, 224
Willamette River characterized as, 7, 101, 103, 224
See also combined sewer overflows (CSOs); Portland sewers; sewage treatment
sewer outfalls
Columbia Slough, 60, 94, 95, 113, 116, 201
Portland Harbor, 40, 58, 60, 92, 113
shipbuilding in Portland during World War II, 102, 113, 231
Sierra Club, 179
Silent Spring (Rachel Carson) compared to *Pollution in Paradise*, 177–79
Simpson Paper mill (West Linn, OR). *See* Crown Zellerbach Paper mill (West Linn, OR)
Simpson, Samuel L., 4, 13, 19–20, 26, 28, 40, 42, 45, 57, 82, 221
Siuslaw National Forest, 228
slime (*sphaerotilus*) in lower Columbia River, 148, 157, 161–64, 233
Smith, C. H., 52
Smith, Jason Scott, 86
Smith, J. Douglas, 218
Smith, Virgil, 114–15
Smith, William Joyce, 84–85, 92, 100–101, 109, 116, 172, 233, 234
Smurfit Newsprint mill (Newberg, OR). *See* Publishers' Paper mill (Newberg, OR)
Snell, Earl, 113, 188
soda wood pulping process, 123, 124
solid wastes, *165*, 201

SP Newsprint mill (Newberg, OR). *See* Publishers' Paper mill (Newberg, OR)
Spaulding Pulp and Paper mill (Newberg, OR). *See* Publishers' Paper mill (Newberg, OR)
Spies, Kenneth H., 101, 144, 170, 172, 176, 182, 195, 200, 234
Sprague, Charles, 87, 185, 186–88, 199
Springfield, OR, 83, 168
pulp mill in, 123, 181, 241
St. Helens Pulp and Paper mill (St. Helens, OR). *See* Crown Zellerbach Paper mill (St. Helens, OR)
St. Helens, OR, pulp mill in, 44, 76, 123, 230, 240
St. Johns Bridge, 229
St. Johns neighborhood, 209
Starbird, Ethel, 7–10, 14–15, 205–206, 233–34
Straub, Robert, 184, 200
Streeter, H. W., 91
Stricker, Frederick, 55, 78, 87
Stricklin, Charles E., 87
Stroud, Ellen, 296–297
sulfate wood pulping process, 89, 123–26, 169, 202, 230
sulfite wood pulping process, 89, 118, 122, 123–26, 129, 135–40, 142–43, 145, *146*, 162, 169, 202–203, 230, 236
Sulphite Pulp Manufacturers' Research League, 121
sulphite wood pulping process. *See* sulfite wood pulping process
Sunday Oregonian. *See* *Oregonian*
Superfund. *See* Comprehensive Environmental Response, Compensation and Liability Act (CERCLA) of 1980
Swan Island, OR, 22, 110, 175, 226
Sweetland, Monroe, 73, 97–98, 108

Tarr, Joel, 31
Technical Association of the Pulp and Paper Industry, 129
television. *See* films and television
tertiary waste treatment, 127, 181, 214
The Dalles, OR, aluminum plant in, 174

Thrush, Coll, 8–9
Total Maximum Daily Load (TMDL),
    217–22
Towne, W. W., 182
Troutdale Rod and Gun Club, 85
Truman, Harry S., 145, 156, 161
Tryon Creek neighborhood, 170, 172
Tualatin River, OR, 22, 23, 218–19
typhoid. *See* bacteria

Udall, Stuart, 184
unemployment relief, 63–65, 69, 80, 86,
    97–98, 104
United Kingdom, 30–31, 43
University of Oregon, 73, 81
    Commonwealth Conferences at, 54–
        55
US Army Corps of Engineers (USACE)
    dredging Willamette and lower
        Columbia rivers by, 13, 34, 59
    water pollution abatement
        involvement by, 36, 91, 182–83
    Willamette River dams studied by,
        158–60
US Biological Survey, 48,
US Bureau of Fisheries, 38, 48, 183
    Willamette River water quality
        research sponsored by, 76, 78
US Bureau of Mines, 34, 37, 129
US Department of Health, Education,
    and Welfare, 213
US Department of the Interior, 184, 200
US Environmental Protection Agency,
    206, 211–15, 218–20, 225–27, 229–
    31
US Fish and Wildlife Service, 162
US Forest Service, 35, 228
US Government Accounting Office
    (GAO), 206
US Public Health Service (USPHS), 48,
    78, 87, 175, 180, 182
    established, 34
    sponsorship of regional water quality
        symposia, 157–58, 161, 163
    water quality laboratory (Cincinnati),
        34, 91, 154
    water quality responsibilities of, 154,
        157, 163

USS *Oregon* (BB-3), 185–86

Van Strum, Carol, 229–30
Vancouver, George, 21
Vancouver, WA, 113, 149
    pulp mill in, 44, 240
Vanport, OR, 102
Veatch, John C., 51, 73, 87, 99, 101, 136,
    138, 234
Velz, Clarence J., 151–53, 217
video. *See* films and television
Vietnam War, 227–28
Vinson-Barkley Bill of 1938 (HR 2711),
    74

wálamt, 25, 27
Waldport, OR, 229
Wallace, Lew, 102–103
Walth, Brent, 204, 235, 297–98
Waltonians. *See* Izaak Walton League of
    America
Ward, Henry B., 38–39
Wasco, OR, 94
Washburn, Sherman, 172, 173
Washington County, OR, 149, 169, 199,
    218
Washington State, 24, 55, 70, 72, 86, 118,
    145, 157, 186
    pulp and paper industry in, 125–26
Washington State Pollution Commission
    (WPCC), 93, 141, 157, 162–63, 245–
    46
"waste sink," 5, 11, 31, 41, 130, 135, 143,
    152, 164, 177
water quality limiting designation, 217,
    220
water pollution, qualitative evidence of,
    15, 53, 90–91, 148, 214
    from rafting on Willamette River, 101
    from workers along Columbia Slough,
        110
water pollution, quantitative evidence of,
    52, 89, 90, 148, 150, 177, 186, 236,
    238
    OSSA and, 91, 144, 148, 172
    role of pulp and paper industry
        research in, 148, 151–53, 217,
        219, 222

role of university research in, 56, 70, 219, 222,
role of David Charlton in, 119, 139, 142
See also bacteria; biochemical oxygen demand (BOD); dissolved oxygen (DO); population equivalent
Water Purification and Prevention of Pollution Bill (1938), 78, 82, 186
water quality
    regulated river flow as critical to, 96, 144, 148, 158–62, 167, 173, 175, 181–84, 210, 216, 221, 237
    See also water pollution, qualitative evidence of; water pollution, quantitative evidence of; pollution abatement
Water Resources Committee, 70–71, 78, 81
Watershed Health Initiative, 219
Wauna, OR, 230
Weber, Louise Palmer, 109
    Wendel, Harold F., 88, 167, 109, 181, 200, 205, 211, 225, 234
    biography of, 87
    cooperative pragmatic approach of, 88, 196–197
    death of, 196
    industrial polluters and, 136–43, 166
    OSSA chair, named as, 87
    Portland sewers and, 98, 102–103, 105, 116
West Linn Paper mill (West Linn, OR). *See* Crown Zellerbach Paper mill (West Linn, OR)
West Linn, OR, 29, 137
    pulp mill in, 1–2, 29, 44, 123, 135, 136, 137, 143, 145, 180, 240
West Slope Sanitary District, 169
Western Kraft mill (Albany, OR). See Weyerhaeuser mill (Albany, OR)
Weyerhaeuser mill (Albany, OR), 123, 180, 241
Weyerhaeuser mill (Longview, WA), 135, 139, 140

Weyerhaeuser Timber mill (Springfield, OR). *See* International Paper mill (Springfield, OR)
White, Richard, 13–14
whites. *See* European Americans
Willamette Falls Locks, OR, 137, 137
Willamette Falls Pulp and Paper mill (West Linn, OR). *See* Crown Zellerbach Paper mill (West Linn, OR)
Willamette Falls, OR, 23, 25, 25, 26, 29, 49, 84, 137, 137, 152, 153, 175, 182, 238
    fish and, 50, 68, 183
    Oregon City pulp mill in, 1–2, 44, 50, 123, 137, 145–46, 163, 180, 239
    Pacific Ocean tidal reach to, 23, 58, 153
    West Linn pulp mill in, 1–2, 29, 44, 123, 135, 136, 137, 143, 145, 180, 240
    See also Native Americans; Oregon City, OR; West Linn, OR; Willamette Falls Locks, OR,
Willamette Pulp and Paper mill (Lebanon, OR). *See* Crown Zellerbach Paper mill (Lebanon, OR)
Willamette Pulp and Paper mill (West Linn, OR). *See* Crown Zellerbach Paper mill (West Linn, OR)
Willamette River Basin Committee, 158
Willamette River Greenway, 195
Willamette River watershed, 3–4, 23
    description of, 13, 22–25
    empirical data about, 57, 59, 127, 134, 148, 169, 221
    hydrology of, 96, 131, 151, 238
    industrialization of, 2–5, 80
    pollution of, 5–7, 10, 44, 50, 56–57, 92–93, 125, 146, 225
    Portland as largest city in, 11, 80
    proposal to divert water from, 181–82
    See also demographics as contributor to pollution; Multnomah Channel, OR; Native Americans; Pacific Ocean, tidal influence on Willamette River;

pollution abatement; Pulp and paper industry; sewage treatment plants in Oregon; Total Maximum Daily Load (TMDL); Watershed Health Initiative; Willamette River, OR; Willamette Valley Project

Willamette River, OR
among the most polluted rivers in the US, 7, 92–93, 93
dams in tributaries of. See Willamette Valley Project
flooding of, 1, 13, 24, 39, 57, 84, 158–60
national exemplar of pollution abatement, 7–10, 14–15, 205–206, 233–34
proposal to divert water from, 181–82
sewer, characterization as, 7, 101, 103, 224
swimming and baptisms in, 6, 33, 40, 41, 170, 224–25
tributaries of, 22, 23
water quality studies of. See water pollution, quantitative evidence of
"working river," characterization as, 11, 42, 57, 234
See also Willamette River watershed; Willamette Valley Project
Willamette Riverkeeper, 227
Willamette Slough, St. Helens, OR, 76
Willamette Slough, Salem, OR, 203
Willamette Valley
romanticizing of by white settlers, 19–20, 26
See also European Americans; Willamette River; Willamette River watershed
Willamette Valley City Engineer's Association, 49
Willamette Valley Project, 159
planning for, 158–60
pollution abatement and, 96, 144, 148, 158–62, 167, 173, 175, 181–84, 210, 216, 221, 237
See also individual dams
Wilsonville, OR, 222

Wisconsin, 37, 51, 72, 85, 121, 122, 129, 140, 145
Wolman, Abel
assimilative capacity and, 162–64, 217
1938 election campaign and, 81, 92, 94
New Deal-era water quality research and, 71, 92
"Wolman Report," author of, 92, 94–96, 99, 102, 105, 106, 110, 112, 113, 116
"Wolman Report," 95–96, 113
women
involvement in pollution abatement of, 77–78, 100, 109, 155, 172, 195, 221, 227
See also Cummins, Mrs. C. D.; Federation of Women's Clubs; Finley, Irene; League of Women Voters; McCullough Lee, Dorothy; Weber, Louise Palmer
Woodward, William F., 66, 68
Works Progress Administration (WPA), 5, 62, 69, 94
World War II, 5, 6, 87, 99, 102, 104–105, 110, 113, 136, 154, 188, 228, 234, 236

Yamhill River, OR, 22, 23
Young, F. H. (Frederic Harold)
biography of, 73
Oregon Business & Investors, Inc. and, 77, 78, 82, 87, 107–108
pollution abatement stances of, 73–74, 76, 78, 87, 98, 107–108